Richard H. Groshong, Jr.

3-D Structural Geology

A Practical Guide to Quantitative Surface
and Subsurface Map Interpretation

Second Edition

Richard H. Groshong, Jr.

3-D Structural Geology

A Practical Guide to Quantitative Surface and Subsurface Map Interpretation

Second Edition

With 453 Figures and a CD-ROM

 Springer

Author

Richard H. Groshong, Jr.

University of Alabama
and
3-D Structure Research
10641 Dee Hamner Rd.
Northport, AL 35475
USA

E-mail: rhgroshong@cs.com

Library of Congress Control Number: 2005937627

Additional material to this book can be downloaded from http://extras.springer.com

ISBN-10   3-540-31054-1 Springer Berlin Heidelberg New York
ISBN-13   978-3-540-31054-9 Springer Berlin Heidelberg New York

ISBN   3-540-65422-4 (first edition) Springer Berlin Heidelberg New York

Springer is a part of Springer Science+Business Media
springer.com
© Springer-Verlag Berlin Heidelberg 1999, 2006
Printed in The Netherlands

Cover design: Erich Kirchner, Heidelberg
Typesetting: Büro Stasch, Bayreuth (stasch@stasch.com)
Production: Almas Schimmel
Printing: Krips bv, Meppel
Binding: Stürtz GmbH, Würzburg

Printed on acid-free paper   32/3141/as – 5 4 3 2 1 0

Preface

Geological structures are three dimensional, yet are typically represented by, and interpreted from, outcrop maps and structure contour maps, both of which are curved two-dimensional surfaces. Maps plus serial sections, called 2½-D, provide a closer approach to three dimensionality. Computer technology now makes it possible for geological interpretations to be developed from the beginning in a fully three dimensional environment. Fully 3-D geological models allow significantly better interpretations and interpretations that are much easier to share with other geologists and with the general public. This book provides an overview of techniques for constructing structural interpretations in 2-D, 2½-D and 3-D environments; for interpolating between and extrapolating beyond the control points; and for validating the final interpretation. The underlying philosophy is that structures are three-dimensional solid bodies and that data from throughout the structure, whether in 2-D or 3-D format, should be integrated into an internally consistent 3-D interpretation.

It is assumed that most users of this book will do their work on a computer. Consequently, the book provides quantitative structural methods and techniques that are designed for use with spreadsheets, mapping software, and three-dimensional computer-graphics programs. The book is also intended to provide the background for understanding what interpretive software, for example, a computer contouring program, does automatically. Most techniques are presented in both a traditional format appropriate for paper, pencil, and a pocket calculator, and in quantitative format for use with spreadsheets and computer-graphics or computer-aided-design programs.

The methods are designed for interpretations based on outcrop measurements and subsurface information of the type derived primarily from well logs and two-dimensional seismic reflection profiles. These data sets all present a similar interpretive problem, which is to define the complete geometry from isolated and discontinuous observations. The techniques are drawn from the methods of both surface and subsurface geology and provide a single methodology appropriate for both. The focus is on the interpretation of layered sediments and rocks for which bedding surfaces provide reference horizons.

The presentation is directed toward geoscience professionals and advanced students who require practical and efficient techniques for the quantitative interpretation of real-world structural geometries at the map scale. The techniques are designed to help identify and develop the best interpretation from incomplete data and to provide unbiased quality control techniques for recognizing and correcting erroneous data and erroneous interpretations. The second edition has been reorganized to more nearly follow the typical interpretation workflow. Several topics that were previously distributed across several chapters now have their own chapters. A significant amount of new material has been added, in particular numerous examples of 3-D models and techniques for using kinematic models to predict fault and ramp-anticline geometry.

Recognizing that not all users of this book will have had a recent course in structural geology, Chap. 1 provides a short review of the elements of structural geology, including the basic definitions and concepts needed for interpretation. The mechanical interpretation of folds and faults and the relationships between the geometry and mechanics are emphasized. Even with abundant data, structural interpretation requires inferences, and the best inferences are based on both the hard data and on mechanical principles.

Chapter 2 covers the fundamental building blocks of structural interpretation: the locations of observation points in 3-D and the orientations of lines and planes. Both analytical solutions and graphical representations of lines and planes on stereograms and tangent diagrams are provided.

Structure contours form the primary means for representing the geometry of surfaces in three dimensions. In Chap. 3, techniques for effective hand and computer contouring are described and discussed. This chapter also contains discussions of building structure contour maps from cross sections and for improving the maps by using the additional information obtained from dip measurements, fluid-flow barriers, and multiple marker surfaces.

Accurate thickness information is as important to structural interpretation as it is to stratigraphic interpretation. Chapter 4 covers the multiple definitions of thickness, thickness measurements, and the interpretation of isopach and isocore maps.

Chapter 5 covers the geometry of folds, including finding the fold trend and the recognition of cylindrical and conical folds on tangent diagrams; using the fold trend in mapping; dip-domain fold geometry and the importance of axial surfaces; the recognition and use of minor folds; and growth folding.

Cross sections are used both to illustrate map interpretations and to predict the geometry from sparse data by interpolation and extrapolation. Construction techniques for both illustrative and predictive cross sections are given in Chap. 6, including techniques for the projection of data onto the line of section. Also in this chapter is a discussion of constructing maps from serial cross sections.

Chapter 7 discusses the recognition of faults and unconformities; calculating heave and throw from stratigraphic separation; and the geometric properties of faults, including associated growth stratigraphy. The correlation of separate observations into mappable faults is treated here.

Chapter 8 completes the basic steps required to build internally consistent 3-D structural interpretations. Techniques are provided for constructing structure contour maps of faulted surfaces, for constructing and interpreting fault cutoff maps (Allan diagrams) and for interpreting faults from isopach maps. Also in this chapter are discussions of the geometry of overlapping, intersecting, and cross-cutting faults.

Dip-sequence analysis of both folds and faults is treated in Chap. 9. Also known as SCAT analysis, the methodology provides a systematic approach to interpreting the structure found along dip traverses in the field and from dipmeters in wells.

Chapter 10, quality control, is a discussion of methods for recognizing problem areas or mistakes in completed maps and cross sections. Quality control problems range from simple data-input errors, to contouring artifacts, to geometrically impossible maps. Corrective strategies are suggested for common problems.

Chapter 11 is a discussion of concepts and techniques for structural validation, restoration, and prediction. The area-depth relationship is treated first because it is a validation and prediction technique that does not require a kinematic model or require restoration. A structure that is restorable to the geometry it had before deformation is considered to be valid. Because restoration techniques are based on models for the kinematic evolution of the geometry, they are inherently predictive of both the geometry and the evolution. The generally applicable kinematic models for predicting fault geometry from hangingwall geometry and hangingwall geometry from fault geometry are presented here along with discussions of the best choice of method for a given structural style.

Vector geometry is often the most efficient approach to deriving the equations needed in 3-D structural interpretation. Chapter 12 provides derivations of the basic equations of vector geometry, the results of which are used in several of the previous chapters. In addition this chapter includes suggestions about how other useful relationships can be derived.

Numerous worked examples are presented throughout the text in order to explain and illustrate the techniques being discussed. Exercises are provided at the ends of Chap. 2 through 11. Many of the map interpretation exercises provide just enough information to allow a solution. It is instructive to see what answers may be obtained by deleting a small amount of the information from the well or the map or by deliberately introducing erroneous data of a type commonly encountered, for example by transposing numbers in a measurement, reversing a dip direction, or by mislocating a contact. For additional practice, use the questions provided at the ends of the chapters to interpret other geologic maps and cross sections.

This edition includes a CD which supplements the text in several ways. Color, shading, and transparency all communicate important information in 3-D models. The CD contains a complete copy of the text with the model-based figures in color. The 3-D models presented here were constructed using the software program Tecplot (www.amtec.com). The CD contains representative Tecplot files that can be viewed or modified as desired. For those interested in working exercises in mapping software, *xyz* input files are provided for many of the map-based exercises. Spreadsheet templates are are included for some of the key calculations, the area-depth relationship, and for SCAT analysis, including the tangent diagram. Answers to a number of exercises are also on the CD.

The first edition benefited from the helpful suggestions of a number of University of Alabama graduate students, especially Bryan Cherry, Diahn Johnson, and Saiwei Wang, whose thesis work has been utilized in some of the examples. I am extremely grateful to Denny Bearce, Lucian Platt, John Spang and Hongwei Yin for their reviews and for their suggestions which have led to significant improvements in the presentation. Additional helpful suggestions have been made by Jean-Luc Epard, Gary Hooks, Jack Pashin, George Davis, Jiafu Qi, Jorge Urdaneta, and the University of Alabama Advanced Map Interpretation class of 1997.

The second edition owes a great debt to Richard H. Groshong, III, who redrafted many of the figures and who constructed all the otherwise unreferenced Tecplot models. Without his help this edition would not have been possible. Alabama graduate students Roger Brewer, Baolong Chai, Mike Cox, Guohai Jin, Carrie Maher, and Marcella McIntyre have provided insights and examples for which I thank them. I have benefited from helpful discussions with Jim Morse, Jim Tucker and Bruce Wrightson about the SCAT analysis of dipmeters. I began assembling the material on restoration and prediction in Chap. 11 as a visiting professor at l'Université de Lausanne in Switzerland, and I am extremely grateful to Professor Henri Masson for making it possible. Collaboration with Dr. Jiafu Qi, Director, Key Laboratory for Hydrocarbon Accumulation Mechanism, China Petroleum University, P.R. China, partially funded by the National Natural Science Foundation of China Contract No. 40372072 and the Ministry of Education Contract No. 200303, have helped advance the work on several of the topics presented in Chap. 11. Finally, I would like to thank the numerous students in my OGCI/Petroskills classes for their comments, questions, and suggestions which, I hope, have helped to make this edition clearer and more complete.

Tuscaloosa, Alabama   Richard H. Groshong, Jr.
January 2006

Contents

Chapter 1

Elements of Map-Scale Structure

1.1
Introduction

The primary objective of structural map making and map interpretation is to develop an internally consistent three-dimensional picture of the structure that agrees with all the data. This can be difficult or ambiguous because the complete structure is usually undersampled. Thus an interpretation of the complete geometry will probably require a significant number of inferences, as, for example, in the interpolation of a folded surface between the observation points. Constraints on the interpretation are both topological and mechanical. The basic elements of map-scale structure are the geometries of folds and faults, the shapes and thicknesses of units, and the contact types. This chapter provides a short review of the basic elements of the structural and stratigraphic geometries that will be interpreted in later chapters, reviews some of the primary mechanical factors that control the geometry of map-scale folds and faults, and examines the typical sources of data for structural interpretation and their inherent errors.

1.2
Representation of a Structure in Three Dimensions

A structure is part of a three-dimensional solid volume that probably contains numerous beds and perhaps faults and intrusions (Fig. 1.1). An interpreter strives to develop

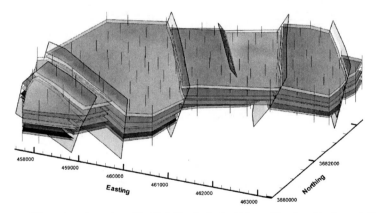

Fig. 1.1. 3-D oblique view of a portion of the Black Warrior Basin, Alabama (data from Groshong et al. 2003b), viewed to the NW. *Thin vertical lines* are wells, semi-transparent surfaces outlined in *black* are faults

a mental and physical picture of the structure in three dimensions. The best interpretations utilize the constraints provided by all the data in three dimensions. The most complete interpretation would be as a three-dimensional solid, an approach possible with 3-D computer graphics programs. Two-dimensional representations of structures by means of maps and cross sections remain major interpretation and presentation tools. When the geometry of the structure is represented in two dimensions on a map or cross section, it must be remembered that the structure of an individual horizon or a single cross section must be compatible with those around it. This book presents methods for extracting the most three-dimensional interpretive information out of local observations and for using this information to build a three-dimensional interpretation of the whole structure.

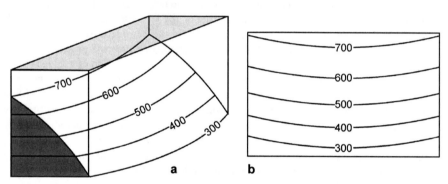

Fig. 1.2. Structure contours. **a** Lines of equal elevation on the surface of a map unit. **b** Lines of equal elevation projected onto a horizontal surface to make a structure contour map

Fig. 1.3.
Structure contour map of the faulted upper horizon from Fig. 1.1. Contours are at 50 ft intervals, with negative elevations being below sea level. Faults are indicated by gaps where the horizon is missing

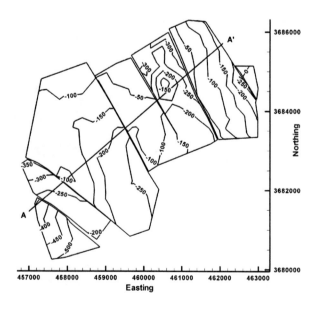

1.2.1
Structure Contour Map

A structure contour is the trace of a horizontal line on a surface (e.g., on a formation top or a fault). A structure contour map represents a topographic map of the surface of a geological horizon (Figs. 1.2, 1.3). The dip direction of the surface is perpendicular to the contour lines and the dip amount is proportional to the spacing between the contours. Structure contours provide an effective method for representing the three-dimensional form of a surface in two dimensions. Structure contours on a faulted horizon (Fig. 1.3) are truncated at the fault.

1.2.2
Triangulated Irregular Network

A triangulated irregular network (TIN; Fig. 1.4) is an array of points joined by straight lines that define a surface. In a TIN network, the nearest-neighbor points are connected to form triangles that form the surface (Banks 1991; Jones and Nelson 1992). If the triangles in the network are shaded, the three-dimensional character of the surface can be illustrated. This is an effective method for the rendering of surfaces by computer. The TIN can be contoured to make a structure contour map.

1.2.3
Cross Section

Even though a structure contour map or TIN represents the geometry of a surface in three dimensions, it is only two-dimensional because it has no thickness. To completely represent a structure in three dimensions, the relationship between different horizons must be illustrated. A cross section of the geometry that would be seen on the face of a slice through the volume is the simplest representation of the relationship between

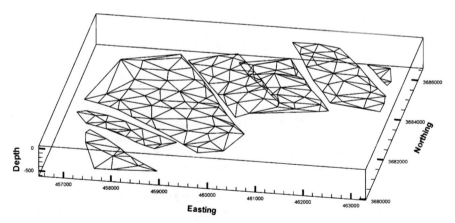

Fig. 1.4. Triangulated irregular network (TIN) of points used to form the upper map horizon in Figs. 1.1 and 1.3. 3-D perspective view to the NW, 3× vertical exaggeration. Vertical scale in ft, horizontal scale in meters

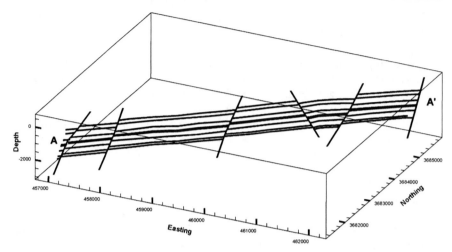

Fig. 1.5. East-west cross section across the structure in Fig. 1.1 produced by taking a vertical slice through the 3-D model. Line of section shown on Fig. 1.3

horizons. In this book cross sections will be assumed to be vertical unless it is stated otherwise. A cross section through multiple surfaces illustrates their individual geometries and defines the relationships between the surfaces (Fig. 1.5). The geometry of each surface provides constraints on the geometry of the adjacent surfaces. The relationships between surfaces forms the foundation for many of the techniques of structural interpretation that will be discussed.

1.3
Map Units and Contact Types

The primary concern of this book is the mapping and map interpretation of geologic contacts and geologic units. A contact is the surface where two different kinds of rocks come together. A unit is a closed volume between two or more contacts. The geometry of a structure is represented by the shape of the contacts between adjacent units. Dips or layering within a unit, such as in a crossbedded sandstone, are not necessarily parallel to the contacts between map units. Geological maps are made for a variety of purposes and the purpose typically dictates the nature of the map units. It is important to consider the nature of the units and the contact types in order to distinguish between geometries produced by deposition and those produced by deformation. Units may be either right side up or overturned. A stratigraphic horizon is said to face in the direction toward which the beds get younger. If possible, the contacts to be used for structural interpretation should be parallel and have a known paleogeographic shape. This will allow the use of a number of powerful rules in the construction and validation of map surfaces, in the construction of cross sections, and should result in geometries that can be restored to their original shapes as part of the structural validation process. Contacts that were originally horizontal are preferred. Even if a restoration is not actually done, the concept that the map units were originally horizontal is implicit in many structural interpretations.

1.3.1
Depositional Contacts

A depositional contact is produced by the accumulation of material adjacent to the contact (after Bates and Jackson 1987). Sediments, igneous or sedimentary extrusions, and air-fall igneous rocks have a depositional lower contact which is parallel to the pre-existing surface. The upper surface of such units is usually, but not always, close to horizontal. A conformable contact is one in which the strata are in unbroken sequence and in which the layers are formed one above the other in parallel order, representing the uninterrupted deposition of the same general type of material, e.g., sedimentary or volcanic (after Bates and Jackson 1987).

Lithologic boundaries that represent lateral facies transitions (Fig. 1.6a), were probably not horizontal to begin with. Certain sedimentary deposits drape over pre-existing topography (Fig. 1.6b) while others are deposited with primary depositional slopes (Fig. 1.6c). The importance of the lack of original horizontality depends on the scale of the map relative to the magnitude of the primary dip of the contact. Contacts that dip only a few degrees might be treated as originally horizontal in the interpretation of a local map area, but the depositional contact between a reef and the adjacent basin sediments may be close to vertical (Fig. 1.7), for example, and could not be considered as originally horizontal at any scale. Depositional contacts that had significant original topographic relief (Fig. 1.7) should be restored to their original depositional geometry, not to the horizontal.

Fig. 1.6.
Cross sections showing primary depositional lithologic contacts that are not horizontal. **a** Laterally equivalent deposits of sandstone and shale. The depositional surface is a time line, not the lithologic boundary. **b** Draped deposition parallel to a topographic slope. **c** Primary topography associated with clinoform deposition

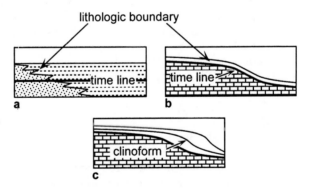

Fig. 1.7.
Cross sections showing primary sedimentary facies relationships and maximum flooding surface. All time lines are horizontal in this example

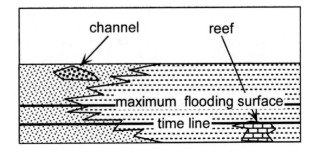

Fig. 1.8.
Unconformity types. The unconformity (*heavy line*) is the
contact between the older, underlying *shaded* units and the
younger, overlying *unshaded*
units. **a** Angular unconformity.
b Buttress or onlap unconformity. **c** Disconformity. **d** Nonconformity. The patterned unit
may be plutonic or metamorphic rock

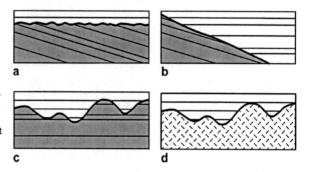

1.3.2
Unconformities

An unconformity is a surface of erosion or nondeposition that separates younger strata
from older strata. An angular unconformity (Fig. 1.8a) is an unconformity between
two groups of rocks whose bedding planes are not parallel. An angular unconformity
with a low angle of discordance is likely to appear conformable at a local scale. Distin-
guishing between conformable contacts and low-angle unconformities is difficult but
can be extremely important to the correct interpretation of a map. A progressive or buttress unconformity (Fig. 1.8b; Bates and Jackson 1987) is a surface on which onlapping
strata abut against a steep topographic scarp of regional extent. A disconformity (Fig. 1.8c)
is an unconformity in which the bedding planes above and below the break are essentially parallel, indicating a significant interruption in the orderly sequence of sedimentary rocks, generally by an interval of erosion (or sometimes of nondeposition),
and usually marked by a visible and irregular or uneven erosion surface of appreciable
relief. A nonconformity (Fig. 1.8d) is an unconformity developed between sedimentary rocks and older plutonic or massive metamorphic rocks that had been exposed to
erosion before being covered by the overlying sediment.

1.3.3
Time-Equivalent Boundaries

The best map-unit boundaries for regional structural and stratigraphic interpretation
are time-equivalent across the map area. Time-equivalent boundaries are normally
established using fossils or radiometric age dates and may cross lithologic boundaries.
Volcanic ash fall deposits, which become bentonites after diagenesis, are excellent time
markers. Because an ash fall drapes the topography and is relatively independent of
the depositional environment, it can be used for regional correlation and to determine
the depositional topography (Asquith 1970). It can be difficult to establish time-equivalent map horizons because of the absence or inadequate resolution of the paleontologic or radiometric data, lithologic and paleontologic heterogeneity in the depositional environment, and because of the occurrence of time-equivalent nondeposition
or erosion in adjacent areas. Time-equivalent map-unit boundaries may be based on
certain aspects of the physical stratigraphy. A *sequence* is a conformable succession of

genetically related strata bounded by unconformities and their correlative conformities (Mitchum 1977; Van Wagoner et al. 1988). A parasequence is a subunit within a sequence that is bounded by marine flooding surfaces (Van Wagoner et al. 1988) and the approximate time equivalence along flooding surfaces makes them suitable for structural mapping. A maximum flooding surface (Fig. 1.7; Galloway 1989) can be the best for regional correlation because the deepest-water deposits can be correlated across lithologic boundaries. At the time of maximum flooding, the sediment input is at a minimum and the associated sedimentary deposits are typically condensed sections, seen as radioactive shales or thin, very fossiliferous carbonates.

1.3.4
Welds

A weld joins strata originally separated by a depleted or withdrawn unit (after Jackson 1995). Welds are best known where a salt bed has been depleted by substratal dissolution or by flow (Fig. 1.9). If the depleted unit was deposited as part of a stratigraphically conformable sequence, the welded contact will resemble a disconformity. If the depleted unit was originally an intrusion, like a salt sill, the welded contact will return to its original stratigraphic configuration. A welded contact may be recognized from remnants of the missing unit along the contact. Lateral displacement may occur across the weld before or during the depletion of the missing unit (Fig. 1.9b). Welded contacts may crosscut bedding in the country rock if the depleted unit was originally crosscutting, as, for example, salt diapir that is later depleted.

1.3.5
Intrusive Contacts and Veins

An intrusion is a rock, magma, or sediment mass that has been emplaced into another distinct unit. Intrusions (Fig. 1.10) may form concordant contacts that are parallel to the layering in the country rock, or discordant contacts that crosscut the layering in the country rock. A single intrusion may have contacts that are locally concordant and discordant.

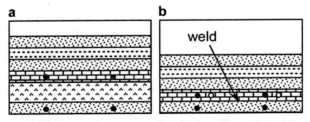

Fig. 1.9. Cross sections illustrating the formation of a welded contact. *Solid dots* are fixed material points above and below the unit which will be depleted. **a** Sequence prior to depletion. **b** Sequence after depletion: *solid dots* represent the final positions of original points in a hangingwall without lateral displacement; *open circles* represent final positions of original points in a hangingwall with lateral displacement

Fig. 1.10.
Cross sections of intrusions.
Intrusive material is *patterned*
and *lines* represent layering in
the country rock. **a** Concordant. **b** Discordant

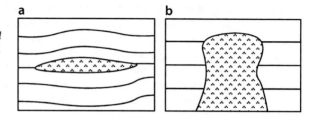

A vein is a relatively thin, normally tabular, rock mass of distinctive lithologic character, usually crosscutting the structure of the host rock. Many veins are depositional
and represent the filling of a fracture, whereas others are the result of replacement of
the country rock. Veins are mentioned here with intrusions because the contact relationships and unit geometries may be similar to those of some intrusions.

1.3.6
Other Boundaries

Many other attributes of the rock units and their contained fluids can be mapped, for
example, the porosity, the oil-water contact, or the grades of mineral deposits. Most of
the mapping techniques to be discussed will apply to any type of unit or contact. Some
interpretation techniques, particularly those for fold interpretation, depend on the
contacts being originally planar boundaries and so those methods may not apply to
nonstratigraphic boundaries.

1.4
Thickness

The thickness of a unit is the perpendicular distance between its bounding surfaces
(Fig. 1.11a). The true thickness does not depend on the orientation of the bounding
surfaces. If a unit has variable thickness, various alternative measurements might be
used, such as the shortest distance between upper and lower surfaces or the distance
measured perpendicular to either the upper or lower surface. The definition used here
is based on the premise that if the unit was deposited with a horizontal surface but a
variable thickness, then the logical measurement direction would be the thickness
measured perpendicular to the upper surface, regardless of the structural dip of the
surface (Fig. 1.11b).

Thickness variations can be due to a variety of stratigraphic and structural causes.
Growth of a structure during the deposition of sediment typically results in thinner
stratigraphy on the structural highs and thicker stratigraphy in the lows. Both growth
folds and growth faults occur. A sedimentary package with its thickness influenced by
an active structure is known as a growth unit or growth sequence. The high part of a
growth structure may be erosional at the same time that the lower parts are depositional. Thickness variations may be the result of differential compaction during and
after deposition. If the composition of a unit undergoes a facies change from relatively
uncompactable (i.e., sand) to relatively compactable (i.e., shale) then after burial and

compaction the unit thickness will vary as a function of lithology. Deformation-related thickness changes are usually accompanied by folding, faulting, or both within the unit being mapped. Deformation-related thickness changes are likely to correlate to position within a structure or to structural dip.

1.5
Folds

A fold is a bend due to deformation of the original shape of a surface. An antiform is convex upward; an anticline is convex upward with older beds in the center. A synform is concave upward; a syncline is concave upward with younger beds in the center. Original curves in a surface, for example grooves or primary thickness changes, are not considered here to be anticlines or synclines.

1.5.1
Styles

Folds may be characterized by domains of uniform dip, by a uniform variation of the dip around a single center of curvature, or may combine regions of both styles. Regions of uniform dip (Fig. 1.12a) are called dip domains (Groshong and Usdansky 1988). Dip domains are separated from one another by axial surfaces or faults. Dip domains have also been referred to as kink bands, but the term kink band has mechanical

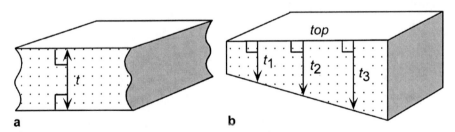

Fig. 1.11. Thickness (*t*). **a** Unit of constant thickness. **b** Unit of variable thickness

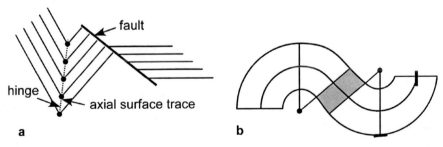

Fig. 1.12. Regions of uniform dip properties. **a** Dip domains. **b** Concentric domains separated by a planar dip domain (*shaded*)

implications that are not necessarily appropriate for every structure. An axial surface (Fig. 1.12a) is a surface that connects fold hinge lines, where a hinge line is a line of maximum curvature on the surface of a bed (Dennis 1967). A hinge (Fig. 1.12a) is the intersection of a hinge line with the cross section. A circular domain (Fig. 1.12b) is defined here as a region in which beds approximate a portion of a circular arc. If multiple surfaces in a circular domain have the same center of curvature, the fold is concentric. Dips in a concentric domain are everywhere perpendicular to a radius through the center of curvature. The center of curvature is determined as the intersection point of lines drawn perpendicular to the dips (Busk 1929; Reches et al. 1981). A circular curvature domain does not possess a line of maximum curvature and thus does not strictly have an axial surface. Dips within a domain may vary from the average values. If the dips are measured by a hand-held clinometer, variations of a few degrees are to be expected due to the natural variation of bedding surfaces and the imprecision of the measurements.

As a generality, the structural style is controlled by the mechanical stratigraphy and the directions of the applied forces. Mechanical stratigraphy is the stratigraphy described in terms of its physical properties. The mechanical properties that control the fold geometry are the stiffness contrasts between layers, the presence or absence of layer-parallel slip, and the relative layer thicknesses. Stiff (also known as competent) lithologies (for example, limestone, dolomite, cemented sandstone) tend to maintain constant bed thickness, and soft (incompetent) lithologies (for example, evaporites, overpressured shale, shale) tend to change bed thickness as a result of deformation (Fig. 1.13). Very stiff and brittle units like dolomite may fail by pervasive fracturing, however, and then change thickness as a unit by cataclastic flow. Deformed sedimen-

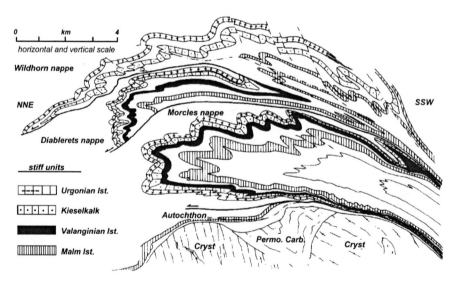

Fig. 1.13. Cross section of the Helvetic Alps, central Switzerland. Mechanical stratigraphy consists of thick carbonate units (stiff) separated by very thick shale units (soft). The folds, especially at the hinges, are circular in style. (After Ramsay 1981)

tary rocks tend to maintain relatively constant bed thickness, although the thickness changes that do occur can be very important. If large thickness changes are observed in deformed sedimentary rocks other than evaporites or overpressured shale, primary stratigraphic variations should be considered as a strong possibility.

In cross section, folds may be harmonic, with all the layers nearly parallel to one another (Fig. 1.14), or disharmonic, with significant changes in the geometry between different units in the plane of the section (Figs. 1.13, 1.15). The fold geometry is controlled by the thickest and stiffest layers (or multilayers) called the dominant members (Currie et al. 1962). A stratigraphic interval characterized by a dominant (geometry-controlling) member between two boundary zones is a structural-lithic unit (Fig. 1.15;

Fig. 1.14.
Dip-domain style folds in an experimental model having a closely spaced multilayer stratigraphy. The *black* and *white layers* are plasticene and are separated by grease to facilitate layer-parallel slip. The *black layers* are slightly stiffer than the *white layers*. (After Ghosh 1968)

0   cm   1

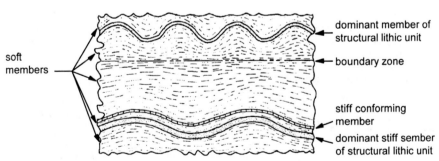

soft members

dominant member of structural lithic unit

boundary zone

stiff conforming member

dominant stiff sember of structural lithic unit

Fig. 1.15. Structural-lithic units. (After Currie et al. 1962)

Currie et al. 1962). A short-wavelength structural lithic unit may form inside the boundary of a larger-wavelength unit and be folded by it, in which case the longer-wavelength unit is termed the dominant structural-lithic unit and the shorter-wavelength unit is a conforming structural-lithic unit (Currie et al. 1962).

The dip changes within structural-lithic units may obscure the map-scale geometry. For example, the regional or map-scale dip in Figure 1.15 is horizontal, although few dips of this attitude could be measured. Where small-scale folds exist, the map-scale geometry may be better described by the orientation of the median surface or the enveloping surface (Fig. 1.16; Turner and Weiss 1963; Ramsay 1967). The median surface (median line in two dimensions) is the surface connecting the inflection points of a folded layer. The inflection points are located in the central region of the fold limbs where the fold curvature changes from anticlinal to synclinal. An enveloping surface is the surface that bounds the crests (high points) of the upper surface or the troughs (low points) of the lower surface on a single unit. Figure 1.16 shows that the dip observed at a single location need not correspond to the dip of either the median surface, the enveloping surface, or to the trace of a formation boundary. The dip of the median surface may be more representative of the dip required for map-scale interpretation than the locally observed bedding dips.

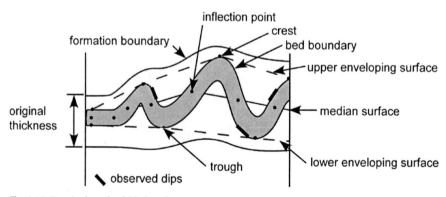

Fig. 1.16. Terminology for folded surfaces

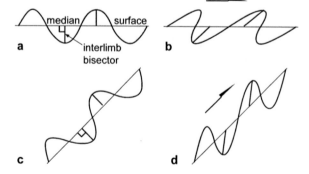

Fig. 1.17.
Fold symmetry. **a** Symmetrical, upright. **b** Asymmetrical, overturned. **c** Symmetrical, overturned. **d** Asymmetrical, upright. *Arrow* gives direction of vergence

The symmetry of a fold is determined by the angle between the plane bisecting the interlimb angle and the median surface (Fig. 1.17a; Ramsay 1967). The angle is close to 90° in a symmetrical fold (Fig. 1.17a,c) and noticeably different from 90° in an asymmetrical fold (Fig. 1.17b,d). An essential property of an asymmetrical fold is that the limbs are unequal in length. Fold asymmetry is not related to the relative dips of the limbs. The folds in Fig. 1.17b,c have overturned steep limbs and right-way-up gentle limbs, but only the folds in Fig. 1.17b are asymmetric. This is a point of possible confusion, because in casual usage a fold with unequal limb dips (Fig. 1.17b,c) may be referred to as being asymmetrical. Folds may occur as regular periodic waveforms as shown (Fig. 1.17) or may be non-periodic with wavelengths that change along the median surface.

The vergence of an asymmetrical fold is the rotation direction that would rotate the axial surface of an antiform from an original position perpendicular to the median surface to its observed position at a lower angle to the median surface. The vergence of the folds in Fig. 1.17b,d is to the right.

1.5.2
Three-Dimensional Geometry

A cylindrical fold is defined by the locus of points generated by a straight line, called the fold axis, that is moved parallel to itself in space (Fig. 1.18a). In other words, a cylindrical fold has the shape of a portion of a cylinder. In a cylindrical fold every straight line on the folded surface is parallel to the axis. The geometry of a cylindrical fold persists unchanged along the axis as long as the axis remains straight. A conical fold is generated by a straight line rotated through a fixed point called the vertex (Fig. 1.18b). The cone axis is not parallel to any line on the cone itself. A conical fold changes geometry and terminates along the trend of the cone axis.

The crest line is the trace of the line which joins the highest points on successive cross sections through a folded surface (Figs. 1.18, 1.19a; Dennis 1967). A trough line is the trace of the lowest elevation on cross sections through a horizon. The plunge of a cylindrical fold is parallel to the orientation of its axis or a hinge line (Fig. 1.19b). The most useful measure of the plunge of a conical fold is the orientation of its crest line or trough line (Bengtson 1980).

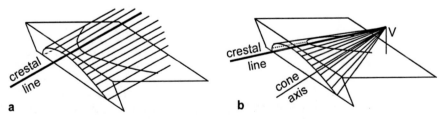

Fig. 1.18. Three-dimensional fold types. **a** Cylindrical. All *straight lines* on the cylinder surface are parallel to the fold axis and to the crestal line. **b** Conical. *V* vertex of the cone. *Straight lines* on the cone surface are not parallel to the cone axis

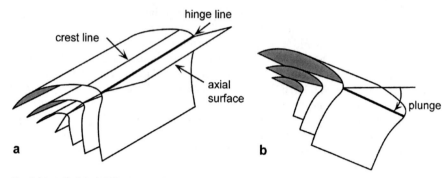

Fig. 1.19. Cylindrical folds. **a** Non-plunging. **b** Plunging

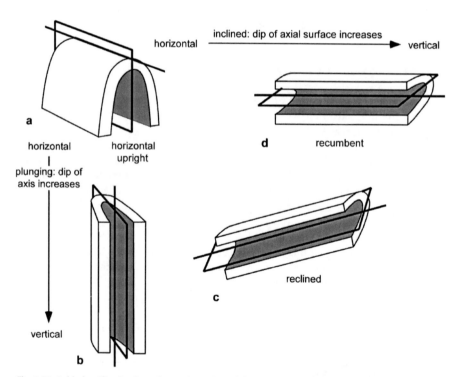

Fig. 1.20. Fold classification based on orientation of the axis and axial surface. **a** *Horizontal upright:* horizontal axis and vertical axial surface. **b** *Vertical:* vertical axis and vertical axial surface. **c** *Reclined:* inclined axis and axial surface. **d** *Recumbent:* horizontal axis and axial surface. (After Fleuty 1964)

The complete orientation of a fold requires the specification of the orientation of both the fold axis and the axial surface (Fig. 1.20). In the case of a conical fold, the orientation can be specified by the orientation of the axial surface and the orientation of the crestal line on a particular horizon. Common map symbols for folds are given in Fig. 1.21.

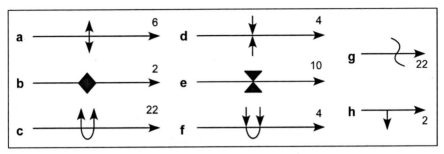

Fig. 1.21. Common map symbols for folds. Fold trend is indicated by the *long line*, plunge by the *arrowhead*, with amount of plunge given. The fold trend may be the axial trace, crest or trough line or a hinge line. **a, b** Anticline. **c** Overturned anticline; both limbs dip away from the core. **d, e** Syncline. **f** Overturned syncline; both limbs dip toward the core. **g** Minor folds showing trend and plunge of axis. **h** Plunging monocline with only one dipping limb

1.5.3
Mechanical Origins

The fundamental mechanical types of folding are based on the direction of the causative forces relative to layering (Ramberg 1963; Gzovsky et al. 1973; Groshong 1975), namely longitudinal contraction, transverse contraction, and longitudinal extension (Fig. 1.22). If the stratigraphy is without mechanical contrasts, forces parallel to layering produce either uniform shortening and thickening or uniform extension and thinning. If some shape irregularity is pre-existing, then it is amplified by layer-parallel shortening to give a passive fold. If the stratigraphy has significant mechanical contrasts, then a mechanical instability can occur that leads to buckle folding in contraction and pinch-and-swell structure (boudinage) in extension. If the forces are not equal vertically, then a forced fold is produced, regardless of the mechanical stratigraphy. Longitudinal contraction, transverse contraction, and longitudinal extension are end-member boundary conditions; they may be combined to produce folds with combined properties.

Buckle folds normally form with the fold axes perpendicular to the maximum principal compressive stress, σ_1. The folds are long and relatively unchanging in geometry parallel to the fold axis but highly variable in cross section. Buckle folds are characterized by the presence of a regular wavelength that is proportional to the thickness of the stiff unit(s). A single-layer buckle fold consists of a stiff layer in a surrounding confining medium. The dip variations associated with a given stiff layer die out into the regional dip within the softer units at a distance of about one-half arc length away from the layer (Fig. 1.15). In the author's experience, buckle folds in sedimentary rocks typically have arc-length to thickness ratios of 5 to 30, with common values in the range of 6 to 10.

As buckled stiff layers become more closely spaced, the wavelengths begin to interfere (Fig. 1.23) resulting in disharmonic folds. Once the layers are sufficiently closely spaced, they fold together as a multilayer unit. A multilayer unit has a much lower buckling stress and an appreciably shorter wavelength than a single layer of same thickness (Currie et al. 1962). Stiff units, either single layers or multilayers, tend to

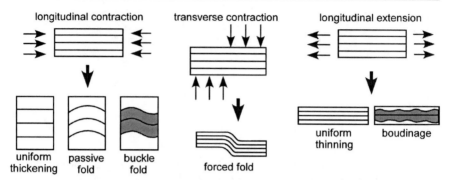

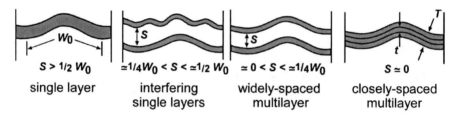

Fig. 1.22. End-member displacement boundary conditions showing responses related to the mechanical stratigraphy. *Shaded* beds are stiff lithologies, *unshaded* beds are soft lithologies

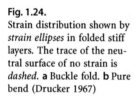

Fig. 1.23. Transition from single layer folding to multilayer folding as space between stiff layers decreases. The stiff layers are *shaded*

Fig. 1.24.
Strain distribution shown by *strain ellipses* in folded stiff layers. The trace of the neutral surface of no strain is *dashed*. **a** Buckle fold. **b** Pure bend (Drucker 1967)

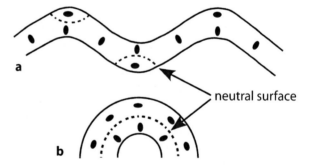

produce circular to sinusoidal folds if enclosed in a sufficient thickness of softer material (Fig. 1.13). Stratigraphic sections made up of relatively thin-bedded multilayer units result in folds that tend to have planar dip-domain styles (Fig. 1.14).

The strain distribution in the stiff layer of a buckle fold (Fig. 1.24a) is approximately layer-parallel shortening throughout most of the fold. The neutral surface separates regions of layer-parallel extension from regions of layer-parallel contraction. Only in a pure bend (Fig. 1.24b) are the areas of extension and contraction about equal and the neutral surface in the middle of the layer. The strain in thick soft layers between stiff layers is shortening approximately perpendicular to the axial surface. The strain in thin soft layers between stiff layers may be close to layer-parallel shear strain. The pure bending strain distribution is usually more closely approached in transverse contraction folds, for example above a salt dome, than in buckle folds.

Cleavage planes and tectonic stylolites in a fold can indicate the mechanical origin of the fold because they form approximately perpendicular to the maximum shortening direction by processes that range from grain rotation to pressure solution (Groshong 1988). Cleavage in a buckle fold is typically at a high angle to bedding (Fig. 1.25), being more nearly perpendicular to bedding in stiff units and more nearly parallel to the axial surface in soft units. Cleavage that is approximately perpendicular to bedding produces a cleavage fan across the fold. The line of the cleavage-bedding intersection is approximately or exactly parallel to the fold axis and can be used to help determine the axis.

Folds produced by an unequal distribution of forces in transverse contraction (Fig. 1.22) are termed forced folds (Stearns 1978). Forced folds tend to be round to blocky or irregular in map view. The major control on the form of the fold is the rheology of the forcing member (Fig. 1.26). A stiff and brittle forcing member (i.e., crystalline basement) leads to narrow fault boundaries at the base of the structure and strain that is highly localized in the zone above the basement fault. A soft unit between a stiff forcing member and the cover sequence will cause the deformation to be disharmonic. A soft forcing member (like salt) typically produces round to elliptical structures with deformation widely distributed across the uplift.

Little strain need occur in the uplifted or downdropped blocks associated with a stiff forcing member. Nearly all the strain is localized in the fault zone between the

Fig. 1.25.
Cleavage pattern in a buckle fold. The gently dipping surfaces are bedding and the steeply dipping surfaces represent cleavage or stylolites. *Arrows* show directions of the boundary displacements

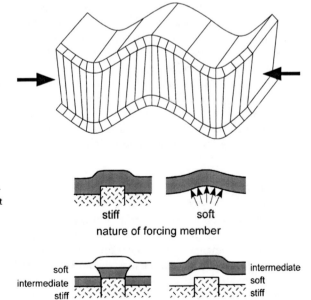

Fig. 1.26.
Effect of mechanical stratigraphy on drape folds. The lowest unit is the forcing member

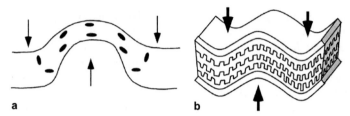

Fig. 1.27. Strain and cleavage patterns in transverse contraction folds produced by differential vertical displacement. **a** Strain distribution above a model salt dome (after Dixon 1975). **b** Cleavage or stylolites parallel to bedding. *Arrows* show directions of the boundary displacements

Fig. 1.28.
Veins due to outer-arc bending stresses

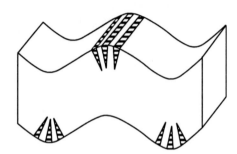

blocks or in the steep limb of the drape fold over the fault zone. A soft forcing member (i.e., salt) will distribute the curvature and strain widely over the uplifted region (Fig. 1.27a). Because cleavage and stylolites form perpendicular to the shortening direction, in folds produced by displacements at a high angle to bedding, the expected cleavage and stylolite direction is parallel to bedding (Fig. 1.27b). In highly deformed rocks, cleavage parallel to bedding might be the result of deformation caused by a large amount of layer-parallel slip or by isoclinal refolding of an earlier axial-plane cleavage.

Extension fractures and veins may form due to the bending stresses in the outer arc of a fold (Fig. 1.28). Such features should become narrower and die out toward the neutral surface. The fracture plane is expected to be approximately parallel to the axis of the fold and the fracture-bedding line of intersection should be parallel to the fold axis. Bending fractures might occur in any type of fold.

1.6
Faults

A fault (Fig. 1.29) is a surface or narrow zone across which there has been relative displacement of the two sides parallel to the zone (after Bates and Jackson 1987). The term displacement is the general term for the relative movement of the two sides of the fault, measured in any chosen direction. A shear zone is a general term for a relatively narrow zone with subparallel boundaries in which shear strain is concentrated (Mitra and Marshak 1988). As the terms are usually applied, a bed, foliation trend, or other marker horizon maintains continuity across a shear zone but is broken and displaced

Fig. 1.29.
General terminology for a
surface (*patterned*) offset by a
fault. *Heavy lines* are hangingwall and footwall cutoff lines

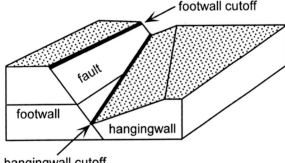

Fig. 1.30.
Fault slip is the displacement
of points (*open circles*) that
were originally in contact
across the fault. Here the correlated points represent the
intersection line of a dike and
a bed surface at the fault plane

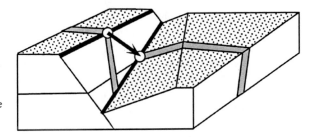

across a fault. It may be difficult to distinguish between a shear zone and a fault zone
on the basis of observations at the map scale, and so here the term fault will be understood to include both faults and shear zones.

The term hangingwall refers to the strata originally above the fault and the term
footwall to strata originally below the fault (Fig. 1.29). Because of the frequent repetition of the terms, hangingwall and footwall will often be abbreviated as HW and FW,
respectively. A cutoff line is the line of intersection of a fault and a displaced horizon
(Fig. 1.29). The HW and FW cutoff lines of a single horizon were in contact across a
fault plane prior to displacement. Across a fault zone of finite thickness or across a
shear zone, the HW and FW cutoffs were originally separated by some width of the
offset horizon that is now in the zone.

1.6.1
Slip

Fault slip is the relative displacement of formerly adjacent points on opposite sides of
the fault, measured along the fault surface (Fig. 1.30; Dennis 1967). Slip can be subdivided into horizontal and vertical components, the strike slip and dip slip components,
respectively. A fault in which the slip direction is parallel to the trace of the cutoff line
of bedding can be called a trace-slip fault. In horizontal beds a trace-slip fault is a
strike-slip fault. Measurement of the slip requires the identification of the piercing
points of displaced linear features on opposite sides of the fault. Suitable linear features at the map scale might be a channel sand, a facies boundary line, a fold hinge line,

Fig. 1.31.
Fault separation terminology

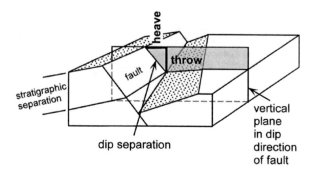

or the intersection line between bedding and a dike (Fig. 1.30). The displacement of dipping beds on faults oblique to the strike of bedding leads to complex relationships between the displacement and the slip in a specific cross-section direction, such as parallel to the dip of bedding. A strike-slip component of displacement is never visible on a vertical cross section.

1.6.2
Separation

Fault separation is the distance between any two parts of an index plane (e.g., bed or vein) disrupted by a fault, measured in any specified direction (Dennis 1967). The separation directions commonly important in mapping are parallel to fault strike, parallel to fault dip, horizontal, vertical and perpendicular to bedding. It should be noted that the definitions of the terms for fault separation and the components of separation are not always used consistently in the literature. Stratigraphic separation (Fig. 1.31) is the thickness of strata that originally separated two beds brought into contact at a fault (Bates and Jackson 1987) and is the stratigraphic thickness missing or repeated at the point, called the fault cut (Tearpock and Bischke 2003), where the fault is intersected. The amount of the fault cut is always a stratigraphic thickness.

Throw and heave (Fig. 1.31) are the components of fault separation most obvious on a structure contour map. Both are measured in a vertical plane in the dip direction of the fault. Throw is the vertical component of the dip separation measured in a vertical plane (Dennis 1967). Stratigraphic separation is not equal to the fault throw unless the marker horizons are horizontal (see Sect. 5.5.3). Heave is the horizontal component of the dip separation measured in a vertical plane normal to the strike of the fault (Dennis 1967).

1.6.3
Geometrical Classifications

A fault is termed normal or reverse on the basis of the relative displacement of the hangingwall with respect to the footwall (Fig. 1.32). For a normal fault, the hangingwall is displaced down with respect to the footwall, and for a reverse fault the hangingwall is displaced up with respect to the footwall. The relative displacement may be either a slip or a separation and the use of the term should so indicate, for example, a *normal-*

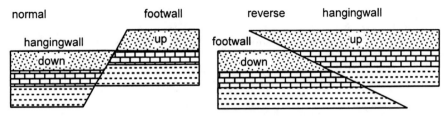

Fig. 1.32. Vertical cross section showing the relative fault displacement terminology with horizontal as the reference plane

Fig. 1.33.
Vertical cross section showing the relative fault displacement terminology with bedding as the reference plane

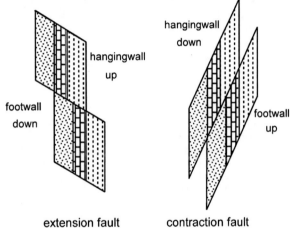

separation fault. Using the horizontal as the plane of reference (i.e., originally horizontal bedding), a normal-separation fault extends a line parallel to bedding and a reverse-separation fault shortens the line.

Using bedding as the frame of reference is not the same as using a horizontal plane, as illustrated by Fig. 1.33 which shows the faults from Fig. 1.32 after a 90° rotation. With bedding vertical, a reverse displacement (Fig. 1.33) extends the bedding while shortening a horizontal line. The fault might have been caused by reverse slip on a fault formed after the beds were rotated to vertical or by the rotation of a normal fault. Using bedding as the frame of reference (Norris 1958), an extension fault extends the bedding, regardless of the dip of bedding, and a contraction fault shortens the bedding.

A fault cut is the point at which a well crosses a fault. A fault with a component of dip separation has the effect of omitting or repeating stratigraphy across the fault at the fault cut (Fig. 1.34). With respect to a vertical line or a vertical well, a normal fault causes the omission of stratigraphic units (Fig. 1.34a) and a reverse fault causes the repetition of units (Fig. 1.34b). Opposite-sense omissions or repetitions may occur in a well that is not vertical (Mulvany 1992). For example, a well drilled from the footwall to the hangingwall of a normal fault will show repeated section down the well (Fig. 1.34a) and will show missing section down the well if the fault is reverse (Fig. 1.34b).

Fig. 1.34.
Effect of well orientation on occurrence of missing or repeated section. All units are right side up and cross sections are vertical. **a** Wells penetrating a normal fault. **b** Wells penetrating a reverse fault

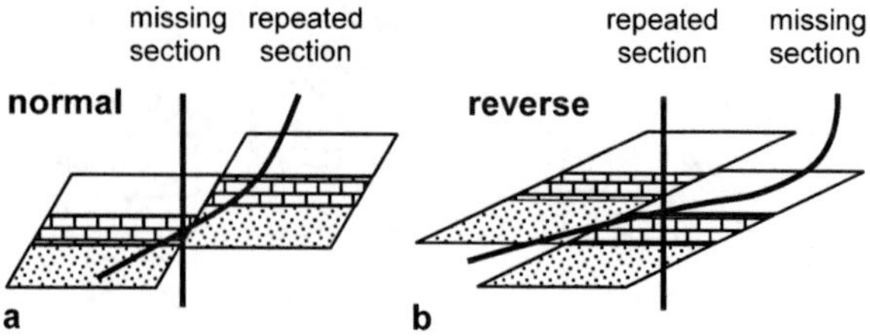

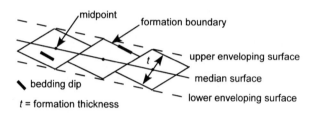

Fig. 1.35.
Regional dip of faulted bedding surface, indicated by enveloping surfaces is different than bedding dip

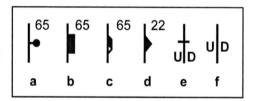

Fig. 1.36.
Map symbols for faults indicating separation. **a–c** Normal separation, *symbol* on hangingwall. **d** Reverse separation, *triangle* on hangingwall. **e** Vertical fault, vertical separation indicated (*U* up; *D* down). **f** Fault of unknown dip, vertical separation indicated

The median surface of a faulted unit (median line in two dimensions) is the surface connecting the midpoints of the blocks in the middle of the reference unit (Fig. 1.35). An enveloping surface is the surface that bounds the high corners or the low corners on a single unit. Dips within the fault blocks may all be different from the dip of either the median surface or the enveloping surface. Within a fault block the original thickness may remain unaltered by the deformation (Fig. 1.35), although the entire unit has been thickened or thinned as indicated by the changed thickness between the enveloping surfaces. Common map symbols for faults are given in Fig. 1.36.

1.6.4
Mechanical Origins

Faults commonly initiate in conjugate pairs (Fig. 1.37). Conjugate faults form at essentially the same time under the same stress state. This geometry has been produced in countless experiments (Griggs and Handin 1960). The acute angle between the two conjugate faults is the dihedral angle which is usually in the range of 30 to 60° but may be significantly smaller if the least principal stress is tensile (Ramsey and Chester 2004). In experiments the maximum principal compressive stress, σ_1, bisects the dihedral angle. The least principal compressive stress, σ_3, bisects the obtuse angle,

Fig. 1.37.
Conjugate pair of faults related to orientation of principal stresses. *D:* dihedral
angle. The principal stresses
are σ_1, σ_2, and σ_3, in order of
greatest to least compressive
stress. *R* right-lateral fault
plane; *L* left-lateral fault plane;
half-arrows show the sense of
shear on each plane

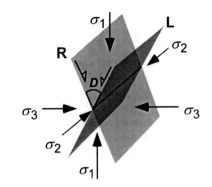

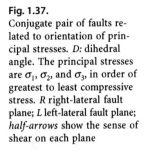

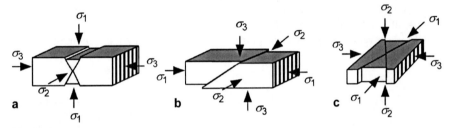

Fig. 1.38. Fault orientations at the surface of the earth predicted from Andersonian stress theory.
a Normal. **b** Reverse. **c** Strike slip

and the intermediate principal stress, σ_2, is parallel to the line of intersection of the
two faults. The slip directions are directly related to the orientation of the principal
stresses (Fig. 1.37), with one set being right lateral (dextral) and the other set left
lateral (sinestral).

The surface of the ground is a plane of zero shear stress and therefore one of the
principal stresses is perpendicular to the surface and the other two principal stresses
lie in the plane of the surface (Anderson 1905, 1951). From the experimental relationship between fault geometry and stress (Fig. 1.37), this leads to a prediction of the
three most common fault types and their dips (Fig. 1.38). Relative to the horizontal,
normal faults typically dip 60°, reverse faults average 30°, and strike-slip faults are
vertical.

The predicted dips in Fig. 1.38 are good for a first approximation, but there are
many exceptions. Fault orientations may be controlled by lithologic differences,
changes in the orientations of the stress field below the surface of the ground, and by
the presence of pre-existing zones of weakness. True triaxial stress states can result
in the formation of two pairs of conjugate faults having the same dips but slightly
different strikes, forming a rhombohedral pattern of fault blocks (Oertel 1965). Oertel
faults are likely to be arranged in low-angle conjugate pairs that are 10–30° oblique to
each other. Faults will rotate to different dips as the enclosing beds rotate. Even with
all the exceptions, it is still common for faults to have the approximate orientations
given in Fig. 1.38.

1.6.5
Fault-Fold Relationships

A planar fault with constant displacement (Fig. 1.39a) is the only fault geometry that does not require an associated fold as a result of its displacement. Of course, all faults eventually lose displacement and end. A fault that dies out without reaching the surface of the earth is called blind, and a fault that reaches the present erosion surface is emergent, although whether it was emergent at the time it moved may not be known. Where the displacement ends at the tip of a blind fault, a fold must develop (Fig. 1.39b). Displacement on a curved fault will cause the rotation of beds in the hangingwall and perhaps in the footwall and will produce a fold (Fig. 1.39c). A generic term for the fold is a ramp anticline. The fold above a normal fault is commonly called a rollover anticline if the hangingwall beds near the fault dip toward the fault.

Fault dips may be controlled by the mechanical stratigraphy to form ramps and flats, although at the scale of the entire fault, the average dip may be maintained (i.e., 30° for a reverse fault). A flat is approximately parallel to bedding, at an angle of say,

Fig. 1.39.
Relationships between folds and faults. **a** Constant slip on a planar fault does not cause folding. **b** Slip on either a plane or curved fault that dies out produces a fold in the region of the fault tip. **c** Slip on a curved fault causes folding in the hangingwall

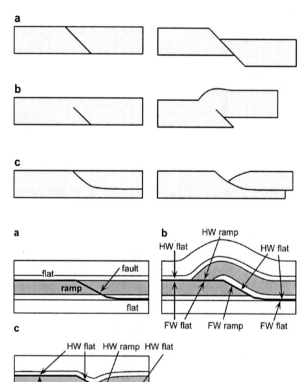

Fig. 1.40.
Ramp-flat fault terminology. *HW* is hangingwall; *FW* is footwall. **a** Before displacement. **b** After reverse displacement. **c** After normal displacement. (After Woodward 1987)

10° or less. A ramp crosses bedding at an angle great enough to cause missing or repeated section at the scale of observation, say 10° or more. Characteristically, but not exclusively, ramps occur in stiff units such as limestone, dolomite, or cemented sandstone, whereas flats occur in soft units, such as shale or salt. Both normal and reverse faults may have segments parallel to bedding and segments oblique to bedding (Fig. 1.40). After displacement, hangingwall ramps and flats no longer necessarily match across the fault (Fig. 1.40b,c).

Another common fault geometry is listric (Fig. 1.41a) for which the dip of the fault changes continuously from steep near the surface of the earth to shallow or horizontal at depth. Both normal and reverse faults may be listric. The lower detachment of a listric fault is typically in a weak unit such as shale, overpressured shale or salt. Faults that flatten upward are comparatively rare, except in strike-slip regimes, and may be termed antilistric. Antilistric reverse faults have been produced by the stresses above a rigid block uplift (Sanford 1959).

In three dimensions, fault ramps are named according to their orientation with respect to the transport direction (Dahlstrom 1970; McClay 1992). Frontal ramps are approximately perpendicular to the transport direction, lateral ramps are approximately parallel to the transport direction and oblique ramps are at an intermediate angle (Fig. 1.42a). Displacement of the hangingwall produces ramp anticlines having orientations that correspond to those of the ramps (Fig. 1.42b). The ramp terminology was developed for thrust-related structures but is equally applicable to normal-fault structures. Lateral and oblique ramps necessarily have a component of strike slip and lateral ramps may be pure strike-slip faults.

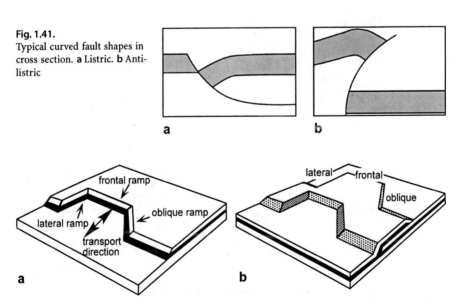

Fig. 1.41.
Typical curved fault shapes in cross section. **a** Listric. **b** Antilistric

a   b

a   b

Fig. 1.42. Ramps and ramp anticlines in three dimensions. **a** Ramps in the footwall. *Arrows* indicate transport direction. **b** Thrust ramp anticlines. (After McClay 1992)

1.7
Sources of Structural Data and Related Uncertainties

The fundamental information generally available for the interpretation of the structure in an area is the attitude of planes and locations of the contacts between units. The primary sources of this information are direct observations of exposures, well logs, and seismic reflection profiles. These data are never complete and may not be correct in terms of the exact locations or attitudes. Before constructing or interpreting a map, it is worth considering the uncertainties inherent in the original data.

1.7.1
Direct Observations

Outcrop- and mine-based maps are constructed from observations of the locations of contacts and the attitudes of planes and lines. Good practice is to show on the working map the areas of exposure at which the observations have been made (Fig. 1.43a). Exposure is rarely complete and so uncertainties typically exist as to the exact locations of contacts. Surface topography is usually directly related to the underlying geology and should be used as a guide to contact locations. Contacts that control topography may be traced with a reasonable degree of confidence, even in the absence of exposure, at least where the structure is simple. The assignment of a particular exposure to a specific stratigraphic unit may be in doubt if diagnostic features are absent. Bedding attitudes measured in small outcrops might come from minor folds, minor fault blocks, or cross beds and not represent the attitude of the formation boundaries. The connectivity of the contacts in the final map is usually an interpretation, not an observation (Fig. 1.43b).

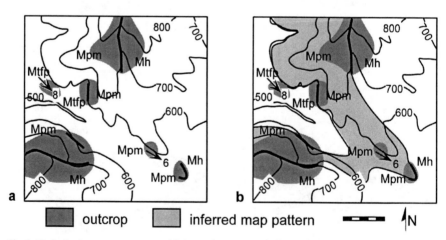

Fig. 1.43. Geologic map on a topographic base. *Contours* are in feet and the scale bar is 1 000 ft. Three formations are present, from oldest to youngest: *Mtfp, Mpm, Mh*. Attitude of bedding is shown by an *arrow* pointing in the dip direction, with the dip amount indicated. **a** Outcrop map showing locations of direct observations (*shaded*). **b** Completed geologic map, contacts *wide* where observed, *thin* where inferred. *Lighter shading* is the interpreted outcrop area of the *Mpm*

1.7.2
Wells

Wells provide subsurface information on the location of formation boundaries and the attitude of planes. Measurements of this information are made by a variety of techniques and recorded on well logs. Sample logs are made from cores or cuttings taken from the well as it is drilled. In wells drilled with a cable tool, cuttings are collected from the bottom of the hole every 5 or 10 feet and provide a sample of the rock penetrated in that depth interval. In wells that are rotary drilled, drilling fluid is circulated down the well and back to remove the cuttings from the bottom of the hole. The drilling fluid is sampled at intervals as it reaches the surface to determine the rock type and fossil content of the cuttings. Depths are calculated from the time required for the fluid to traverse the length of the hole and are not necessarily precise.

A wire-line log is a continuous record of the geophysical properties of the rock and its contained fluids that is generated by instruments lowered down a well. Lithologic units and their contained fluids are defined by their log responses (Asquith and Krygowski 2004; Jorden and Campbell 1986). Two logs widely used to identify different units are the spontaneous potential (SP) and resistivity logs (Fig. 1.44). In general, more permeable units show a larger negative SP value. The resistivity value depends on presence of a pore fluid and its salinity. Rocks with no porosity or porous rocks filled with oil generally have high resistivities and porous rocks containing saltwater have low resistivities. A variety of other log types is also valuable for lithologic interpretation, including gamma-ray, neutron density, sonic and nuclear magnetic resonance logs. The gamma-ray log responds to the natural radioactivity in the rock. Very radioactive (hot) black shales are often widespread and make good markers for correlations between wells. A caliper log measures the hole diameter in two perpendicular directions. Weak lithologies like coal or fractured rock can be recognized on a caliper log by intervals of hole enlargement. In wells drilled with mud, fluid loss into very porous lithologies or open fractures may cause mud cakes that will be recognized on a caliper log by a reduced well-bore size.

Logs from different wells are correlated to establish the positions of equivalent units (Levorsen 1967; Tearpock and Bischke 2003). Geologic contacts may be correlated from well to well to within about 30 ft in a lithologically heterogeneous sequence or to within inches or less on high-resolution logs in laterally homogeneous lithologies. The cable that lowers the logging tool into the well stretches significantly in deep wells. The recorded depth is corrected for the stretch, but the correction may not be exact. Different log runs, or a log and a core, may differ in depth to the same horizon by 20 ft at 10 000 ft. Normally, different log runs will duplicate one of the logs, for example the SP, so that the runs can be accurately correlated with each other.

The orientation of the well bore is measured by a directional survey. Some wells, especially older ones, may be unintentionally deviated from the vertical and lack a directional survey, resulting in spatial mislocation of the boundaries recorded by the well logs (if interpreted as being from a vertical well) which will lead to errors in dip and thickness determinations. The most common effect is for a well to wander down dip with increasing depth.

A dipmeter log is a microresistivity log that simultaneously measures the electrical responses of units along three or more tracks down a well (Schlumberger 1986). The

Fig. 1.44.
Typical example of electric logs used for lithologic and fluid identification. The interpreted lithologic column is in the *center*. Short normal (16 in) and long normal (64 in) refer to spacing between electrodes on the resistivity tool. (After Levorsen 1967)

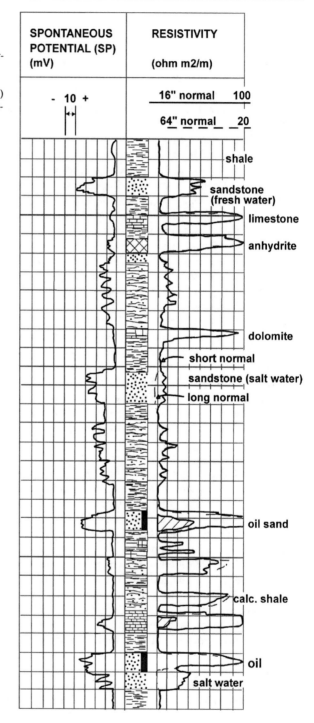

Fig. 1.45.
Representative segment of a dipmeter log. The depth scale could be in feet or meters. *Solid points* indicate the higher quality correlations, *open points* lower quality correlations

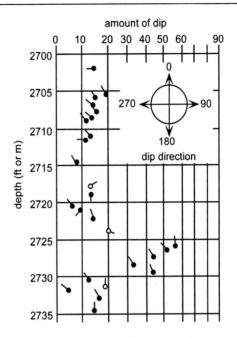

responses are correlated around the borehole, and the dip of the unit is determined by a version of the 3-point method (Sect. 2.4.2). Dips may be calculated for depth intervals as small as 8–16 cm. A typical record (Fig. 1.45) shows the dips as "tadpoles", the heads of which mark the amount of dip and the tails of which point in the direction of dip. The dips presented on the log are corrected for well deviation.

The correlations required to determine the dip for a dipmeter log are not always possible and may not always be correct. On the printed log, solid points (Fig. 1.45) indicate the highest quality correlations and open points indicate lower quality correlations. Sparse data or gaps on the dipmeter record indicate that no correlations were possible, a likely occurrence in a very homogeneous lithology (including fault gouge). Closely spaced dips that are scattered in amount and direction, such as between the depths of 2 715 and 2 725 in Fig. 1.45, suggest miscorrelations or perhaps small-scale bedding features, and are probably not reliable dips for structural purposes. A log may use a special symbol to show dips that are consistent over vertical intervals five or more times that of the minimum correlation interval. These large-interval dips are more likely to represent the structural dip. The correlations in a dipmeter log are made by scanning some distance (the scan angle) up and down the individual tracks to look for correlations. If the angle between the well bore and bedding is small (equivalent to a steep dip in a vertical well), the correlative units may lie outside the search interval. Thus dipmeter logs rarely show dips that are at angles of less than 30–40° from the well bore (50–60° dip in a vertical well) unless they were specifically programmed to look for them. If the dipmeter interpretation program is unable to make good correlations across the well it will probably show either no dips in the interval or may have made false correlations and so show low-quality dips at a high angle to the well bore.

1.7.3
Seismic Reflection Profiles

Many interpretations of subsurface structure are based on seismic reflection profiles. Sound energy generated at or near the earth's surface is reflected by various layer boundaries in the subsurface. The time at which the reflection returns to a recorder at the surface is directly related to the depth of the reflecting horizon and the velocity of sound between the surface and the reflector. Seismic data are commonly displayed as maps or cross sections in which the vertical scale is the two-way travel time (Fig. 1.46).

The geometry of a structure that is even moderately complex displayed in travel time is likely to be significantly different from the true geometry of the reflecting boundaries because of the distortions introduced by steep dips and laterally and vertically varying velocities (Fig. 1.47). Reflections from steeply dipping units may return

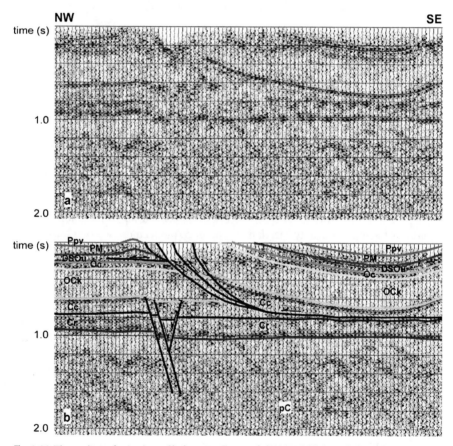

Fig. 1.46. Time-migrated seismic profile from southern Appalachian fold-thrust belt (Maher 2002), displayed with approximately no vertical exaggeration. The vertical scale is two-way travel time in seconds. **a** Uninterpreted. **b** Interpreted

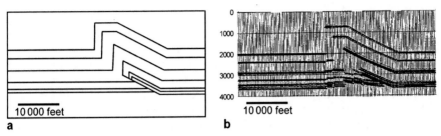

Fig. 1.47. Seismic model of a faulted fold. **a** Geometry of the model, no vertical exaggeration. **b** Model time section based on normal velocity variations with lithology and depth. Vertical scale is two-way travel time in milliseconds. (After Morse et al. 1991)

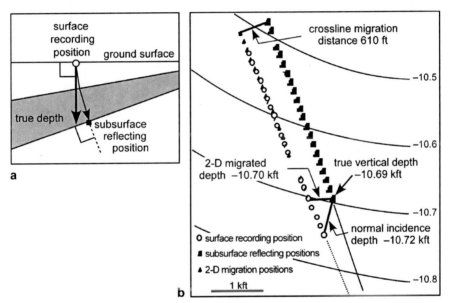

Fig. 1.48. Mislocation of seismic reflection points caused by dip of the reflector. **a** Ray path end points in vertical cross section. **b** Structure contours on a seismic reflector, depths subsea in kilofeet, showing actual and interpreted locations of the reflecting points on a seismic line. (After Oliveros 1989)

to the surface beyond the outer limit of the recording array and so are not represented on the seismic profile. The structural interpretation of seismic reflection data requires the conversion of the travel times to depth. This requires an accurate model for the velocity distribution, something not necessarily well known for a complex structure. The most accurate depth conversion is controlled by velocities measured in nearby wells (Harmon 1991).

If the trend of a seismic line is oblique to the dip of the reflector surface, two-dimensional reflection data have location problems similar to those of unknowingly deviated wells. This is in addition to the location problems associated with the conversion of travel time to depth. Reflections are interpreted to originate along ray paths that are normal to the reflector boundaries (Fig. 1.48a). A normal-incidence seismic ray is deflected up the dip (Fig. 1.48a). The true location of the reflecting point is up the dip and at a shallower depth than a point directly below the surface recording position. When the locations of reflecting points are plotted as if they were vertically below the surface recording stations, there will be a decrease in the calculated depth (or two-way travel time) relative to the true depth. Two-dimensional time migration is a standard processing procedure that corrects for the apparent dip of reflectors in the plane of the seismic line, but does not correct for the shift of the reflector positions in the true dip direction. Two-dimensional depth migration may give the correct depth to the reflecting point, but still does not correct for the out-of-plane position shift. For example (Fig. 1.48b), four degrees of oblique dip leads to a 400- to 600-ft shift in the true position of the reflection points on a seismic section at depths of 10 500–10 800 ft.

Chapter 2

Location and Attitude

2.1
Introduction

This chapter covers the basic building blocks required to construct 3-D models of geological structures. Included here are methods for locating points in three dimensions on a map or in a well, for determining the attitude of a plane and the orientation of a line, and for representing planes with structure contours. Both graphical and analytical solutions for finding and displaying lines and planes are presented. Many of the analytical techniques are based on vector geometry, a topic treated separately in Chap. 12. The graphical techniques utilize stereograms and tangent diagrams.

2.2
Location

The locations of data points recorded on maps and well logs must be converted to a single, internally consistent coordinate system in order to be used in the interpretive calculations given in this book. The positions of points in three dimensions will be described in terms of a right-handed Cartesian coordinate system with $+x$ = east, $+y$ = north, and $+z$ = up. Dimensions will be given in feet and kilofeet or meters and kilometers, depending on the units of the original source of the data. Parts of a foot will be expressed as a decimal fraction. Unit conversions are a common source of error which are largely avoided by retaining the original units of the map or well log. The relationships between locations on a topographic map or well log and the xyz coordinate system are given next.

2.2.1
Map Coordinate Systems, Scale, Accuracy

The true locations of points on or in the earth are given in the spherical coordinate system of latitude, longitude, and the position along an earth radius. Maps are converted from the spherical coordinate system to a plane Cartesian coordinate system by projection. Some distortion of lengths or angles or both is inherent in every projection technique, the amount of which depends on the type of projection and the scale of the map (Greenhood 1964; Robinson and Sale 1969; Bolstad 2002) but is not significant at the scale of normal field mapping.

In many regions, maps are based on the transverse Mercator or polar stereographic projections and contain a superimposed rectangular grid called the Universal Transverse Mercator (UTM) or the Universal Stereographic Projection (USP) grid (Robinson

Fig. 2.1.
UTM grid (NAD-27) in north-central Alabama as given on a USGS topographic map. Each square block is 1 000 m on a side. Block 1 954 is *shaded*

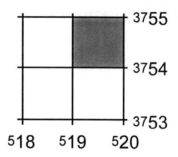

and Sale 1969; Snyder 1987). The UTM grid is used at lower latitudes and the USP grid is used in the polar regions. The UTM grid divides the earth into $6 \times 8°$ quadrilaterals that are identified by reference numbers and letters. All UTM coordinates are with respect to a survey datum that should be specified. In North America, for example, UTMs are referenced to either the North American Datum of 1927 (NAD-27) or the datum of 1983 (NAD-83) (Bolstad 2002).

UTM coordinates along the margin of a United States Geological Survey (USGS) map are given at intervals of 1 000 m (Fig. 2.1). On a topographic quadrangle map, the first digit or digits of the UTM coordinates are shown as a superscript and the last three zeros are usually omitted. Locations within the grid are given as the coordinates of the southwest corner of a block within the grid system. The x value is called the easting, and the y value is called the northing. For example, the lower left coordinate in Fig. 2.1, ^{5}18, is 518 000 m east of the origin. Any block can be subdivided into tenths in both the x and y directions, adding one significant digit to the coordinates of the sub-block. Sub-blocks may be similarly subdivided. UTM coordinates are commonly written as a single number, easting first, then the northing, with the superscript and the trailing zeros omitted. For example, a grid reference of 196 542 in the map of Fig. 2.1 represents a block 100 m on a side with its southwest corner at $x = 196, y = 542$. A grid reference of 19 605 420 represents a block 10 m on a side with its southwest corner at $x = 1 960, y = 5 420$. This coordinate system is internally consistent over large areas and is convenient for maintaining large databases. The United Kingdom uses a similar metric system called the National Grid (Maltman 1990).

The surface locations of wells are commonly recorded according to the coordinates given on a cadastral map, which is a map for officially recording property boundaries, land ownership, political subdivisions, etc. In much of the United States the cadastral system is based on the Land Office grid of Townships and Ranges (Fig. 2.2). The grid is aligned with latitude at base lines and with longitude at guide meridians. Every 24 miles the grid is readjusted to maintain the 6 mile dimensions of the blocks. An individual township is located according to its east-west coordinate (Township) and its north-south coordinate (Range). The township that has been subdivided (Fig. 2.2) is T.2N., R.3E. A township is subdivided into 36 sections, each 1 mile on a side and numbered from 1 to 36 as shown in Fig. 2.2. For a more precise location, each section is subdivided into quarter sections (may be called corners) as in the northeast quarter of sect. 7, abbreviated NE¼ sect. 7. Quarter sections may themselves be divided into quarters, as in the northwest quarter of the northeast quarter of sect. 7, abbreviated NW¼NE¼ sect. 7, T.2N., R.3E. Locations within

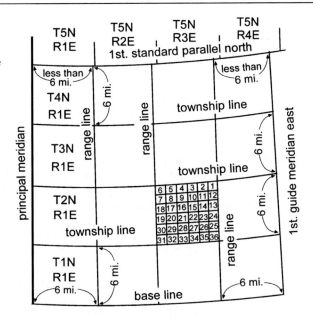

Fig. 2.2.
The Township-Range grid system. The basic unit is a 24-mile block of 6-square-mile townships. Township T2N, R3E is divided into sections. (After Greenhood 1964)

a quarter are given in feet measured from a point that is specified. The surveys were not always done perfectly and were sometimes forced into irregular shapes by the topography. It is necessary to see the local survey map to be certain of the locations.

Map scales are expressed as a ratio in which the first number is the length of one unit on the map and the second number is the number of units of the same length on the ground. The larger the scale of the map, the smaller the second number in the ratio. Geological maps suitable for detailed interpretation are typically published at scales ranging from 1 : 63 300 (1 in to the mile: 1 in = 63 300 in = 1 mile) to 1 : 24 000 (1 in = 24 000 inches, the 7.5 minute quadrangles of the USGS). Larger scales are useful for making very detailed maps. Base maps should always contain bar scales which make it easy to enlarge or reduce the map while preserving the correct scale.

Most governmentally produced base maps have an accuracy equivalent to the U.S. Class 1 map standard (Fowler 1997). This standard states that the horizontal position of 90% of the points must be within 0.5 mm of their true location at the scale of the map. For example, at the 1 : 24 000 scale, a point on the map must be within 0.5 mm × 24 000 = 12 m (39.4 ft) of its true location. The vertical position of 90% of all contours must be within half the contour interval in open areas and spot elevations must be within one-quarter of a contour of their true elevation. A 20-ft contour (characteristic of 1 : 24 000 USGS topographic maps) can be expected to be accurate to ±10 ft (3 m). For comparison, the width of the thinnest contour line is about 0.01 in on a 1 : 24 000 USGS topographic map and represents 20 ft (6.01 m) at the scale of the map. The accuracy of an enlarged (or reduced) map is no better than it was at the original scale. Ground surveys of point locations such as wells can be expected to be accurate to within 1 m and satellite surveys can be accurate to 0.1 m within the survey area (Aitken 1994), significantly greater accuracies than that of many base maps.

2.2.2
Geologic Mapping in 3-D

Widely available computer technology now makes it possible to construct and interpret geological maps in 3-D. Three-dimensional geologic outcrop mapping usually begins with a digital elevation model (DEM). A DEM is a grid of elevation values and their corresponding xy coordinates (Bolstad 2002). The DEM is contoured (typically by triangulation because it is computationally fast and fits the control points, see Chap. 3) to produce a computer visualization of the topographic surface (Fig. 2.3). The accuracy of the most widely available DEMs is not as great as the corresponding topographic map, which may be important in critical applications. For DEMs produced at the scale of a 7.5-minute quadrangle and derived from a photogrammetric source, 90 percent have a vertical accuracy of 7-meter root mean square error or better and 10 percent are in the 8- to 15-meter range (source USGS EROS Data Center).

Geological contact lines can be constructed as three-dimensional traces on the surface of the contoured DEM. This can be achieved by extracting lines from the 3-D topographic surface within 3-D software (Fig. 2.4), or by mapping 3-D lines in outcrop using a Global Positioning System (GPS) receiver, which provides a digital file that can be used with the computer map (Bolstad 2002). Extracting lines from the contoured DEM usually is done over a superimposed image of the hard-copy geological map. Geographic information system (GIS) software allows the image of a geological map to be draped over the DEM surface, giving a realistic view of the map surface. Images used for this purpose need to be ortho rectified, aligned and georeferenced to fit the DEM. Once the

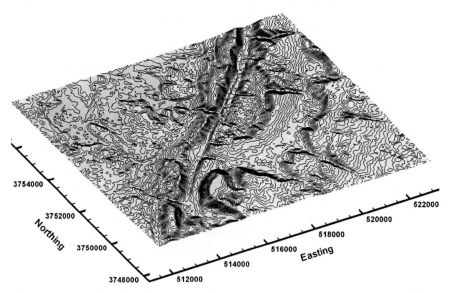

Fig. 2.3. Oblique view to the NE of the Blount Springs area, Alabama, 50 ft topographic contours derived from a 30 m Digital Elevation Model (NAD-27). Horizontal coordinates are UTMs in meters, no vertical exaggeration

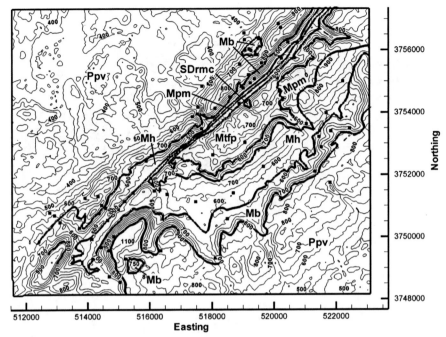

Fig. 2.4. Geologic map of a portion of the southern Sequatchie anticline in the vicinity of Blount Springs, Alabama. Base map is a DEM. The units are, from oldest to youngest, Silurian Red Mountain Formation and Devonian Chattanooga Shale (*SDrmc*), Mississippian Tuscumbia Limestone and Fort Payne Formation (*Mtfp*), Mississippian Pride Mountain Formation (*Mpm*), Mississippian Hartselle Sandstone (*Mh*), Mississippian Bangor Limestone (*Mb*), Pennsylvanian Pottsville Formation (*Ppv*). *Thick lines* are geologic contacts, *thin lines* are topographic contours (50 ft interval). Horizontal scale is UTM in meters. *Solid squares* are locations where bedding attitudes have been measured. (After Cherry 1990; Thomas 1986)

contacts are digitized, they can be visualized as part of the 3-D model (Fig. 2.5), edited and otherwise manipulated digitally.

The geologic map of the Blount Springs area (Figs. 2.4, 2.5) will provide the data for an ongoing example of the process of creating and validating a structure contour map. The map area is located along the Sequatchie anticline at the southern end of the Appalachian fold-thrust belt and is the frontal anticline of the fold-thrust belt.

2.2.3
Wells

The location of points in a well are measured in well logs with respect to the elevation of the wellhead and are usually given as positive numbers. Depths in oil and gas wells are usually measured from the Kelly bushing (Fig. 2.6a). The elevation of the Kelly bushing (KB) is given in a surveyor's report included as part of the well-log header information. Alternatively, depths may be measured from ground level (GL) or the derrick floor (DF).

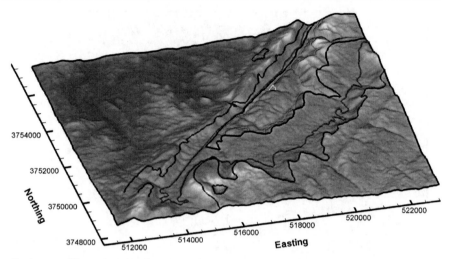

Fig. 2.5. 3-D oblique view of geologic map of Blount Springs area, from Fig. 2.4. Topography is shown as shaded relief map, without vertical exaggeration

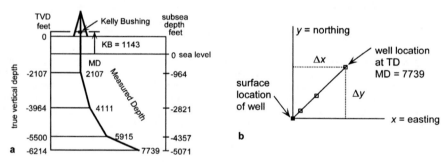

Fig. 2.6. Location in a deviated well. **a** Vertical section through a deviated well, in the northeast-southwest direction. Depths in the well are measured downward from the Kelly bushing (*KB*). True vertical depths (*TVD*) are calculated from the borehole deviation survey. **b** Map view of the deviated well. Locations of points down hole are given by their distance from the surface location. *TD* total depth; *MD* measured depth

2.2.3.1
Datum

The coordinates of points in a well need to be corrected to a common datum elevation, normally sea level. The depths should be adjusted so that they are positive above sea level and negative below. In a vertical well the log depths are converted to a sea-level datum with the following equation:

$$SD = KB - MD \quad , \tag{2.1}$$

where SD = subsea depth, KB = elevation of Kelly bushing or other measurement of surface elevation, MD = measured depth on well log.

2.2.3.2
Deviated Well

Many wells are purposely drilled to deviate from the vertical. This means that points in the well are not directly below the surface location. The shape of the well is determined by a deviation survey, the terminology of which is given in Fig. 2.6. The primary information from a deviation survey is the azimuth and inclination of the borehole and the downhole position of the measurement, for a number of points down the well. This information is converted by some form of smoothing calculation into the *xyz* coordinates of selected points in well-log coordinates, known as true vertical depth (TVD, Fig. 2.6a) and is given in the log of the survey. The TVD must be corrected for the elevation of the Kelly bushing to give the locations of points with respect to the datum. In a deviated well the log depths are converted to a sea-level datum with the following equation:

$$SD = KB - TVD \ , \tag{2.2}$$

where SD = subsea depth, KB = elevation of Kelly bushing or other measurement of surface elevation, TVD = true vertical depth from the deviation survey.

The calculated position of the points determined from a deviation survey depends on the spacing between the measurement points and the particular smoothing calculation used to give the TVD locations. If the measurement points are spaced tens of meters or a hundred or more feet apart, the positions of points toward the bottom of a 3 000-m (10 000-ft) well might be uncertain by tens of meters or a hundred feet or so. This is because the absolute location of a point depends on the accuracy of the location of all the points above it in the well. Small errors accumulate. The relative positions of points spaced a small distance apart along the well should be fairly accurate. Points will be accurately located if the deviation survey is based on points spaced only a meter or less apart.

Locations in a deviated well are commonly given as the *xyz* coordinates of points in the well relative to the surface location, for example, P_1 and P_2 in Fig. 2.7a. Here *z* is the vertical axis, positive upward, the subscript "1" will always denote the upper point and the subscript "2" will denote the lower point. A unit boundary is likely to be located between control points, making it necessary to calculate its location.

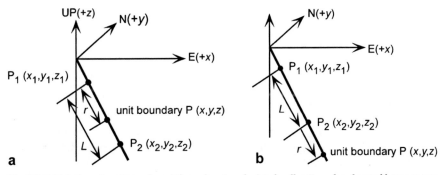

Fig. 2.7. Calculation of position of a unit boundary in a deviated well. **a** Boundary located between two control points. **b** Boundary located below the control points

To find the coordinates of an intermediate point, P, between P_1 and P_2, let r be the distance along the well from the upper control point to the point P. In a right-handed xyz coordinate system (Fig. 2.7) with z positive upward, positive x = east and positive y = north, the point P (x, y, z) between P_1 (x_1, y_1, z_1) and P_2 (x_2, y_2, z_2) at a location given by the ratio r/s (where $r + s = L$), has the coordinates (Eves 1984)

$$x = (rx_2 + sx_1)/(r + s) \ , \tag{2.3a}$$

$$y = (ry_2 + sy_1)/(r + s) \ , \tag{2.3b}$$

$$z = (rz_2 + sz_1)/(r + s) \ . \tag{2.3c}$$

L is found from the Pythagorean theorem:

$$L = [(x_2 - x_1)^2 + (y_2 - y_1)^2 + (z_2 - z_1)^2]^{1/2} \ . \tag{2.4}$$

Substitute $s = L - r$ into Eq. 2.3 to obtain

$$x = (rx_2 - rx_1 + Lx_1)/L \ , \tag{2.5a}$$

$$y = (ry_2 - ry_1 + Ly_1)/L \ , \tag{2.5b}$$

$$z = (rz_2 - rz_1 + Lz_1)/L \ . \tag{2.5c}$$

If the well is straight between the upper and lower points, L will be equal to the log distance. If L is not equal to the log distance, then the well is not straight and the calculated value of L will give the more internally consistent answer. More precise location of the point will require definition of the curvature of the well between the control points.

It may be required to locate a contact below the last control point given by the deviation survey (Fig. 2.7b). A simple linear extrapolation can be used, based on the assumption that the well continues in a straight line below the last control point with the same orientation it had between the last two control points. The subscripts "1" and "2" again represent the upper and lower control points. Let $\Delta x = (x_2 - x_1)$, $\Delta y = (y_2 - y_1)$, and $\Delta z = (z_2 - z_1)$, then

$$x = x_2 + r\Delta x/L \ , \tag{2.6a}$$

$$y = y_2 + r\Delta y/L \ , \tag{2.6b}$$

$$z = z_2 + r\Delta z/L \ , \tag{2.6c}$$

where x, y, and z = the coordinates of a point at a well-log distance r below the last control point, P_2, having the coordinates (x_2, y_2, z_2), and L (Eq. 2.4) = the straight-line distance between the last two control points.

As an example of the location of points in a deviated well, consider the following information from a well (KB at 660 ft) on which the formation boundary of interest is at a log depth of 1 225 ft. The deviation survey gives the coordinates of points above and below the boundary as 1 200 ft TVD, 50 ft northing, –1 050 ft easting and 1 400 ft TVD, 150 ft northing, –1 150 ft easting. What are the coordinates of the formation boundary? The subsea depths of the upper and lower points, found by subtracting the log depths from the elevation of the Kelly bushing are P_1: z = –540 ft and P_2: z = –740 ft, respectively. From Eq. 2.4, the straight-line distance between the two points is L = 225 ft. From Eq. 2.3, the coordinates of a point r = 25 ft down the well from the upper point are –560 ft subsea, 60 ft northing, –1 060 ft easting. Note that negative northing is to the south and negative easting is to the west.

2.3
Orientations of Lines and Planes

The basic structural measurements at a point are the orientations of lines and planes. The attitude of a plane is its orientation in three dimensions. The attitude may be given as the strike and dip (Fig. 2.8). Strike is the orientation of a horizontal line on the plane and the dip is the angle between the horizontal and the plane, measured perpendicular to the strike in the downward direction. Compass directions will be given here as the *trend*, which is the azimuth on a 360° compass (Ragan 1985), and will be indicated by numbers always containing three digits. Strike and dip are written in text form as strike, dip, dip direction; for example 340, 22NE and represented by the map symbols of Fig. 2.9a–d. The degree symbol may be written after each angle (i.e., 340°, 22° NE) or may be omitted for the sake of simplicity, as will be done here (Rowland and Dueben-dorfer 1994). The alternative form on a quadrant compass is the *bearing* and the dip (Ragan 1985), for which the same attitude would be written as N20W, 22NE. In subsurface geology, where the frame of reference may be a vertical shaft or well bore, the inclination of a plane may be given as the hade, which is the angle from a downward-

Fig. 2.8.
Attitude of the *shaded* plane can be given by its strike and dip, dip vector, or pole

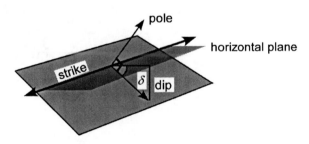

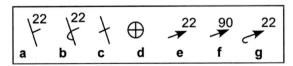

Fig. 2.9. Map symbols for the attitude of a plane. **a** Strike and dip. **b** Strike and dip of overturned bed. **c** Strike of vertical bed. **d** Horizontal bed. **e** Azimuth and plunge of dip. **f** Facing (stratigraphic up) direction of vertical bed. **g** Azimuth and plunge of dip, overturned bed

pointing vertical line and the plane. The hade angle is the complement of the dip ($90° - δ$). The attitude of a plane may also be given as the azimuth and plunge of the dip vector, written as dip amount, dip direction. By this convention, the previous attitude is given as 22, 070, indicated by the map symbol of Fig. 2.9e. This is the form that is usually used in this book because it is short and convenient for numerical calculations. The dip vector will occasionally be written in short form as **δ**, where the bold type indicates a vector. The attitude of a plane may also be represented by the orientation of its pole, a line perpendicular to the plane (Fig. 2.8).

The orientation of a line is given by its trend and plunge. The trend is the angle, $β$, between north and the vertical projection of the line onto a horizontal plane (Fig. 2.10). The plunge is the angle, p, in the vertical plane between the line and the horizontal. The orientation of a line is written as plunge amount, azimuth of trend, for example 30, 060 is a line plunging 30° toward the azimuth 060°. On a map the orientation of a line is represented by the same type of symbol as the azimuth and plunge of the dip (Fig. 2.9b) which is also a line. The two symbols could be differentiated on a map by means of line weight.

Given the *xyz* coordinates of two points, it is possible to calculate the trend and plunge of the line between them (Fig. 2.11). If subscript 1 represents the higher of the two points, the plunge of the line determined from the following equations will be downward. The preliminary value of the azimuth $θ'$ and plunge $δ$ of the line between points 1 and 2 are (from Eqs. 12.11 and 12.12)

$$θ' = \arctan(Δx / Δy) \ , \tag{2.7}$$

$$δ = \arcsin[(z_2 - z_1)/L] \ , \tag{2.8}$$

where

Fig. 2.10.
Trend and plunge of a line in the *shaded* surface

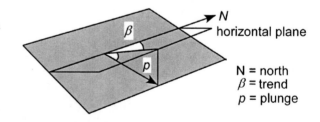

N = north
$β$ = trend
p = plunge

Fig. 2.11.
Trend and plunge of a line between two points

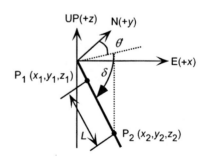

$$\Delta x = (x_2 - x_1) \ , \tag{2.9a}$$

$$\Delta y = (y_2 - y_1) \ , \tag{2.9b}$$

and L is given by Eq. 2.4. Division by zero is not allowed in Eq. 2.7. The preliminary azimuth, θ', calculated from Eq. 2.7, is always in the range of 000 to 090°. The angle must be converted to the true azimuth, θ, in the range of 000 to 360° using Table 2.1.

As an example of the calculation of the orientation of a line, find the trend and plunge of the well between points P_1 and P_2 from the example in Sect. 2.2.3.2. The locations of the two points are found from the deviation survey to be P_1: $z = -540$ ft, 50 ft northing, $-1\,050$ ft easting and P_2: $z = -740$ ft, 150 ft northing, $-1\,150$ ft easting. The preliminary plunge and trend of the line, from Eqs. 2.7 and 2.8, is 55, -045. The sign of Δx is negative ($\Delta x = (-1\,150 - (-1\,050)) = -100$) and Δy is positive ($\Delta y = 150 - 50 = +100$), and so from Table 2.1, the true azimuth is $\theta = 360 + (-45) = 315$.

The orientation of a line can also be represented by its pitch or rake, r, which is the angle measured in a specific plane between the line and the strike of the plane (Fig. 2.12). Both the rake and the attitude of the plane must be specified. This form is convenient for recording lines that are attached to important planes, like striations on a fault plane. Alternatively the orientation of a line in a plane can be represented by its apparent dip and bearing (Fig. 2.12). Apparent dip, δ', is the orientation of a line that lies in a plane in some direction other than the true dip (Fig. 2.12). An apparent dip can be written as a plunge and trend or as a rake in the plane. Apparent dips are important in drawing a cross section that is in a direction oblique to the true dip direction of bedding.

It is possible to represent the three-dimensional orientations of lines and planes quantitatively on graphs called stereograms and tangent diagrams. The graphical techniques aid in the visualization of geometric relationships and allow for the rapid solution of many three-dimensional problems involving lines and planes.

Table 2.1.
True azimuth, θ, determined from preliminary azimuth, θ' based on the signs of Δx and Δy (Eq. 2.9) or $\cos \alpha$ and $\cos \beta$ (Eq. 2.14)

Azimuth (deg)	$\cos \alpha; \Delta x$	$\cos \beta; \Delta y$	θ
000+ to 090	+	+	θ'
090+ to 180	+	−	$180 + \theta'$
180+ to 270	−	−	$180 + \theta'$
270+ to 360	−	+	$360 + \theta'$

Fig. 2.12.
True dip, δ, apparent dip, δ', and rake (or pitch), r

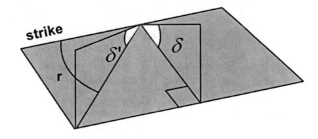

2.3.1
Stereogram

A stereogram (Dennis 1967; Lisle and Leyshon 2004) is the projection of the latitude and longitude lines of a hemisphere onto a circular graph. Two different projections are commonly used in structural interpretation, the equal-area net (Schmidt or Lambert net) and the equal-angle stereographic net (Wulff net). The different projection techniques result in the preservation of different properties of the original sphere (Greenhood 1964). The equal-area net is required if points are to be contoured into spatially meaningful concentrations. An equal-angle net produces false concentrations from an evenly spaced distribution of points, although the angular relationships are correct. The lower-hemisphere equal-area stereograms (Fig. 2.13a) will be used throughout this book (an enlarged copy for use in working problems is given as Fig. 2.28 at the end of the chapter).

The outer or primitive circle of the equal-area stereogram is the projection of a horizontal plane (Fig. 2.13a). The compass directions are marked around the edge of the primitive circle. The projections of longitude lines form circular arcs that join the north and south poles of the graph and represent great circles. The trace of a plane is a great circle. A line plots as a point. The projections of latitude lines are elliptical curves, concentric about the north and south poles and represent small circles. The center of the graph is a vertical line. Planes and lines plotted on a stereogram can be visualized as if they were intersecting the surface of a hemispherical bowl (Fig. 2.13b).

A stereogram is used with a transparent overlay on which the data are plotted and which can be rotated about the center of the graph. To begin using a stereogram, on the overlay, mark the center and the north, east, south, and west directions, and the primitive circle (Fig. 2.14a). To plot the attitude of a plane from the strike and dip (for example, 60, 32SE), mark the strike on the primitive circle. Then rotate the overlay so that the strike direction lies on the N-S axis (Fig. 2.14b), find the great circle corresponding to the dip by counting down (inward) along the E-W axis from the primitive circle (zero dip) the dip amount in the dip direction. Draw a line along the great circle and

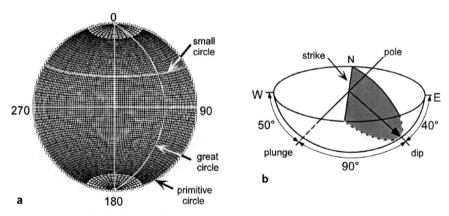

Fig. 2.13. Stereogram. **a** Equal-area stereogram (Schmidt or Lambert net), lower hemisphere projection. **b** Visualization of a plane and its pole in a lower hemisphere. (After Rowland and Duebendorfer 1994)

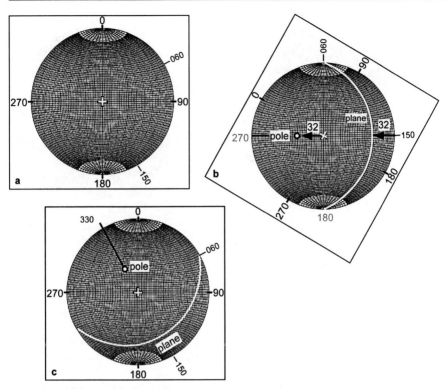

Fig. 2.14. Plotting the orientation of the plane and its pole on an equal-area lower-hemisphere stereogram. The plane has a dip of 32SE (150), and a strike of 060. **a** NESW compass directions and strike and dip directions are marked on the overlay (*square*). **b** Overlay rotated to bring the strike to north and the dip to east. Trace of plane and pole to plane marked on overlay. **c** Overlay returned to its original position. The plunge and trend of the pole is 58, 330

then return the overlay to its original position to see the plane in its correct orientation (Fig. 2.14c). Find the pole to the plane by placing the overlay in the starting position for drawing bedding, that is, with the strike direction N-S. Along the E-W axis, count in from the primitive circle an amount equal to the dip, then count another 90° to find the pole (Fig. 2.14b) or, equivalently, count the dip amount up (outward) from the center point of the diagram. Mark the position of the E-W line on the primitive circle and return the overlay to its original position to find the trend of the pole (330°). The plunge of the pole is 90° minus the dip. To plot planes and poles from the dip and dip azimuth of the plane (the plane previously plotted is $\delta = 32, 150$) mark the dip direction on the primitive circle (Fig. 2.14a), rotate the overlay to bring this direction to E-W, and count the dip amount inward to find the point that represents the orientation of the dip vector. The great circle projection of the bed goes through this point. Plot 90° plus the dip along E-W to find the pole (Fig. 2.14b). Return the overlay to its original position to see the result (Fig. 2.14c). If the plunge and trend to be plotted are those of a line, follow the same steps as in plotting a dip vector.

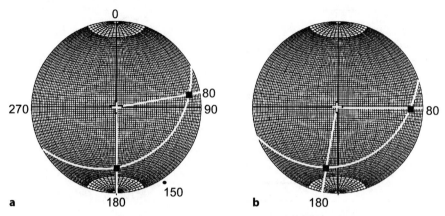

Fig. 2.15. Apparent dip in the plane given by the dip vector 32, 150 on an equal-area, lower-hemisphere stereogram. **a** Plane plotted with apparent dips shown as *black squares*. The south-trending apparent dip is 28, 180. **b** Overlay rotated 10° to bring the 080° trending apparent dip into measurement position; apparent dip is 15, 080

Plotting poles and lines can be done more quickly on a stereogram using a Biemsderfer plotter (Wise 2005), a calibrated dip scale that rotates on the center of the net to allow plotting of points without rotating the overlay. The attitudes of lines such as dip vectors and lineations are also quickly and easily plotted on a tangent diagram, described in Sect. 2.3.3.

Apparent dips are quickly determined on a stereogram as the orientation of the point of intersection between a line in the direction of the apparent dip and the great-circle trace of a plane. For example, find the apparent dips along the azimuths 080 and 180 for the plane plotted previously ($\delta = 32, 150$). Plot lines from the center of the graph in the azimuth directions (Fig. 2.15a). The intersections of the azimuth lines with the great circle are the apparent dips. Dips can be read from the N-S axis as well as the E-W and so the 180° azimuth is in measurement direction. The angle measured inward from the primitive circle to the intersection is the apparent dip, 28°. The overlay is rotated into measurement position for the 080° azimuth (Fig. 2.15b) to find the apparent dip of 15°.

Apparent dip problems can be worked backwards to find the true dip from two apparent dips. Plot the points representing the apparent dips, then rotate the overlay until both points fall on the same great circle. This great circle is the true dip plane.

2.3.2
Natural Variation of Dip and Measurement Error

The effect of measurement error or of the natural irregularity of the measured surface on the determination of the attitude of a plane is readily visualized on a stereogram. The plane is represented by its pole (Fig. 2.16). Irregularities of the measured surface and measurement errors should produce a circular distribution of error around this pole (Cruden and Charlesworth 1976). An error of 4° around the true dip is probably

Fig. 2.16.
Equal-area, lower-hemisphere
stereogram. Measurement
error of 4° (radius of circles)
around true bedding poles
(points *a* and *b*)

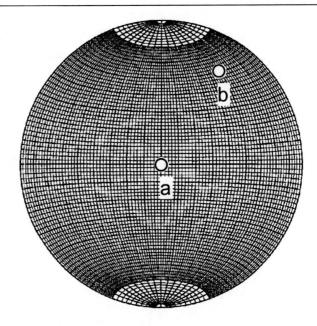

the maximum expected for routine field measurements on a normally smooth bed surface. Measurements on a rough surface may show an even greater variability. A good average attitude from a rough surface can be obtained by making several measurements and then separately averaging the strikes and dips or the trends and plunges of the dips. A good field measurement procedure is to lie a flat field notebook or square of rigid plastic on a rough bed surface to average out the irregularities.

The effect of the error is related to the attitude of the plane. If the bed is horizontal, the pole is vertical (Fig. 2.16, point a) and the error means that the azimuth of the dip could be in any direction, even though the true three-dimensional orientation of the plane is rather well constrained. Small irregularities on a bed surface have the same effect (Woodcock 1976; Ragan 1985). On a steeply dipping plane (Fig. 2.16, point b), the same amount of error causes little variation in either the azimuth or the dip. Conversely, the measurement of the trend of a line on a gently dipping surface is accurate to within a few degrees, but the direction measured on a steeply dipping plane may show significantly greater error (Woodcock 1976). A precision of about 2° is about normal for calculations done using a stereogram. For greater precision, the calculations should be done analytically by methods that will be presented later in the chapter.

2.3.3
Tangent Diagram

The other useful diagram for representing the attitudes of planes is the tangent diagram (Fig. 2.17; an enlarged copy is given at the end of the chapter as Fig. 2.29). Developed by Hubbert (1931), it has been popularized by Bengtson (1980, 1981a,b) in the context of dipmeter interpretation. It is particularly valuable in the determina-

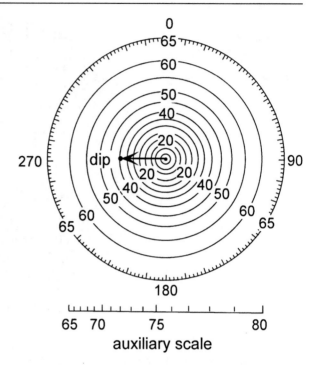

Fig. 2.17.
Tangent diagram. *Arrow* represents a plane dipping 40°
to the west (40, 270). (After
Bengtson 1980)

tion of the crest lines of cylindrical and conical folds (Chap. 5). The concentric circles
on the diagram represent the dip magnitude and their spacing is proportional to the
tangent of the dip, hence the name tangent diagram. The center of the diagram is
zero dip. The azimuth is marked around the margin of the outer circle. The attitude
of a plane is represented by a vector from the origin in the direction of the azimuth
and having a length equal to the dip amount (Fig. 2.17). The attitude can be shown
with the complete vector or as a point plotted at the location of the tip of the vector.
See Sect. 2.8 for how to plot dip vector points on a tangent diagram using a spreadsheet. A major convenience of the tangent diagram is that no overlay is required and
that certain problems are solved very quickly and without the rotations that are required with the stereogram. The drawback of the tangent diagram is that very steep
dips require a very large diagram. The diagram in Fig. 2.17 extends to a dip of 65°. A
calibrated scale that can be used to plot dips from 65° to 80° is given at the bottom of
this figure. To use it, plot the vector along the appropriate radius and use the auxil-
iary scale to find the added length of the vector beyond the outer circle of the diagram. The diagram is not practical if a significant percentage of the dips are over 70°,
for which a stereogram is more suitable. The tangent diagram can be used as a circular histogram, even for steep dips, by plotting the steep dips with their correct azimuths along the outer circle.

A tangent diagram is a convenient tool for finding the true or apparent dip. The
apparent dip in a given direction (Fig. 2.18a) is the vector in the appropriate direction.

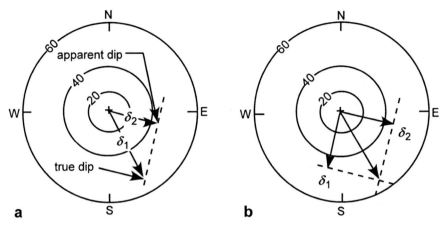

Fig. 2.18. Apparent dip on a tangent diagram. **a** Apparent dip from true dip. **b** True dip from two apparent dips. (After Bengtson 1980)

For example, given a dip vector of a bed δ_1, find the apparent dip in the direction of δ_2. Project the tip of δ_2 onto the direction of δ_1. The projection is along a line perpendicular to the apparent dip. The length of the projected bedding vector in the direction of the apparent dip is the amount of the apparent dip. The true dip can be found from two apparent dips. The perpendiculars from the apparent dips (δ_1 and δ_2) intersect at the tip of the true dip vector (Fig. 2.18b).

2.4
Finding the Orientations of Planes

The attitude of a plane measured by hand with a compass or given by a dipmeter log is effectively the value at a single point. Measured over such a small area, the attitude is very sensitive to small measurement errors, surface irregularities, and the presence of small-scale structures. The following two sections describe how to find the attitude from three points that can be widely separated and from structure contours. Both methods provide an average attitude representative of the map-scale structure.

The farther the points depart from a straight line, the more reliable the expected result because small irregularities or location errors will have less influence on the result. If more than three points are available to find the attitude of the plane, a best-fit can be determined using planar regression or moment of inertia analysis (Fernández 2005). A typical situation for this application would be a series of points along the outcrop trace of a plane (e.g., Fig. 2.4). On the other hand, the farther the points are from one another, the greater the possibility that they no longer fall on a single plane. A high-quality calculated dip will be compatible with the surrounding data as demonstrated with a structure contour map (Chap. 3) and cross section (Chap. 6). If the unit thickness is known (Chap. 4), a very powerful test of the quality of a calculated dip is to show that both the top and base of the unit fit their respective outcrop traces.

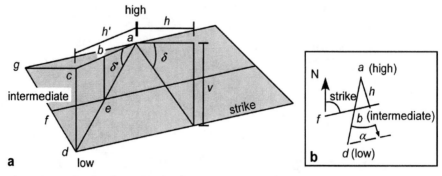

Fig. 2.19. True dip, δ, and apparent dip, δ'. **a** Perspective view. **b** Map view. *N:* north. For explanation of *a–h* and *v*, see text

2.4.1
Graphical Three-Point Problem

The attitude of a plane can be uniquely determined from three points that are not on a straight line. Let the highest elevation be point a and the lowest be point d (Fig. 2.19). The intermediate elevation, *f*, must also occur along the line joining a and d as point e. The line fe is the strike line. The horizontal (map) distance from a to e by linear interploation is ab, where

$$ab = (ac \times be) / cd \ . \tag{2.10}$$

Plot the length ab on the map (Fig. 2.19b) and join point f and b to obtain the strike line. The dip vector lies along the perpendicular to the strike, directed from the high point to the intermediate elevation along the strike line. The dip amount is

$$\delta = \arctan(v/h) \ , \tag{2.11}$$

where *v* = the elevation difference between the highest and the lowest points and *h* = the horizontal (map) distance between the highest point and a strike-parallel line through d. The azimuth of the dip is measured directly from the map direction of the dip.

A typical example of a 3-point problem is seen on the map of Fig. 2.20a. The map shows the elevations of three locations identified in the field as being on the same contact (a, f, d). These points could just as easily be the elevations of a formation boundary identified in three wells. To find the attitude of the contact, draw a line between the highest and lowest points (a–d, Fig. 2.20b) and measure its length. Use Eq. 2.10 to find the distance along the line from the high point to the level of the intermediate elevation (e). Connect the two intermediate elevations to find the strike line (Fig. 2.20b,c). Draw a perpendicular from the strike line (e–f) to the lowest point (d, Fig. 2.20d). The horizontal length of the line is *h* and the elevation change is *v*. Determine the dip from Eq. 2.11. The azimuth of the dip is measured from the map. The dip vector in this example is 22, 125.

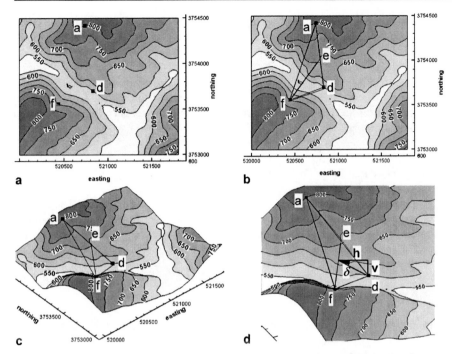

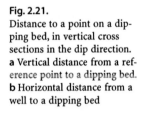

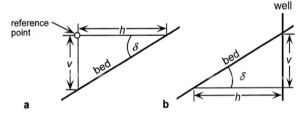

Fig. 2.20. Attitude determination from three points on a topographic map. Horizontal scale in km, vertical scale in ft. **a** Three points (*solid squares*) on a contact. **b** Determination of the strike line (e–f). **c** The same three points in oblique 3-D view to NE. **d** Enlarged 3-D view of three-point solution

Fig. 2.21.
Distance to a point on a dipping bed, in vertical cross sections in the dip direction.
a Vertical distance from a reference point to a dipping bed.
b Horizontal distance from a well to a dipping bed

A dip can be converted from degrees into feet/mile or meter/kilometer by solving Eq. 2.11 for v and letting h be the reference length (5 280 ft for ft/mile or 1 000 m for m/km):

$$v = h \tan \delta \, , \tag{2.12}$$

where v = the vertical elevation change, h = reference length, and δ = dip. The same relationship can be used to determine the vertical distance from a reference point to a dipping horizon seen in a nearby outcrop (Fig. 2.21a).

Another useful application of Eq. 2.11 is to find the distance to the intersection between a horizontal plane (such as an oil-water contact) and a dipping plane (such as the top of the unit containing the contact) which are separated by a known distance in the well. Solve Eq. 2.11 for h (Fig. 2.21b):

$$h = v / \tan \delta \ , \tag{2.13}$$

where h = distance from the well bore to the intersection with a dipping bed and v = vertical distance in the well between the intersection of the dipping horizon and the horizontal horizon.

2.4.2
Analytical Three-Point Problem

The attitude of a plane is given by the trend and plunge of the dip vector (Fig. 2.22). The dip vector can be determined analytically from the xyz coordinates of three non-colinear points (derived in Sect. 12.3). The preliminary trend and plunge of the dip vector is

$$\theta' = \arctan(A / B) \ , \tag{2.14}$$

$$\delta = \arcsin \{-\cos[90 + \arccos(C / E)]\} \ , \tag{2.15}$$

where θ' = the preliminary azimuth of the dip, δ = the amount of the dip, and

$$A = y_1 z_2 + z_1 y_3 + y_2 z_3 - z_2 y_3 - z_3 y_1 - z_1 y_2 \ , \tag{2.16a}$$

$$B = z_2 x_3 + z_3 x_1 + z_1 x_2 - x_1 z_2 - z_1 x_3 - x_2 z_3 \ , \tag{2.16b}$$

$$C = x_1 y_2 + y_1 x_3 + x_2 y_3 - y_2 x_3 - y_3 x_1 - y_1 x_2 \ , \tag{2.16c}$$

$$D = z_1 y_2 x_3 + z_2 y_3 x_1 + z_3 y_1 x_2 - x_1 y_2 z_3 - y_1 z_2 x_3 - z_1 x_2 y_3 \ , \tag{2.16d}$$

$$E = (A^2 + B^2 + C^2)^{1/2} \ . \tag{2.16e}$$

Fig. 2.22.
Three points on a plane and the dip vector of the plane

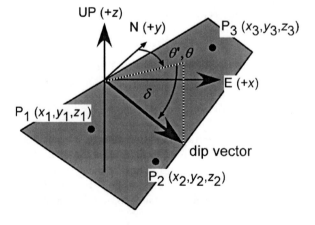

The sign of $E = -\text{sign}\,D$ if $D \neq 0$; $= \text{sign}\,C$ if $D = 0$ and $C \neq 0$; $= \text{sign}\,B$ if $C = D = 0$. Division by zero is not allowed in Eq. 2.14. The value of θ' computed from Eq. 2.7 is always between the values of $000°$ and $090°$ and is equal to $90°$ if $B = 0$. The true azimuth, θ, of the dip in the complete range from $000°$ to $360°$ can be determined from θ' and the signs of $\cos\alpha$ and $\cos\beta$ (Eqs. 2.17 and Table 2.1):

$$\cos\alpha = A/E \; , \tag{2.17a}$$

$$\cos\beta = B/E \; , \tag{2.17b}$$

where A, B, and E are given by Eqs. 2.16a,b,e above.

As an example, find the analytical solution to the 3-point problem in Fig. 2.20. The three points have the coordinates, in xyz order, of 520 739, 3 754 420, 800; 520 438, 3 753 560, 700; 520 833, 3 753 700, 600. From Eqs. 2.14–2.17, the dip is 22° at an azimuth of 125°. The leading UTM digits of the x and y coordinates are identical and need not be included in the calculation.

2.5
Apparent Dip

Apparent dip, δ', is the angle in a plane between the horizontal and some direction other than the true dip (Fig. 2.19a). To find the apparent dip, let the horizontal angle between the true and apparent dip be α, then

$$\delta' = \arctan\left(\tan\delta\,\cos\alpha\right) \; . \tag{2.18}$$

On a completed structure-contour map, the apparent dip in a given direction is found from

$$\delta' = \arctan\left(I/h'\right) \; , \tag{2.19}$$

where $I =$ the contour interval and $h' =$ the horizontal distance on the map between the contours in the direction of interest (Fig. 2.23). If the direction perpendicular to the contours is selected, then the apparent dip is the true dip. If the strike direction is selected, the apparent dip is zero.

Fig. 2.23.
Relationships between structure contours, point elevations, and dip. Points a and b are at the same elevation; c and d are at different elevations. C_1 and C_2 are structure contours. For explanation of other symbols, see text

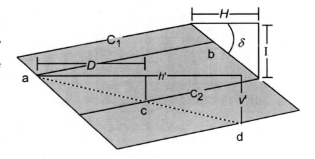

2.6
Structure Contours

Structure contours provide an efficient and effective representation of the attitude of a surface. The contours are strike lines and the dip direction is perpendicular to the contours (Fig. 2.23). The simplest way to construct structure contours from a geologic map is to connect points on a contact that lie at the same elevation. If two points on a contact have the same elevation, for example a and b in Fig. 2.23, then a line joining them is a strike line or structure contour. Finished maps normally show contours at even increments of elevation.

Structure contours are the primary method for illustrating the shape of structures in the subsurface. Generated from outcrop maps, structure contours provide an effective method for validating the outcrop pattern and for smoothing out local dip variations caused by bed roughness. If a map horizon is planar, the contours on it are parallel and uniformly spaced. Folding produces curved contours that usually are not far from parallel at a local scale. Abrupt changes in direction indicate fold hinges, faults, or mistakes. Unless radical thickness changes occur, the structure contours on adjacent horizons are approximately parallel.

The following sections describe the creation of straight-line structure contours from the elevations of three points and the interpretation of contours in the vicinity of a dip measurement. The generation of curved structure contours from multiple control points is discussed in Chap. 3.

2.6.1
Structure Contours from Point Elevations

Structure contours are usually generated from point elevations, especially in subsurface work. Except for the unlikely circumstance where all the points happen to fall on the selected contour elevations, contouring will require placing contours between control points that have different elevations. This means that contour values must

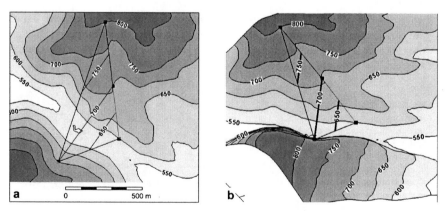

Fig. 2.24. Structure contours on a dipping plane (from Fig. 2.20). The 650-ft and 750-ft contours are interpolated between the four control points. **a** Plan view. **b** 3-D oblique view to NE

be interpolated between the control points. Using linear interpolation (more complex forms of interpolation are considered in Chap. 3), an interpolated even-elevation contour lies at point c (Fig. 2.23) on the straight line between the two points at known elevations *a* and *d*. Following the method of Eq. 2.10, the elevation of the upper control point is *a*, the horizontal distance of the desired contour from the highest control point is *D*, the elevation of the desired contour is C_2, the map distance between the two control points is *h'*, and the vertical distance between the two control points is *v'*, giving

$$D = h' \, (a - C_2) / v' \; . \tag{2.20}$$

As an example, the strike line constructed in Fig. 2.20b is itself a structure contour line. Two additional structure contour lines are shown in Fig. 2.24.

2.6.2
Structure Contours from Attitude

If the dip is known, then the spacing between structure contours is found from the method of Eq. 2.11 (Fig. 2.23) to be

$$H = I / \tan \delta \; , \tag{2.21}$$

where *H* = horizontal spacing between structure contours measured perpendicular to the contour strike, *I* = contour interval, and δ = dip. The trend of the structure contours is perpendicular to the dip direction.

2.6.3
Dip from Structure Contours

If the attitude is given by structure contours, the dip is found by solving Eq. 2.21:

$$\tan \delta = I / H \; , \tag{2.22}$$

where δ = dip, *I* = contour interval, and *H* = spacing between contours. The dip direction is perpendicular to the contour, in the direction toward the lower contour.

2.7
Intersecting Contoured Surfaces

The trace of a contact on a geologic map represents the intersection of the topographic surface with the boundaries of the geologic units. Finding the intersection between two surfaces is a fundamental technique in three-dimensional interpretation. In Sect. 2.6.1, three points along the outcrop trace of a planar geological contact were used to generate a structure-contour map. Conversely, the outcrop trace of a contact can be determined from a structure contour map. The following procedure works for both plane and curved structure contour surfaces. The structure contour map is superimposed on the topographic map (Fig. 2.25a). All intersection points where both surfaces have

Fig. 2.25.
Intersection between topographic contours and a planar marker horizon. Contour interval 20 ft. **a** Structure contours on a geologic horizon (*within shaded rectangle*) and topographic contours. The outcrop trace of the contact is the *thick line*. **b** Oblique 3-D view to *E*. **c** Geologic map of the outcrop trace (*thick line*) with the structure contours of the plane removed

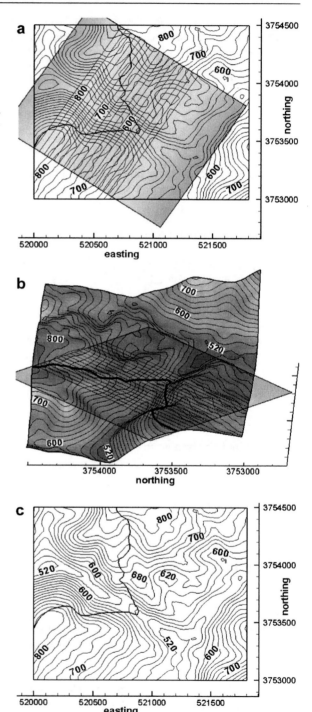

identical elevations are marked. Additional topographic and structural contours can be interpolated to add additional control. Marking the intersection points is most convenient if the topographic map is on a transparent overlay that can be placed above the structure contour map, something that is easily done in computer drafting programs. A 3-D view of the intersecting surfaces (Fig. 2.25b) helps in understanding the procedure. The intersection points are connected to obtain the outcrop pattern of the horizon (Fig. 2.25c). The inferred outcrop pattern becomes a working hypothesis for the location of the horizon. Exactly the same procedure is used to find the intersection between a fault and a stratigraphic marker or the line of intersection between two faults.

2.8
Derivation: Tangent Diagram on a Spreadsheet

To plot a tangent diagram with a spreadsheet it may be necessary to shift the origin for the azimuth (Az) from zero to the east to zero to the north using

$$Az = Az - 90 \ . \tag{2.23}$$

Change the dip to the tangent of the dip,

$$r = \tan(\text{dip}) \ , \tag{2.24}$$

and change from polar to Cartesian coordinates with

$$x = r \cos(Az) \ , \tag{2.25a}$$

$$y = r \sin(Az) \ . \tag{2.25b}$$

The native angle format in spreadsheet trigonometric functions is radians and so where angles in degrees are used in trigonometric expressions they must always be converted to degrees from radians. Plot the transformed points as an xy graph in the spreadsheet.

2.9
Exercises

2.9.1
Interpretation of Data from an Oil Well

The Appleton oil field is located near the northern rim of the Gulf of Mexico basin. Use the data from well 4835-B in this field (Table 2.2) to solve the following problems. Find the subsea depths of the stratigraphic markers. Write a spreadsheet program using Eqs. 2.3 and 2.4 to find the coordinates of a point between two known points. Find the true vertical depths and the total rectangular coordinates of the formation contacts in the well. What is the orientation of the well where it penetrates the Smackover? Write a spreadsheet program to solve this problem based on Eq. 2.4. Find the true vertical depths and the total rectangular coordinates of the oil-water contact (OWC) and the

Table 2.2. Data from well 4835-B, Appleton field, Alabama. Kelly Bushing: 244 ft

Measured depth = log depth (ft)		True vertical depth (ft)	Depth subsea (ft)	Total rectangular coordinates (ft)	
0		0		0.00 N	0.00 E
928		928		1.72 N	1.93 E
2308		2308		4.53 N	9.51 W
3282		3282		6.37 N	6.29 W
4150	Eutaw				
4400	U. Tuscaloosa				
4811		4811		13.38 N	5.99 W
4890	M. Tuscaloosa				
5150	L. Tuscaloosa				
5525	L. Cretaceous				
6198		6198		30.82 N	17.30 W
7696		7695		61.48 N	7.50 W
8328		8327		74.90 N	1.25 E
8482		8480		74.59 N	18.82 E
8661		8655		71.22 N	53.19 E
8935		8922		66.16 N	117.74 E
9186		9161		62.25 N	194.12 E
9556		9512		54.50 N	310.78 E
10006		9940		34.61 N	447.09 E
10266	Cotton Valley				
10431		10346		7.63 N	568.29 E
11103		10982		55.68 S	777.65 E
11621		11470		116.46 S	938.37 E
11960		11791		142.28 S	1045.58 E
12370	Haynesville				
12450		12253		173.48 S	1206.13 E
12641		12432		185.03 S	1271.62 E
13020	Buckner				
13072	Smackover[a]				
13086		12851		210.98 S	1418.79 E
13190	OWC[b]				
13286	Base Smack				

[a] Attitude from dipmeter 12,056. [b] Oil-water contact.

base of the Smackover. Plot the locations of the stratigraphic tops relative to the surface location of the well on the map (Fig. 2.26). Connect the points to show the shape of the well in map view. How far from the well (horizontally and in which direction) would you expect to first find the intersection of the oil-water contact with the top of the Smackover Formation, assuming constant dip for the Smackover?

2.9.2
Attitude

Given the dip vector of the plane 35, 240, show on an overlay of a stereogram the trace of the plane, its dip vector, and its pole. Show the dip vector 25, 240 on a tangent diagram. What is the apparent dip along a line that trends 260?

2.9.3
Attitude from Map

Use the map of the Blount Springs area (Fig. 2.27) to answer the following questions. Find the attitude of the Mpm in its southeastern outcrop belt using the 3-point method. Draw structure contours for the Mtfp, Mpm, and Mh on the eastern side of the map. Are the upper and lower contacts of each unit parallel to each other? Do you think the contacts are mapped correctly? Determine the attitude of the eastern contact between the Mpm and the Mtfp by the 3-point method and from the structure contours. Are they the same? If they are different, discuss which answer is better. What would be the apparent dip of the Mpm in a north-south roadcut through the northwestern limb of the anticline?

Fig. 2.26.
Map grid for plotting
points in a deviated well

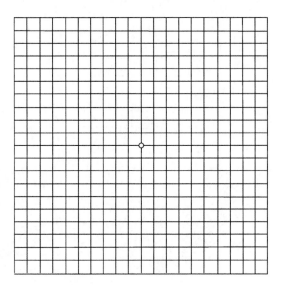

Fig. 2.27.
Geological map of the northeast corner of the Blount
Springs area, southern Appalachian fold-thrust belt. *Thin
lines* are topographic contours
(elevations in feet). *Thick lines*
are geologic contacts. *Arrows*
are dip directions; *numbers*
give the amount of dip

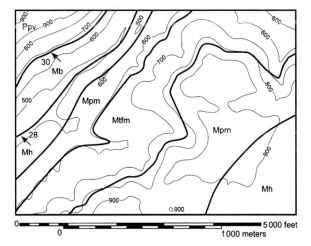

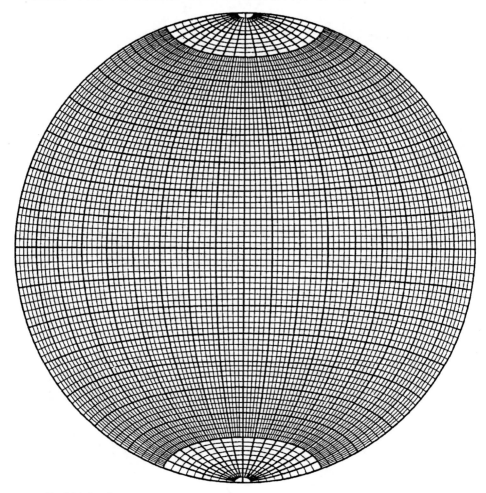

Fig. 2.28. Equal-area stereogram

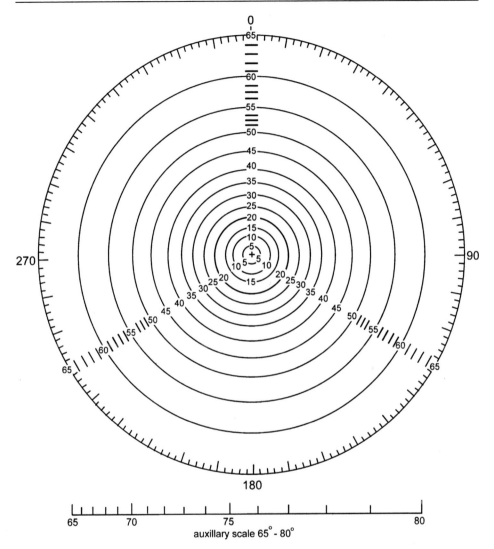

Fig. 2.29. Tangent diagram

Chapter 3

Structure Contouring

3.1
Introduction

This chapter covers the basic techniques for contouring continuous surfaces and for the construction of composite-surface maps. The contouring of faults and faulted surfaces is treated in Chap. 7.

3.2
Structure Contouring

A structure contour map is one of the most important tools for three-dimensional structural interpretation because it represents the full three-dimensional form of a map horizon. The mapping techniques to be discussed are equally applicable in surface and subsurface interpretation. The usual steps required to produce a structure contour map are:

1. Plot the points to be mapped.
2. Determine an appropriate contour interval.
3. Interpolate the locations of the contour elevations between the control points. There are several techniques for doing this and they may give very different results when only a small amount of data is available.

A structure contour map is constructed from the information at a number of observation points. The observations may be either the *xyz* positions of points on the surface, the attitude of the surface, or both. A relatively even distribution of points is desirable and which, in addition, includes the local maximum and minimum values of the elevation. If the data are from a geologic map or from 2-D seismic-reflection profiles, a very large number of closely spaced points may be available along widely spaced lines that represent the traces of outcrops or seismic lines, with little or no data between the lines. The number of points in a data set of this type will probably need to be reduced to make it more interpretable. Even if the contouring is to be done by computer, it is possible to have too much information (Jones et al. 1986). This is because the first step in contouring is always the identification of the neighboring points in all directions from a given data point and it is difficult and usually ambiguous to choose the neighbors between widely spaced lines of closely spaced points. Before contouring, the lines of data points may need to be resampled on a larger interval that is still small enough to preserve the form of the surface.

A series of rules for contouring has been developed over the years to produce visually acceptable maps that reasonably represent the surface geometry. The reference frame for a contour map is defined by specifying a datum plane, such as sea level. Elevations are customarily positive above sea level and negative below sea level. The contour interval selected depends on the range of elevations to be depicted and the number of control points. There should be more contours where more data are available. The contour interval should be greater than the limits of error involved. Errors of 20 ft (7 m) are typical in correlating well logs and as a result of minor deviations of a well from vertical. Uncertainties in locations in surface mapping are likely to produce errors of similar magnitude. The contour interval should be small enough to show the structures of interest. The map can be read more easily if every fifth or tenth contour is heavier. The map should always have a scale. A bar scale is best because if the map is enlarged or reduced, the scale will remain correct.

The contour interval on a map is usually constant; however, the interval may be changed with the steepness of the dips. A smaller interval can be used for low dips. If the interval is changed in a specific area, make the new interval a simple multiple (or fraction) of the original interval and clearly label the contour elevations and/or show the boundaries of the regions where the spacing is changed on the map.

The contours must obey the following rules (modified from Sebring 1958; Badgley 1959; Bishop 1960):

1. Every contour must pass between points of higher and lower elevation.
2. Contour lines should not merge or cross except where the surface is vertical or is repeated due to overturned folding or reverse faulting. The lower set of repeated contours should be dashed.
3. Contour lines should either close within the map area or be truncated by the edge of the map or by a fault. Closed depressions are indicated by hash marks (tic marks) on the low side of the inner bounding contour.
4. Contour lines are repeated to indicate reversals in the slope direction. Rarely will a contour ever fall exactly on the crest or trough of a structure.
5. Faults cause breaks in a continuous map surface. Normal separation faults cause gaps in the contoured horizon, reverse separation faults cause overlapping contours and vertical faults cause linear discontinuities in elevation. Where beds are repeated by reverse faulting, it will usually be clearer to prepare separate maps for the hangingwall and footwall.
6. The map should honor the trend or trends present in the area. Crestal traces, trough traces, fold hinges and inflection lines usually form straight lines or smooth curves as appropriate for the structural style.

Mapping by hand usually should begin in regions of tightest control and move outward into areas of lesser control. It is usually best to map two or three contours simultaneously in order to obtain a feel for the slope of the surface. The contours will almost certainly be changed as the interpretation is developed; therefore, if drafting by hand, do the original interpretation in pencil. The map should be done on tracing paper or clear film so that it can be overlaid on other maps.

3.3
Structural Style in Contouring

Contouring may be done using different styles, each of which produces its own characteristic pattern (Handley 1954). With a large amount of evenly spaced data, the difference between maps produced by different styles will usually be small. Contouring by any method must be viewed as a preliminary interpretation because unknown structures can always occur between widely spaced control points. The characteristics of the common styles of contouring are summarized next. Contours are usually shown as smooth curves although this should depend on the structural style (Sect. 3.3.4). The following examples (Figs. 3.1–3.2) are based on exactly the same points in order to demonstrate the differences that can be achieved by different methods.

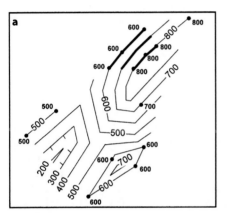 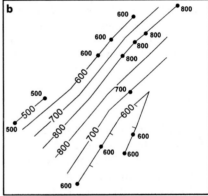

Fig. 3.1. Different contouring methods applied to the same control points (*solid circles*). **a** Equal-spaced contouring. The dip to be maintained (*heavy contours*) is chosen in the region of tightest control. **b** Parallel contouring

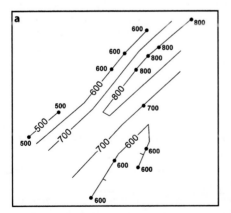 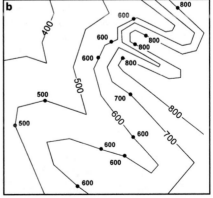

Fig. 3.2. Interpretive contouring; **a** based on interpretation that regional trends are northeast–southwest; **b** based on the interpretation that the regional trends are northwest–southeast

3.3.1
Equal Spacing

Equal-spaced contouring (Fig. 3.1a) is based on the assumption of constant dip magnitude over as much of the map area as possible. In the traditional approach, the dip selected is determined in an area of tightest control. The same dip magnitude is then used over the entire map (Handley 1954; Dennison 1968). Find the dip from the control points with the three-point method (Sect. 2.4), or, if the dip is known, as from an outcrop measurement or a dipmeter, find the contour spacing from the dip (Eq. 2.21). This approach projects dips into areas of no control or areas of flat dip and so will create large numbers of structures that may be artifacts (Dennison 1968). Because of the great potential for producing nonexistent structural closures, this method is usually not preferred (Handley 1954).

Equal-spaced contouring of the Mtfp (Fig. 3.1a) produces multiple closures, two anticlines and a syncline that is much lower than any of the data points. The anticlinal nose defined by the 800-ft contour is reasonable but the closed 700-ft anticline to the south and the syncline bounded by the 200-ft contour on the southwest are forced into regions of low dip or low control.

3.3.2
Parallel

Parallel contouring is based on the assumption that the contours are parallel, in other words, a strong linear trend is present. The contours are drawn to be as parallel as possible and the spacing between contours (the dip) is varied as needed to maintain the parallelism (Dennison 1968). The resulting map may contain cusps and sharp changes in contour direction, but is good for areas with prominent fold trends (Tearpock 1992). The method tends to generate fewer closures between control points than equal-spaced contouring and more than linear interpolation (Tearpock 1992).

The parallel contouring of the top of the Mtfp (Fig. 3.1b) is strongly influenced by the parallelism of the contours that can be drawn through the data points on the northwest limb and on the southeast limb of the major anticline. This method suggests an elongate northeasterly trending anticline in the center of the map and a southwest-plunging syncline on the southeast. The syncline is in the same position as the anticline predicted by the equal-spaced method (Fig. 3.1a).

3.3.3
Interpretive

Interpretive contouring reflects the interpreter's understanding of the geology. The preferred results are usually regular, smooth and consistent with the local structural style and structural grain. Generally, the principle of simplicity is applied and the least complex interpretation that satisfies the data is chosen.

Interpretive contouring (Fig. 3.2a) incorporates the knowledge that the structural grain is northeast–southwest and that the regional plunge is very low. The main anticline seen in each of the other techniques is present and is interpreted to be closed to the southwest, although there is no control as to exactly where the fold nose occurs. The northwestern contours are all interpreted to lie on the limb of a single structure, just as inferred by parallel contouring (Fig. 3.1b). The group of 600-ft contours in the southeast remains a problem. A syncline seems possible.

It is possible to obtain dramatically different results by "highly" interpretive contouring of sparse data (Fig. 3.2b). The two maps in Fig. 3.2 are completely different, although they are derived from exactly the same data. The difference between the two maps reflects the different assumed regional trends. There is no basis for choosing which map is better, given the available information. The simplest and most useful additional information for selecting the best interpretation is the bedding attitude, because the attitudes should indicate the trend direction (Sect. 3.6.1).

3.3.4
Smooth vs. Angular

Structure contours are usually drawn as smooth curves. This is appropriate for circular-arc and other smoothly curved fold styles. Many folds, however, are of the dip-domain style for which the structure contours should be relatively straight between

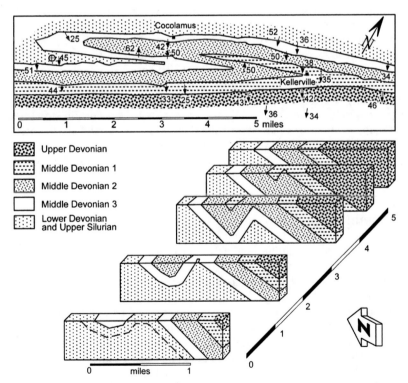

Fig. 3.3. Map and cross sections of dip-domain style folds in the Appalachian fold-thrust belt in Pennsylvania. (After Faill 1969)

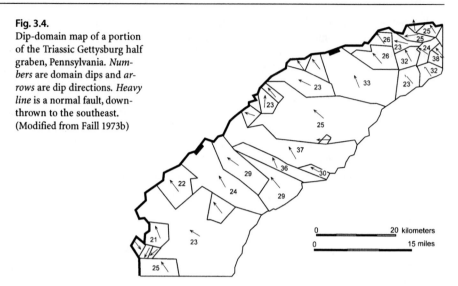

Fig. 3.4.
Dip-domain map of a portion
of the Triassic Gettysburg half
graben, Pennsylvania. *Numbers* are domain dips and *arrows* are dip directions. *Heavy
line* is a normal fault, downthrown to the southeast.
(Modified from Faill 1973b)

sharp hinges and have sharp corners on the map. The characteristic dip-domain geometry is regions of planar dip separated by narrow hinges. A map of dip-domain
compressional folds in the central Appalachian Mountains (Fig. 3.3a) shows long, relatively planar limbs and very narrow, tight hinges. The cross sections (Fig. 3.3b) show
a chevron geometry. Extensional folds may also have a dip-domain geometry. The
extensional dip domains in a portion of the Newark-Gettysburg half graben (Fig. 3.4)
have been synthesized from numerous outcrop measurements. Structure contour maps
with straight lines and sharp bends are appropriate for dip-domain structures.

Computer programs may allow smoothing to be performed after contouring, if
smoother surfaces are desired. For structural interpretation it is recommended that
the unsmoothed contours should always be examined first because they provide more
insight into the interpretation of the data points themselves.

3.4
Contouring Techniques

A given set of points can be contoured into a nearly infinite number of shapes, depending on the methodology followed. There is no absolute best technique for contouring and the overall appearance of the map is not necessarily an indication of its
quality (Davis 1986). Geological interpretation must ultimately be part of the process.
It is usually a good idea to start the interpretation process with a map that is constructed using standardized and reproducible procedures. Data points may either be
contoured directly, using the triangulated irregular network (TIN) method, or can be
interpolated into the elevations at the nodes of a grid (gridding) and then contoured
(Jones and Hamilton 1992). In order to illustrate the process of constructing a structure contour map, the points in Fig. 3.5 will be treated as if they come from locations
where there is no knowledge of the shape of the surface between the points.

Fig. 3.5.
Elevations of the top of a
marker unit that will be contoured using different techniques in Figs. 3.6–3.13

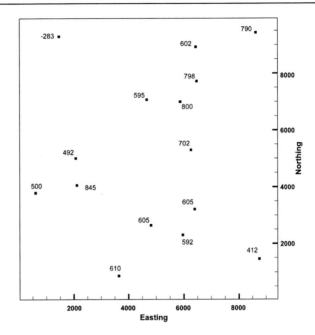

3.4.1
Choosing the Neighboring Points: TIN or Grid?

Drawing a contour between control points requires first deciding which control points
from the complete data set are to be used. This decision is not trivial or simple. The
choice of neighboring points between which the contours are to be drawn has a major
impact on the shape of the final surface. Two procedures are in wide use, triangulation and gridding. Triangulation involves finding the TIN network of nearest neighbors in which the data points form the nodes of the network (Fig. 3.6a). Gridding
involves superimposing a grid on the data (Fig. 3.6b) and interpolating to find the
values at the nodes (intersection points) of the grid. Many different interpolation
methods are used in gridding. Most involve some form of weighted average of points
within a specified distance from each grid node (Hamilton and Jones 1992). Contours developed from either type of network may be smoothed, either as part of the
contouring procedure or afterward.

The first decision is whether the contouring will be based on a TIN or on a grid. The
most direct relationship is to connect adjacent points with straight lines, producing a
TIN. This has long been a preferred approach in hand contouring and is also popular
in computer contouring (Banks 1991; Jones and Nelson 1992). The primary advantages
of the method are that it is very fast, the contoured surface precisely fits the data, and
it is easy to do by hand. For structural interpretation, fitting the data exactly, including
the extreme values, is a valuable property, because the extreme values may provide the
most important information. Plotted in three dimensions, the TIN network alone will
show the approximate shape of the surface. The advantage of gridding is that once the

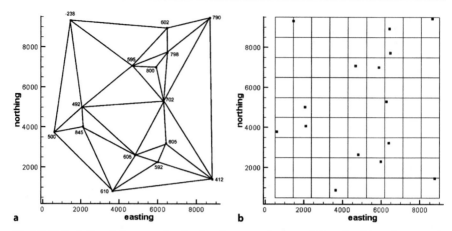

Fig. 3.6. Methods for relating control point elevations to each other. **a** Triangulated irregular network (TIN) with data points at the vertices. **b** Grid superimposed on the data, elevations will be interpolated at the grid nodes

data are on a regular grid, other operations, such as the calculation of the distance between two gridded surfaces, are relatively easy. The major disadvantage is that the contoured surface does not necessarily go through the data points and it may be difficult to make the surface fit the data.

3.4.2
Triangulated Irregular Networks

Creating a TIN requires determining the nearest neighbor points, between which the contours will be located. The possible choices of nearest neighbors are seen by connecting the data points with a series of lines to form triangles (Fig. 3.7a). The points could be connected differently to form different networks. Delauney triangles and greedy contouring are two unbiased approaches to choosing the nearest neighbors. The commonest form of triangulation is in two dimensions and considers only the proximity of the points in a plane, such as x and y but not z. Triangulation in three dimensions considers the xyz location of points and can be generalized to interpret very complex surfaces (Mallet 2002).

3.4.2.1
Delauney Triangles

A Delauney triangle is one for which a circle through the three vertices does not include any other points (Jones and Nelson 1992). In Fig. 3.7b, the solid circle through vertices a, b and c is a Delauney triangle because no other points occur within the circle. The vertices of the triangle are nearest neighbors. The dashed circle through vertices a, b and d (Fig. 3.7b) includes point c and therefore does not define a Delauney triangle. This method is not as practical for use by hand as the next technique.

3.4.2.2
Greedy Triangulation

A triangulation method suitable for use by hand as well as by computer is known as "greedy" triangulation (Watson and Philip 1984; Jones and Nelson 1992). The criterion is that the edge selected is the shortest line between vertices. No candidate edge is included if there is a shorter candidate edge that would intersect it (Jones and Nelson 1992). Figure 3.8a shows a network of candidate edges with the longer edges dashed. The TIN produced by this method is shown in Fig. 3.8b. This method is both logical and convenient for use by hand and is the approach generally used in this book as the first step in the interpretation.

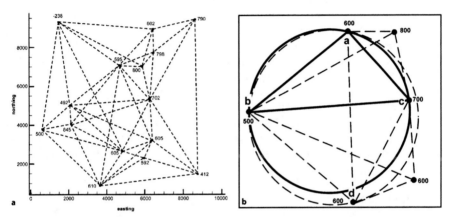

Fig. 3.7. Nearest neighbors in a TIN network. **a** Possible nearest neighbors connected by *dashed lines*. **b** Four potential neighbor points *a–d*. The *solid circle* includes three points (*a–c*) that define a Delauney nearest-neighbor triangle (*heavy lines*). The *dashed circle* through *a*, *b*, and *d* includes one point inside the circle

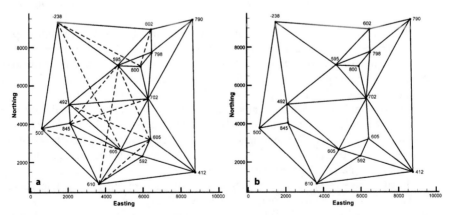

Fig. 3.8. Greedy triangulation. **a** Alternative nearest neighbors. Longer edges *dashed*. **b** Longer edges removed to define nearest neighbors

3.4.3
Interpolation

All contouring methods require interpolation between control points in order to find the structure contours. Discussed here are linear interpolation between nearest neighbor points in a TIN network and interpolation to a grid.

3.4.3.1
Linear Interpolation

Linear interpolation is based on the assumption that slope between the data points is a straight line. As a contouring technique it is also called mechanical contouring (Rettger 1929; Bishop 1960; Dennison 1968). This is a standard approach for producing topographic maps where the high points, low points, and the locations of changes in slope are known, allowing accurate linear interpolation between control points (Dennison 1968). This method may produce unreasonable results in areas of sparse control (Dennison 1968; Tearpock 1992). The resulting map is good in regions of dense control and is the most conservative method in terms of not creating closed contours that represent local culminations or troughs. The method tends to de-emphasize closed structures into noses, is good for gently dipping structures with no prominent fold axes, and is often used in litigation, arbitration and oil-field unitization (Tearpock 1992). This method is applied to the example data in Fig. 3.9. The contours suggest an anticline with two separate culminations.

3.4.3.2
Interpolation to a Grid

Mapping by gridding requires interpolation between and extrapolation beyond the control points to define values at the grid nodes prior to contouring. Gridding variables always include the choice of the grid spacing and the interpolation technique. A typical characteristic of gridded data is that the original control points do not fall on the contoured surface. The reason for this is that the control points are not used to make the final map.

The simplest gridding technique is linear interpolation. The structure contour map in Fig. 10a is the result of linear interpolation to the nodes of a 10×10 grid, then linear-interpolation contouring between the nodes. The best-controlled part of the structure resembles the triangulated map of Fig. 3.9. The contouring algorithm for this technique forces all contours to close within the map area which is not a geologically realistic assumption.

Another simple interpolation technique is the inverse distance method. In this method the value at a grid node is the average of all points within a circle of selected radius around the node, weighted according to distance, such that the farther-away points have less influence on the value. The weighting function is usually an exponential, such as one over the distance squared (Bonham-Carter 1994). Figure 3.10b is a structure contour map produced by inverse-distance interpolation. The contours are significantly more curved than those produced by linear interpolation of the TIN (Fig. 3.9) or the linear interpolated grid (Fig. 3.10a). The side view of the inverse-distance interpretation (Fig. 3.10c) shows that the control points do not all lie on the interpolated surface.

Fig. 3.9.
Structure contour map of the
triangulated data in Fig. 3.8b.
Contours produced by linear
interpolation between nearest
neighbors

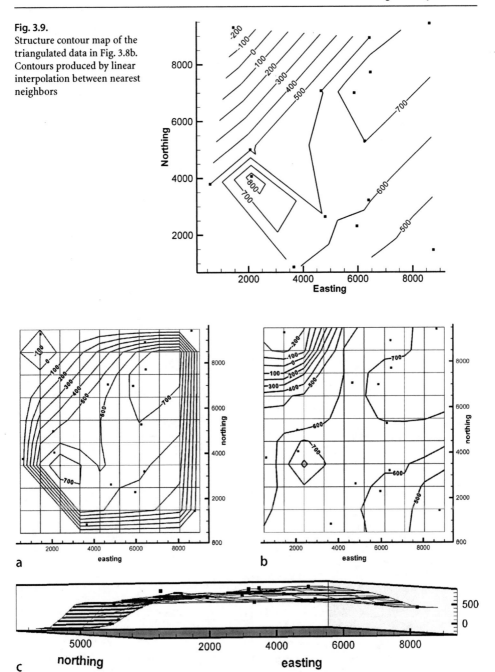

Fig. 3.10. Simple grid-based contouring techniques using a 10 × 10 grid, control points from Fig. 3.5.
Squares are data points. **a** Linear interpolation. **b** Inverse-distance interpolation, weighting exponent 3.5.
c Oblique view to NW of inverse-distance interpolation

Kriging is an interpolation method in which the value at a grid node is a weighted sum of points within a zone of influence, like the inverse-distance method, but with a more complex weighting system (Bonham-Carter 1994). There are several kriging parameters (Davis 1986) that must be set to obtain a result. In the program used to produce the maps below, these parameters are: *range* = distance beyond which the values of the points become insignificant in the average; *drift* = the overall trend of the surface, which can be either zero, linear or quadratic; *zero value* = semi-variance of source points = certainty that the value is correct on a scale of zero to one (zero means the point is exact). Setting the trend to be quadratic trend allows the final surface to be more complex. Larger values of the zero value lead to smoother surfaces.

A range of kriging results as a function of the choices of mapping parameters is illustrated in Fig. 3.11. For each map the zero value is set to zero and the effects of grid spacing and drift are explored. Both surfaces generated with linear drift (Fig. 3.11b,c)

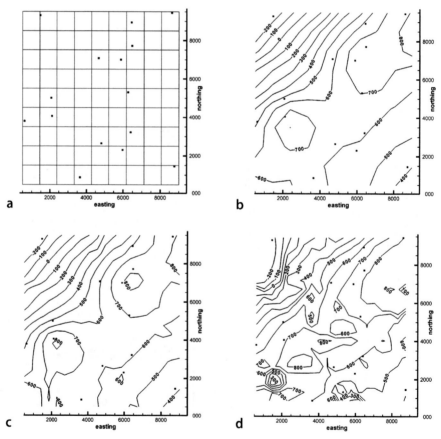

Fig. 3.11. Kriging of the data in Fig. 3.5. *Solid squares* are control points. **a** 10 × 10 grid with control points. **b** Mapped to a 10 × 10 grid, range = 0.3, drift = linear. **c** Mapped to a 20 × 20 grid, range = 0.3, drift = linear. **d** Mapped to a 20 × 20 grid, range = 0, drift = quadratic

indicate an anticline with two local closures and a pronounced saddle between them, similar to the inverse-distance result (Fig. 3.10b). Decreasing the grid spacing from 10×10 (Fig. 3.11b) to 20×20 (Fig. 3.11c) increases the complexity of the surface, but only slightly. Increasing the drift from linear (Figs. 3.11b,c) to quadratic (Fig. 3.11d) greatly increases the complexity of the surface, resulting in numerous small closures on the bigger structure, analogous to those produced by the equal-spaced contouring style. An oblique view (Fig. 3.12) shows that the control points may lie at significant distances from the interpolated surface. For further discussion of working with grid-based computer contouring, see Walters (1969), Jones et al. (1986), and Hamilton and Jones (1992).

3.4.4
Adjusting the Surface Shape

In order to achieve the desired result (interpretive contouring) with computer contouring, it may be necessary to introduce a bias in the choice of nearest neighbors or to introduce pseudopoints. A biased choice of neighbors is used in forming a TIN to control the grain of the final contours or to overcome a poor choice of neighbors that results from inadequate sampling of the surface. Pseudopoints can be used to insure

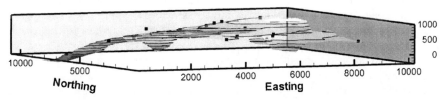

Fig. 3.12. 3-D oblique view to the NE of the kriged surface in Fig. 3.11d showing that some control points (*squares*) lie above or below surface

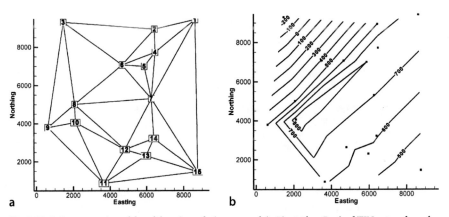

Fig. 3.13. Reinterpretation of the of the triangulation network in Fig. 3.8b. **a** Revised TIN network, nodes are numbered. **b** Linear interpolation contouring of network in **a**

that the surface goes above or below the extreme values of the data points. It is important to carefully label pseudopoints in the data base so that they will not be mistaken for real data.

The relative spacing of grid nodes can be altered to produce a trend. Changing, for example, from a 10×10 grid to a 10×20 grid in the same area will alter the surface. Rotating the grid directions will also have an effect.

A TIN network can be edited to change the nearest neighbors, which will then change the resulting surface. Suppose that the map of the anticline in Fig. 3.9 would be better interpreted without a saddle between separate closures. The control point that creates the saddle should have nearest neighbors on the southeast limb, not the northwest limb of the anticline. The desired result is obtained by re-defining the nearest neighbor network (Fig. 3.13a), resulting in a new map (Fig. 3.13b).

3.5
Mapping from Profiles

Frequently structure contour maps are derived from data distributed along linear traverses, rather than from randomly spaced points. This is particularly true when working with 2-D seismic-reflection profiles, ground penetrating radar profiles, or predictive cross sections (Sect. 6.4). If the profile trend is not parallel or perpendicular to the structural trend, the map may contain apparent structures related to the traverse orientation. The effect is shown by obliquely sampling a cylindrical, sinusoidal fold that has a horizontal axis with a north-south trend (Fig. 3.14). When the traverse data are interpolated by the inverse-distance technique (Fig. 3.15), the correct general form of the anticline is produced but smaller-scale NE and NW trends are superimposed. The oblique trends are most evident in the low-dip region near the crest of the anticline. Triangulation shows even more pronounced oblique trends near the crest (Fig. 3.16a). The triangulation network (Fig. 3.16b) shows the reason for the oblique trends. Nearest neighbors are controlled by the traverse spacing, not the underlying

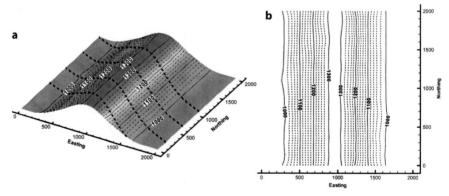

Fig. 3.14. Source data: non-plunging, sinusoidal anticline. **a** Oblique view showing section traces. **b** Structure contour map

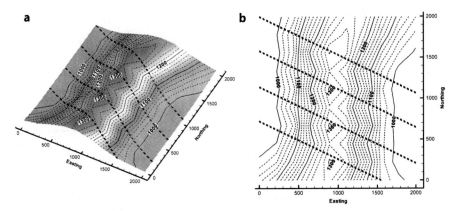

Fig. 3.15. Map made by inverse-distance interpolation (weighting exponent 3.5) of points along cross sections from data in Fig. 3.14. *Lines of black squares* are control points. **a** Oblique view. **b** Structure contour map

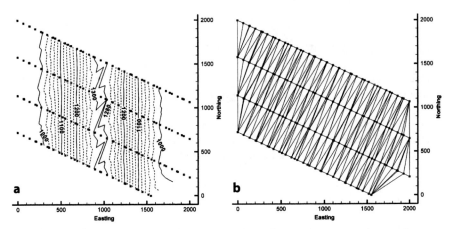

Fig. 3.16. Map made by triangulation of points along cross sections from data in Fig. 3.14. *Lines of black squares* are control points. **a** Structure contour map. **b** TIN network

structural trend. Nearest neighbors are the closest points on adjacent traverses. This problem can be overcome by orienting traverses parallel and/or perpendicular to the structural trend, or by mapping based on the structural trend (Sect. 5.5).

Non-cylindrical folds pose a greater challenge for accurate mapping. The fold in Fig. 3.17a is a simple flat-topped anticline with limbs that converge to the south and disappear, giving a conical geometry. Sampled along three traverses perpendicular to the average crestal trend, the reconstruction does only a fair job of reproducing the original geometry. The fold limbs are reproduced but the flat crest and the plunging nose are misrepresented. Mapping based on 3-D dip domain interpretation (Sect. 6.7) is the most accurate approach for this style of structure.

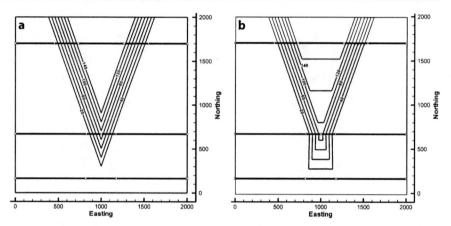

Fig. 3.17. Structure contour maps of a conical, dip-domain anticline. **a** Original map showing three traverses where elevations have been extracted (*heavy EW lines*). **b** Map constructed by triangulating points extracted from the three profiles in **a**

3.6
Adding Information to the Data Base

Structure contour maps can be based on a significant amount of information in addition to the elevations on a single horizon. The shape of the contoured surface can be controlled using the attitudes of bedding, data from multiple stratigraphic horizons, and from pore-fluid behavior such as different groundwater levels or the presence or absence of hydrocarbon traps as indicated by shows of oil and gas in wells. Finding the structural trend and mapping based on trend are discussed in Chap. 5 and 6.

3.6.1
Bedding Attitude

If the bedding attitude is known from outcrop or dipmeter measurements (Fig. 3.18a), it can be incorporated into the contouring. Attitudes at the well-bore or outcrop scale can give insight into the shape of the surface but are subject to influence by small-scale structures. The contours are not required to have the same dip everywhere on the map. The contour spacing will change as the dip changes. Structure contours are perpendicular to the bedding dip. The distance between the contours is given by Eq. 2.21. In the example of Fig. 3.18, for a contour interval of 100 and dip of 25°, the spacing between the contours at map scale is 214, giving the structure contour map in the vicinity of the control point shown in Fig. 3.18b.

If mapping is customarily done at a standard scale, then a map-spacing ruler (Fig. 3.19) can be useful. A contour-spacing ruler shows the spacing between contours for a variety of dips, given the map scale and the contour interval. It is constructed using Eq. 2.21. The ruler is oriented perpendicular to the structure-contour (strike) trend and the contour spacing for a given dip is easily plotted or the dip determined from the contour spacing.

Fig. 3.18.
Structure contour direction and spacing from an attitude measurement. **a** Dip vector at a point having an elevation of 800. **b** Structure contours in the vicinity of the point. For $I = 100$, $H = 214$

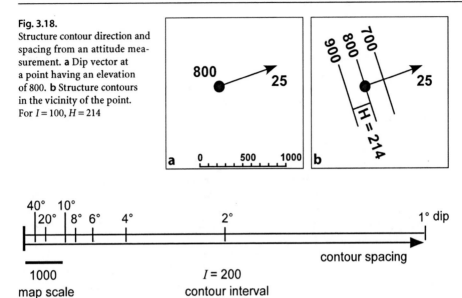

Fig. 3.19. Contour-spacing ruler

3.6.2
Projected and Composite Surfaces

A projected surface is a structure contour map derived entirely by projecting data from other stratigraphic levels. A projected marker is sometimes called a ghost horizon. Usually a projected surface is below the lowest control points. A typical use is to project the subsurface location of an aquifer or an oil reservoir from outcrop or shallow subsurface information.

A composite surface is a structure contour map derived using data from multiple stratigraphic horizons, including the horizon being mapped. One horizon is selected as the reference surface and data from other stratigraphic horizons are projected upward or downward to this horizon, using the known stratigraphic thicknesses. The best choice of a reference horizon is one for which there is already a significant amount of control and that minimizes the projection distance. Usually a reference horizon that is stratigraphically in the middle of the best-controlled units should be selected. The data from multiple horizons provide increased control on the interpretation of the shape of the reference horizon. This type of map is particularly useful in the interpretation of outcrop data because the locations of all formation boundaries can be used to provide control points, greatly increasing the areal distribution of data.

An elevation on a marker surface is transformed into an elevation on a projected surface by adding or subtracting the vertical distance between the two (Fig. 3.20). The projection is made from a point where the elevation of the marker is known. This may be the location of a surface outcrop (as in Fig. 3.20) or the elevation of a contact in a well. The distance to the projected horizon is derived from the thickness of the unit by

$$d = t / \cos \delta \, , \tag{3.1}$$

where d = vertical distance between the surfaces, t = true thickness, and δ = true dip (Badgley 1959). The projection can be either up or down from the known point, that is, from the marker horizon in Fig. 3.20 to either a or b. Be sure to use the same datum (i.e., sea level) for all measurements. Projections from a surface map need to use the topographic elevation to find the elevation with respect to sea level. Projections above the surface of the earth are as valid as projections below the surface; it is not necessary that the reference surface be confined to the subsurface.

If regional thickness variations are present, the thickness used for projection must be adjusted according to the location. An isopach map (Chap. 4) provides the information necessary to determine the thickness at specific points. In regions of low dip, the difference between the vertical distance and the true thickness is small. In this situation an approximate projected surface can be derived by simply adding or subtracting the thickness between the units to or from the elevation of the marker to obtain the projected surface (Handley 1954; Jones et al. 1986; Banks 1993).

Projected data can greatly augment the information on a single horizon and can lead to a significant improvement in the interpreted geometry of the structure. The increase in data available for contouring may significantly improve the map on the reference horizon. Inconsistent data on different horizons can be more easily recognized when all data are projected to the same surface. Accurate projection requires accurate knowledge of the unit thickness and the dip, both of which are likely to contain uncertainties and so a certain amount of "noise" is to be expected in the projected data set. The interpreted surface and the data will be iteratively improved as the inconsistencies are eliminated.

The creation of a composite surface map allows utilization of stratigraphic markers that are not formation boundaries. The location of any marker horizon separated from the reference surface by a known stratigraphic interval can be converted to an elevation on the reference horizon. Even if the marker is not usually mapped, it will provide important information.

The construction of a projected or composite-surface includes assumptions that must be considered in each application. Projection with Eq. 3.1 requires that the dip and

Fig. 3.20.
Vertical cross section showing the projected distance from a point (*small circle*) on a marker horizon to reference surface. Projections may be done either upward (to *a*) or downward (to *b*). The region below ground level is *patterned*; *d*: vertical distance between surfaces; *t*: true thickness; *δ*: true dip

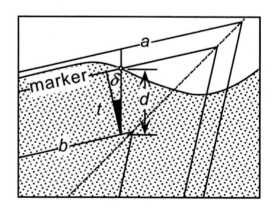

Fig. 3.21.
Cross sections showing directions of constant thickness. The dips are identical in each cross section. **a** Constant bed thickness (t). **b** Constant vertical thickness (v). **c** Constant apparent thickness (t') in an inclined direction

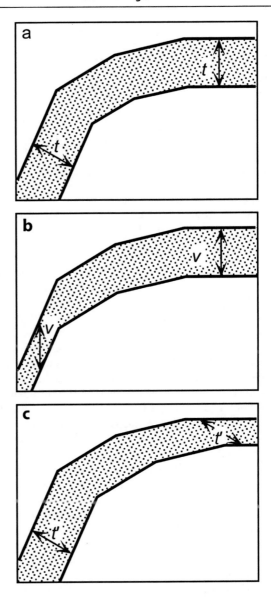

thickness remain constant and the units unfaulted over the projection distance. The most definitive check on the validity of a projected or composite surface is to construct a cross section that shows all the horizons from which data have been obtained (Chap. 6), as in Figs. 3.21 and 3.22. Any projection problems should be reasonably obvious on the cross section.

Folding may produce thickness changes that are a function of position within the fold and the mechanical stratigraphy. Equation 3.1 is based on the assumption that

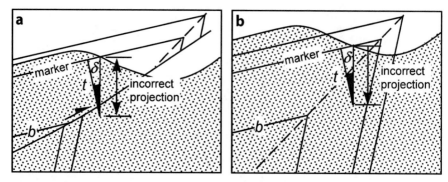

Fig. 3.22. Vertical cross sections showing incorrect projections across discontinuities. **a** Projection across a fault. **b** Projection across an axial surface. The region below ground level is *patterned.* δ: dip; t: thickness of the interval being projected

bed thickness remains constant throughout the fold for all units being projected (Fig. 3.21a), in other words, that all the horizons are parallel. This is called parallel folding, and is only one of the possible fold styles. Deformation can change the thicknesses, especially the thicknesses of thick soft units between stiffer units. The direction of constant thickness is an element of the fold style. A similar fold maintains constant thickness parallel to the axial surfaces. Constant vertical thickness projection (Handley 1954; Banks 1993) is strictly appropriate only for similar folds that have vertical axial surfaces (Fig. 3.21b). The resulting thinning on steep limbs is a common feature of compressional folds, even in those that maintain constant bed thickness elsewhere. If the axial surfaces of a similar fold are inclined, the direction of constant thickness is inclined to the vertical (Fig. 3.21c). Projection of surfaces is probably best restricted to situations in which bed thicknesses are approximately constant (Fig. 3.21a).

Projection of thickness is based on the further assumption that the stratigraphy between the projection point and the composite surface is an unbroken sequence of uniform dip. If the vertical line of projection crosses a fault (Fig. 3.22a) or an axial surface (Fig. 3.22b), then the projection will be incorrect. The shorter the projection distance, the less likely these problems are to occur.

The value of a composite-surface map in structural interpretation is shown by the composite surface of the Mtfp (Fig. 3.23) in the Blount Springs map area. The projected points allow a structure contour map to be constructed for the top of a centrally located stratigraphic horizon (top Mtfp). This map was produced by first interpolating the dip values between control points and then projecting all the points on each contact to the top of the Mtfp. Projecting using very steep dips provided some unrealistic results because the long vertical projection distances cross gently dipping axial surfaces, invalidating the result as in Fig. 3.22b. The obviously incorrectly projected points have been removed. The square points in Fig. 3.23a are the locations of dip measurements and approximately indicate the limits of control. The complex structure northwest of the anticline and the flat surface to the southeast are artifacts of the contouring process (kriging) and do not represent real structure.

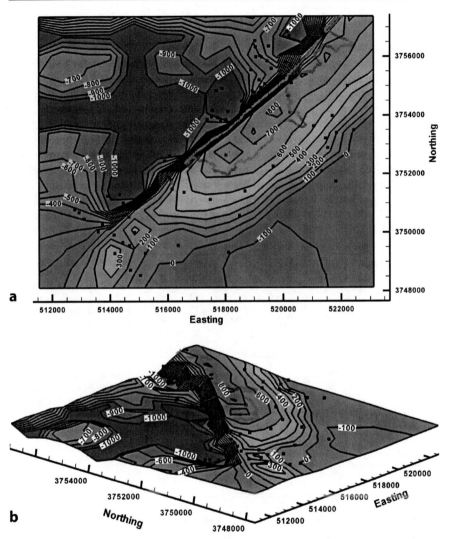

Fig. 3.23. Composite-surface map of the top Mtfp, Blount Springs map area of Fig. 2.4. The outcrop trace of the top Mtfp is a *wide, light gray line.* Thicknesses used for projection are: Mpm = 108 ft and Mh = 105 ft, calculated from the geologic map (Fig. 2.4); Mtfp = 245 ft from a well in the area; Mb = 625 ft from the outcrop just southeast of the map area. Map coordinates are UTM in meters and contour elevations are in feet. *Squares* are locations of dip measurements. **a** Map. **b** Oblique 3-D view to the northeast

The major northeast-southwest trending structure (Fig. 3.23) is the Sequatchie anticline, now clearly shown to be an asymmetric anticline with a steep forelimb on the northwest. The internal consistency of the data appears to be good, confirming the general validity of all the mapped outcrop traces. The flattening and spreading of the anticline at its southwest end is real, and represents a saddle where the fold crest steps to the northwest.

3.6.3
Fluid-Flow Barriers

Fluid movement, or the lack of it, through porous and permeable units can indicate the connectivity or the lack of connectivity between wells. A *show* is a trace of hydrocarbons in a well, and can indicate the presence of a nearby hydrocarbon trap that is otherwise unseen. Different water levels, oil-water contacts, or fluid pressures in nearby wells can indicate a barrier between the wells. The structure contour map in Fig. 3.24a and the corresponding cross section in Fig. 3.25a show four wells that appear to define a region of uniform dip. Suppose, however, that an oil or gas show is present in the downdip well but not in the updip wells. The show suggests proximity to a hydrocarbon trap, yet the map does not indicate a trap. The map must be revised to include some form of barrier because of this additional information (Sebring 1958). Possible alternatives that could produce an oil or gas show in the downdip well include a hydro-

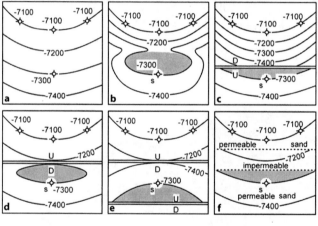

Fig. 3.24.
Alternative maps honoring the same data points. **a** Map based on the four wells being dry holes. **b–f** Maps based on presence of an oil show in the well labeled *s*, implying presence of a barrier between this well and the three updip dry holes. Hydrocarbon accumulations are *shaded*. Figure 3.25 shows the corresponding cross sections. (After Sebring 1958)

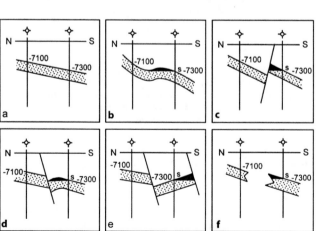

Fig. 3.25.
Alternative cross sections in the dip direction honoring the same data points. **a** Section based on the four wells being dry holes. **b–f** Sections based on presence of an oil show (*s*) in the downdip well, implying a barrier between this well and the updip wells. Hydrocarbon accumulations are *shaded*. Figure 3.24 shows the corresponding structure contour maps. (After Sebring 1958)

carbon-filled structural closure up the dip (Figs. 3.24b, 3.25b), different types of faults between the downdip well and the updip wells (Figs. 3.24c–e, 3.25c–e) or a stratigraphic permeability barrier (Figs. 3.24f, 3.25f). A stratigraphic barrier does not necessarily require the structure to be changed from the original interpretation. The structural configuration is the same in Figs. 3.24f and 3.25f as in Figs. 3.24a and 3.25a.

3.7
Exercises

3.7.1
Contouring Styles

Use the data from the Weasel Roost Formation (Fig. 3.26) to try out different contouring techniques and to see the effect of trend biasing. Use interpretive contouring and assume a surface with no grain. Contour by parallel contouring: *(a)* assuming a northwest-southeast grain; *(b)* assuming a northeast-southwest grain. Draw crestal and trough traces on the structure contour maps just completed. Use interpretive contouring and assume a northeast-southwest grain. Define a TIN using greedy triangulation and contour by linear interpolation.

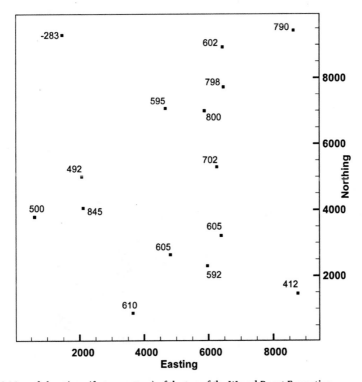

Fig. 3.26. Map of elevations (feet or meters) of the top of the Weasel Roost Formation

3.7.2
Contour Map from Dip and Elevation

Contour the top of the Tuscaloosa sandstone in Fig. 3.27 using the bedding attitudes to help generate the contour orientations and spacings. The elevations are in meters.

3.7.3
Depth to Contact

Find the elevation of the top of the Mtfp below the dot in Fig. 3.28. The thickness of the Mpm is 97 ft and the dip is 04°.

Fig. 3.27.
Map of the top of the porous Tuscaloosa sandstone. Negative elevations are below sea level; azimuth of bedding dip is indicated by *arrows*

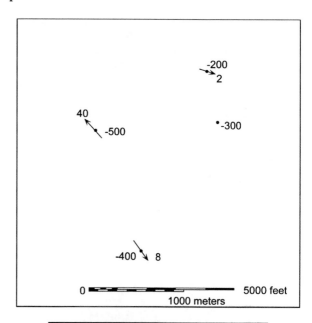

Fig. 3.28.
Geologic map of the Mill Creek area. Topographic elevations are in feet and the scale bar is 1 000 ft

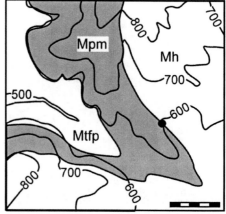

3.7.4
Projected-Surface Map

Use the geologic map of Fig. 3.29 to construct a projected structure contour map of the top of the Fairholme, a potential hydrocarbon reservoir. Use every point where a

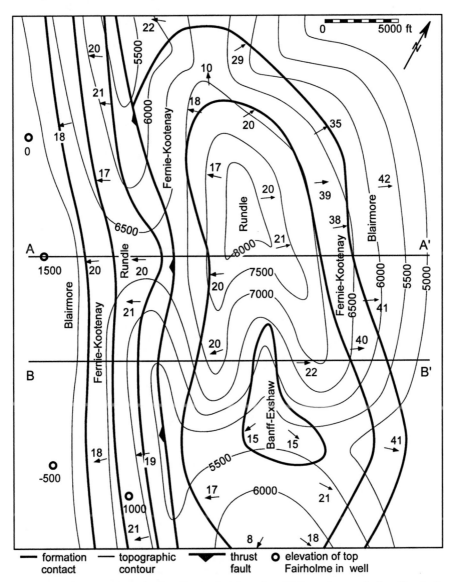

Fig. 3.29. Geologic map from the Canadian Rocky Mountains. All dimensions are in feet. The stratigraphic column (with thickness) from top to base is: Blairmore (2 400), Fernie-Kootenay (700), Rundle (900), Banff-Exshaw (900), Palliser (800), Fairholme (1 200). (After Badgley 1959)

Legend:
— formation contact
— topographic contour
▼ thrust fault
○ elevation of top Fairholme in well

formation boundary crosses a topographic contour. Post all the elevations on your map before contouring. What is the best method for contouring this map? Explain your reasons. Is the geological map correct? Why or why not? Does the projected structure-contour map agree with the drilled depths to the top of the Fairholme? The wells to the Fairholme were drilled to find a hydrocarbon trap but were not successful. What is a structural reason for drilling the wells and what is a structural reason why they were unsuccessful?

Chapter 4

Thickness Measurements and Thickness Maps

4.1
Thickness of Plane Beds

Thickness has multiple definitions, the choice of which depends on the purpose and the data available (Fig. 4.1). The true stratigraphic thickness (TST) is always the distance between the top and base of a unit measured perpendicular to the top. In a completely exposed outcrop, bed thickness can be measured directly across the bed. In a well that is perpendicular to bedding, the measured thickness in the well (MD) is the true stratigraphic thickness. Commonly, however, thicknesses must be determined from oblique traverses across beds or from wells that are not perpendicular to the bed boundaries. The measured thickness in a vertical well (or along a vertical traverse) is the true vertical thickness (TVT). The measured thickness in any other direction is here termed a slant thickness. A "thickness" measurement that is easily derived from well data is the TVD or true vertical depth thickness, and is the difference in elevation between the top and base of a unit in a well log. This "thickness" is more related to the orientation of the well, however, and for a horizontal traverse or a horizontal well, the TVD is zero. In the following sections the true stratigraphic thickness is found first and then other thickness determined from it, as needed.

4.1.1
Universal Thickness Equation

The stratigraphic thickness can be determined from a single equation (Hobson 1942; Charlesworth and Kilby 1981) based on the angle between the direction of the thick-

Fig. 4.1.
Vertical cross section in the dip direction showing measures of thickness: *MD:* measured distance on well log or traverse; *t:* true stratigraphic thickness = *TST*; t_v: true vertical thickness = *TVT*; t_s: slant thickness; *TVD:* true vertical depth thickness = difference in z coordinates between top and base of unit. (Standard well-log terminology from Robert L. Brown of Shell Oil Co.)

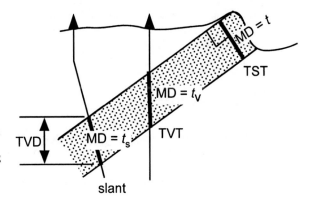

Fig. 4.2.
Data needed to determine
thickness of a unit from the
universal-thickness equation
(Eq. 4.1)

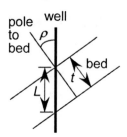

ness measurement and the pole to bedding (Fig. 4.2). The method is convenient for data from a well or a map. It is here called the *universal* thickness equation because it always works, regardless of the direction of the measurement or the dip of the bed. The advantage of this method is that the thickness is given by a single, simple equation, eliminating most potential sources of error. The disadvantage for hand calculation is the need to determine an angle in three dimensions, although this is easily done with a stereogram or spreadsheet. The apparent thickness is measured along the direction L. In the plane defined by the line of measurement and the pole to bedding

$$t = L \cos \rho \ , \tag{4.1}$$

where t = the true stratigraphic thickness and L is the straight-line length between the top and base of the unit (MD), measured along a well or between two points on a map. The angle ρ = the angle between L and the pole to the bed. If the acute angle is used in the equation, the thickness is positive. If the obtuse angle is used the thickness will be correct in magnitude but negative in sign; taking the absolute value gives the correct result for either possibility. If L is vertical, then the angle $\rho = \delta$, the dip of the bed. If the true thickness is known, then the vertical thickness can be found by rewriting Eq. 4.1 as

$$t_v = t / \cos \delta \ , \tag{4.2}$$

where t_v = vertical thickness, t = true thickness, and δ = true dip.

If L is not vertical, its inclination must be determined. For a well, a directional survey will give the azimuth of the deviation direction and the amount of the deviation; the latter which may be reported as the kickout angle, the angle up from the vertical. Alternatively, the deviation of a well may be given as the *xyz* coordinates of points along the well bore. The angle ρ can be found graphically or analytically, as described in the following two sections.

4.1.1.1
Angle between Two Lines, Stereogram

To find the angle ρ with an equal-area stereogram, on an overlay, plot the point representing the pole to bedding by marking the trend of the dip on the overlay, rotating the overlay to bring this mark to the east-west axis, and counting inward from the outer circle (the zero-dip circle) the amount of the dip plus 90°, and mark the point. Return

Fig. 4.3.
Equal-area stereogram showing the angle between two
lines. Angle ρ is the great
circle distance between the
points giving the pole to bedding and the orientation of
the apparent thickness measurement. Lower-hemisphere
projection

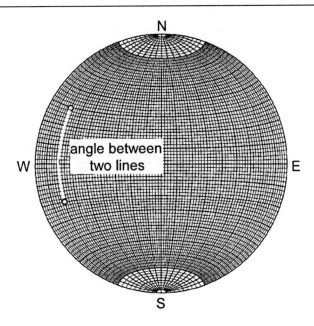

the overlay to its original position. Plot the line of measurement (or well bore) by similarly marking the trend of the measurement on the outer circle, bringing the mark to
the east-west axis and measuring the dip inward from the outer circle if given as a
plunge, or outward from the center of the graph if given as a hade or kickout angle.
Rotate the overlay until the two points fall on the same great circle (Fig. 4.3). The angle ρ
is measured along the great circle between the two points.

As an example of the thickness calculation based on the universal thickness equation, find the true thickness of a bed that is $L = 10$ m thick in a well. The well hades 10°
to 310° and the bed dip vector is 20, 015. Plot the bed on the stereogram and find its
pole. Then plot the well and measure the angle between the two lines ($\rho = 27°$). Equation 4.1 gives $t = 8.9$ m.

4.1.1.2
Angle between Two Lines, Analytical

Method 1. Using bed dip vector and well dip vector
To find the angle between the bed pole and the well, both given as bearing and plunge,
substitute well dip vector (Eq. 12.3) and the bed pole from the dip vector (Eq. 12.13)
into the equation for the angle between two vectors (Eq. 12.25) to obtain

$$\rho = \cos^{-1} \rho = -\cos \delta_w \sin \theta_w \sin \delta_b \sin \theta_b$$
$$- \cos \delta_w \cos \theta_w \sin \delta_b \cos \theta_b + \sin \delta_w \cos \delta_b \, , \tag{4.3}$$

where ρ = angle between bed pole and fault dip vector, δ_w = dip of well, θ_w = azimuth
of well dip, δ_b = dip of bed, θ_b = azimuth of bed dip.

Method 2. Using bed dip vector and points on top and base of unit
If the locations of the unit boundaries are given by their xyz coordinates, and the orientation of bedding by its azimuth and dip, the angle ρ required in Eq. 4.1 may be found by substituting Eq. 12.9 into Eq. 12.25:

$$\rho = \cos^{-1}\{[(x_1 - x_2)/L]\sin\theta_b \sin\delta_b + [(y_1 - y_2)/L]\cos\theta_b \sin\delta_b + [(z_2 - z_1)/L]\cos\delta_b\} , \tag{4.4}$$

where θ_b and δ_b =, respectively, the azimuth and dip of the bed dip vector, and L, the apparent length, is

$$L = [(x_2 - x_1)^2 + (y_2 - y_1)^2 + (z_2 - z_1)^2]^{1/2} . \tag{4.5}$$

Method 3. Using bed dip vector and line on map
If the line of the thickness measurement is defined by its map length, h, change in elevation, v, and orientation given by the azimuth, θ, to the lower end of the line, then the angle ρ required in Eq. 4.1 may be found by substituting Eq. 12.7 into Eq. 12.25:

$$\rho = \cos^{-1}\{\cos\delta_b \sin[\arctan(v/h)] - \sin\delta_b \cos[\arctan(v/h)](\cos\theta_b \cos\theta + \sin\theta_b \sin\theta)\} , \tag{4.6}$$

where θ_b and δ_b =, respectively, the azimuth and dip of the bed dip vector, and L, the apparent length, is

$$L = (v^2 + h^2)^{1/2} , \tag{4.7}$$

where v = vertical distance between end points and h = the horizontal distance between end points. The angle in Eq. 4.6 must be acute and must be changed to $180 - \rho$ if it is obtuse.

4.1.2
Thickness between Structure Contours

The thickness determined between structure contours is straightforward to compute and generally shows much less variability than that determined between individual points on an outcrop map. This approach provides a more reliable value in situations where the attitudes and contact locations are uncertain on a map. Determining the best-fit structure contours uses a large amount of data simultaneously to improve the attitude of bedding and the contact locations. This method requires a structure contour at the top and base of the bed (Figs. 4.4, 4.5a). The width of the unit is always measured in the dip direction, perpendicular to the structure contours. If contours at the same elevation can be constructed on the top and base of the unit, from the geometry of Fig. 4.5b, the thickness of the unit is

$$t = h_c \sin\delta , \tag{4.8}$$

where h_c = horizontal distance between contours at equal elevations on the top and base of the unit, t = true thickness, and δ = true dip. If the contours on the top and

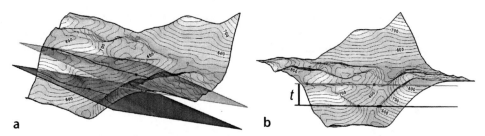

Fig. 4.4. Oblique views of planar unit boundaries cutting a topographic surface. Map view is in Fig. 4.5a. **a** Upper and lower bed surfaces, view toward north. **b** View to northeast parallel to bedding showing thickness, *t*

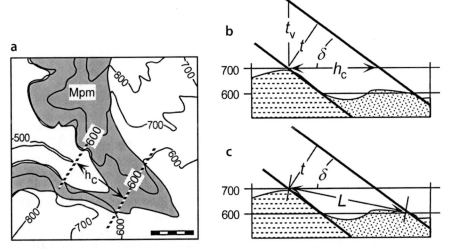

Fig. 4.5. Thickness measured between structure contours. **a** Structure contours at 600-ft elevation on the top and base of a formation; h_c is perpendicular to structure contours (3-D views in Fig. 4.4). **b** Measurement along a constant elevation on a vertical cross section in the dip direction. **c** Measurement between points of different elevations on a vertical cross section in the dip direction. For explanation of *symbols*, see text

base of the unit are at different elevations (Fig. 4.5c), then the line on the map that connects the upper and lower contours has the length *L*, and the thickness can be calculated from Eq. 4.1. Equation 4.1 gives the same result as Eq. 4.8 for the special case where *L* is horizontal.

The vertical thickness can readily be computed by taking the difference in elevation between structure contour maps on the top and base of the unit at a given *xy* point (Fig. 4.5b). Then the true thickness is calculated from Eq. 4.2, rewritten as

$$t = t_v \cos \delta \ , \tag{4.9}$$

where *t* = true stratigraphic thickness, t_v = vertical thickness and δ = dip.

As an example, in Fig. 4.5a, the 600-ft contour has been located on both the top and base of the Mpm, and so the simplest means of thickness determination is with Eq. 4.8. The value of h_c measured from the map is 1 387 ft, and the dip, δ, is 04°, giving a thickness of 97 ft.

4.1.3
Map-Angle Thickness Equations

The calculation of the thickness of a unit based entirely on map distances and directions results in two equations, depending on whether the topographic slope is in the same direction as the dip of the unit or the opposite direction to it. For the bed dip and topographic slope in the same direction, the equation (derived at the end of this chapter as Eqs. 4.18 and 4.21) is

$$t = |h \cos \alpha \sin \delta - v \cos \delta| \ , \tag{4.10}$$

where the notation $|...|$ = the positive value of the expression between the bars. For a bed dip and topographic slope that are in opposite directions:

$$t = h \cos \alpha \sin \delta + v \cos \delta \ , \tag{4.11}$$

where h (Fig. 4.6) = the horizontal distance along a line between the upper and lower contacts on the map, α = the angle between the measurement line and the dip direction, δ = the true dip of the unit, and v = the elevation difference between the end points of the measurement line. If the measurement is in the dip direction, $\alpha = 0$ and $\cos \alpha = 1$; if the base and top of the unit are both at the same elevation, $v = 0$.

As an example, determine the thickness of the Pride Mountain Formation (Mpm) indicated by its outcrop width on the map of Fig. 4.7. Along line a on the map, the horizontal width of the outcrop is $h = 1\,250$ ft, the vertical drop from highest to lowest contact is $v = 100$ ft, the attitude of bedding is $\delta = 07°$ (the average of the 06 and 08° dips mapped) at an azimuth $\theta = 135°$, and the angle between the dip direction and the measurement line is $\alpha = 85°$. The bed and topography slope in the same direction (strike of bedding = 225° and the thickness traverse is along azimuth 220°), making Eq. 4.10 appropriate. The resulting thickness is 86 ft.

Fig. 4.6.
Map data needed to determine the thickness of a unit from outcrop observations using the map-angle thickness Eqs. 4.10 and 4.11

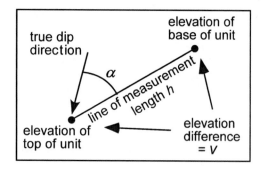

Fig. 4.7.
Line of a thickness measurement (*a*) on a geologic map
on a topographic base

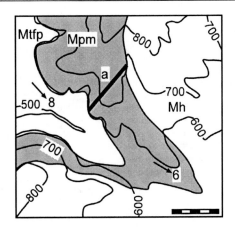

4.1.4
Effect of Measurement and Mapping Errors

In the determination of thickness from measurements on a map, there are three sources of error, the length and the direction measurements, the dip measurement, and the contact locations. Errors in length will result from the finite widths of the contact lines at the scale of the map. The minimum possible length error is approximately equal to the thickness of the geologic contacts as drawn on the map. For example, the geologic contacts in Fig. 4.7 are about 20 ft wide. Thus a thickness error of about ±20 ft is about the best that could be done on this map. Small errors in contact locations may lead to large errors in the calculated thickness, depending on the measurement direction. Uncertainties in the dip amount and dip direction of a degree or two are to be expected. The effects of error in the dip range from very small if the thickness measurement is at a high angle to the bed boundary, to very large if the measurement is at a low angle to the boundary.

The effect of errors in bedding attitude and measurement direction on a thickness determination can be estimated using Eq. 4.1 which places the variables in their simplest form. The true thickness (t) is directly proportional to the apparent thickness (L) and so the erroneous thickness (t_e) is directly proportional to the error in the length of the apparent thickness measurement. The error measure (t_e) is normalized in Fig. 4.8 to remove the effect of the length scale. The true thickness in Eq. 4.1 is proportional to the cosine of the angle between the pole to the bed and the line of the thickness measurement (ρ), leading to a non-linear dependence of the error on the size of the angle (Fig. 4.8). The angle error could be in the orientation of the bed, in the orientation of the apparent thickness, or a combination of both. A combined error in ρ of ±5° from the correct value seems like the upper limit for careful field or well measurements. An angle of $\rho = 0°$ means that the thickness measurement is perpendicular to bedding; this is the most accurate measurement direction. An angle of $\rho = 90°$ means that the measurement direction is parallel to bedding, an impossibility. The sensitivity of a thickness measurement to error in the angle goes up rapidly as ρ increases (Fig. 4.8). The thickness error exceeds ±10% at an angle of 40° and is about 100% at 80 + 5° and

Fig. 4.8.
Effect of attitude and angle
measurement errors on thickness calculation. Normalized
error in thickness measurement, in percent as a function
of the angle ρ (Eq. 4.1) for
angle errors of $\rho = \pm 5°$. t True
thickness of bed; t_e erroneous
thickness given the angle error; ρ angle between the pole
to the bed and direction along
which thickness is measured

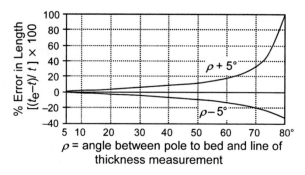

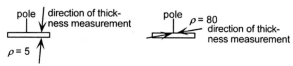

−35% at 80 − 5°. This means that thickness measurements made at a low angle to the
plane of the bed may produce very large errors, with a strong bias toward overestimation. At angles between the measurement direction and bedding of 10° or less ($\rho = 80°$
or greater), very large thickness errors will occur with very small orientation errors
and thickness determinations made from maps should be considered suspect.

From Fig. 4.8, a bed having a true thickness of 100 m measured at an angle of 10°
to bedding ($\rho = 80°$) for which the angle is overestimated by 5° (giving $\rho = 85°$) will
yield a thickness of nearly 200 m. The same measurement with a 5° underestimate in
the angle will give a thickness of 65 m. The average thickness for these two measurements is 132.5 m, still an overestimate. Thickness measurements made nearly perpendicular to bedding ($\rho = 0°$) are rather insensitive to errors in the angle. At $\rho = 20°$
the error is about ±3%; a 100 m thick bed would be measured as being between 97
and 103 m thick.

Thickness calculations between two points on a map are very sensitive to the accuracy of the contact locations and the attitude of bedding, as illustrated by the data
obtained from Fig. 4.9a. The thickness of the Mpm from the seven locations a–g
(Table 4.1) ranges from 84 to 230 ft, as calculated from Eq. 4.1. This is an unreasonably
large variation in thickness at what is nearly a single location at the scale of the map.
What is the probability that the thickness variation is due to small measurement errors? The bedding azimuth of 4, 125 represents the value determined from the structure-contour map (Fig. 4.9b). The difference in dip from 04° on the structure contour
map to 06° from the field measurement is responsible for a large variation in the calculated thickness. For example, at point c, the thickness along a single line is 127 ft for
a 04° dip compared to 230 ft for a 06° dip (Table 4.1). However, the substantial difference in calculated thickness along lines a and g is not the result of uncertainty in the
dip, but must be attributed to the uncertainty both in the location of the lower contact
and in the exact dip direction. Changing the azimuth of the dip from 135 to 127 at
location a increases the thickness from 84 to 108 ft at a bed dip of 08°; at location g the
same change reduces the thickness from 173 to 159 ft. If we say that the thickness of

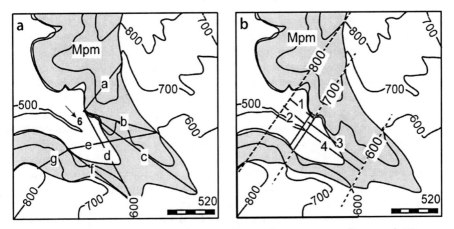

Fig. 4.9. Alternative thickness measurements. **a** Point-to-point measurement lines *a–g*. **b** Measurements *1–4* between structure contours

Table 4.1. Thickness data in map area of Fig. 4.9. All thicknesses calculated with Eq. 4.1

Location	Bed		Azimuth of apparent *t*	*h* (ft)	*v* (ft)	α (deg)	Calculated *t* (ft)
	Dip	Azimuth					
a	7	135	220	1242	100	85	86
	8	135	220	1242	100	85	84
b	7	125	105	1658	0	20	190
	4	125	105	1658	0	20	109
c	6	125	126	2960	80	1	230
	4	125	126	2960	80	1	127
d	6	125	150	2104	0	25	199
	4	125	150	2104	0	25	133
e	7	125	080	1030	0	45	174
	4	125	080	1030	0	45	99
f	7	125	127	1770	0	2	216
	4	125	127	1770	0	2	123
g	7	135	059	741	200	76	177
	8	135	059	741	200	76	173
1 SE	8	127	127	631	200	0	110
2 NW	8	127	127	564	200	0	120
3 SE	4	124	124	1503	0	0	105
4 NW	4	124	124	1387	0	0	97

the Mpm is the average of the values at locations a–g obtained using the observed dips, then the thickness is 181 ft with a range from 84 to 230 ft, a poorly constrained result.

The thicknesses determined at locations 1–4 (Fig. 4.9b) between the structure contours on the top and base of the unit average 108 ft thick and range from 97 to 120 ft

(Table 4.1) or 108 ±12 ft based on the whole range of values. The SE lengths (Table 4.1) are measured to a southeasterly position on the base of the Mpm, at the 600-ft contour, which lies directly beneath the 700-ft contour on the top of the unit; the NW measurements are from the more northwesterly position of the lower contact. Changing the location of the structure contour of the base has only a small effect on the thickness. The average thickness determined from the structure-contour-based measurements falls within the range of the point-to-point measurements, but is much smaller than the average of the point-to-point measurements, as expected from the behavior of the thickness equation (Fig. 4.8). Thickness measurements between two points (Eqs. 4.10 and 4.11, or 4.1) exhibit a non-linear sensitivity to error at low angles between the dip vector and the measurement orientation, leading to a high probability of an artificially high average from multiple measurements. Smoothing of the attitude errors by structure contouring leads to a better average thickness.

Where the thickness is known accurately from a complete exposure or from well-defined contacts in a borehole, the structure contours or bedding attitudes might be adjusted to conform to the thicknesses. The thickness measured between structure contours is the best approach at the map scale where there is uncertainty in the data.

4.2
Thickness of Folded Beds

In a folded bed, the dips of the upper and lower contact are not the same and the previous thickness equations are inappropriate. The fold is likely to approach either the planar dip domain or the circular arc form. Equations for both forms are given in the next two sections. For both methods it is assumed that the thickness is constant between the measurement points and that the line of the thickness measurement and the bedding poles are all in the plane normal to the fold axis. The latter condition is satisfied if the directions of both dips and the measurement direction are the same. If the geometry is more complex than this, then a cross section perpendicular to the fold axis should be constructed to find the thickness and projection may be required, as discussed in Chap. 6.

4.2.1
Circular-Arc Fold

The thickness of a bed that is folded into a circular arc (Fig. 4.10) can be found if the dip direction of the bed and the well or traverse line are coplanar. In this situation the bedding poles intersect at a point. Let ρ_1 be the smaller angle between the well and the pole to bedding, thus always associated with the longer radius, r_1. The thickness, t, is

$$t = r_1 - r_2 \; . \tag{4.12}$$

From the law of sines:

$$r_2 = (L \sin \rho_1) / \sin \gamma \; , \tag{4.13}$$

$$r_1 = (L \sin (180 - \rho_2)) / \sin \gamma \; , \tag{4.14}$$

Fig. 4.10.
Thickness of a bed folded into
a circular arc. The dip of the
bed and the well are co-planar.
C is the center of curvature
where the poles to bedding
intersect

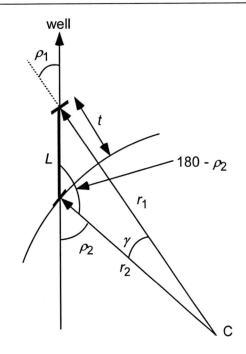

Fig. 4.11.
Example of thickness determination of a circularly folded
bed. **a** Field data: bedding dips
are shown by *heavy lines*, distance between exposures of
upper and lower contacts is
300 ft on a line that plunges 10°.
b Angles required for the thickness calculation

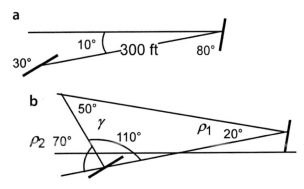

where ρ_2 and $r_2 =$, respectively, the angle between the bed pole and the well and the
radius associated with the larger angle, and $\gamma = \rho_2 - \rho_2$. Substitute Eqs. 4.13 and 4.14
into 4.12 and replace $\sin(180 - \rho_2)$ with $\sin \rho_2$ to obtain the thickness:

$$t = (L / \sin \gamma)\,(\sin \rho_2 - \sin \rho_1) \ . \tag{4.15}$$

A typical data set is shown in Fig. 4.11a. The cross section is in the dip direction.
The angles between the line of measurement and the poles to bedding are determined
as well as the acute angle between the poles (Fig. 4.11b). From Eq. 4.15, the true thickness of the bed is 234 ft.

4.2.2
Dip-Domain Fold

The dip-domain method can be used to find or place bounds on the thickness of a unit that changes dip from its upper to lower contact (Fig. 4.12). For constant bedding thickness, the axial surface bisects the angle of the bend. The total thickness of the bed along the measurement direction, t, is the sum of the thickness in each domain, found from Eq. 4.1 as

$$t = L_1 \cos \rho_1 + (L - L_1) \cos \rho_2 \; , \tag{4.16}$$

where ρ_1 = the angle between the well and the pole to the upper bedding plane, ρ_2 = the angle between the well and the pole to bedding of the lower bedding plane, and L_1 = the apparent thickness of the upper domain. If the position of the dip change can be located, for example with a dipmeter, then it is possible to specify L_1 and find the true thickness. If the location of the axial surface is unknown, the range of possible thicknesses is between the values given by setting $L_1 = 0$ and $L_1 = L$ in Eq. 4.16. The circular-arc thickness (Eq. 4.15) is usually half-way between the extremes that are possible for dip-domain folding.

4.3
Thickness Maps

Thickness maps are valuable for both structural and stratigraphic interpretation purposes. Because multiple measures of thicknesses can be mapped, care is required in the interpretation. The calculated thickness is related to the dip and so uncertainties

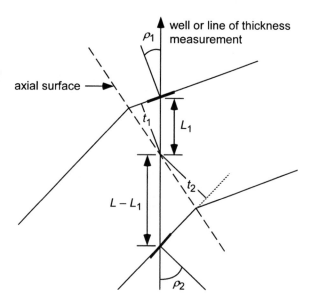

Fig. 4.12.
Thickness of a dip-domain bed that changes dip in the measured interval, in a cross section normal to the fold axis

or errors in the dip may appear as thickness anomalies. An isopach map is a map of the true thickness of the unit (t, Fig. 4.1) measured normal to the unit boundaries (Bates and Jackson 1987). An isocore map is defined as a map of the vertical thickness of a unit (t_v; Fig. 4.1; Bates and Jackson 1987). The drilled thickness in a deviated well (t_s, Fig. 4.1) will usually differ from either the true thickness or the vertical thickness. It is not possible to correct the thickness in a deviated well to the vertical thickness or to the true thickness without knowing the dip of the bed. The thickness differences resulting from the different measurement directions are not large for nearly horizontal beds cut by nearly vertical wells, but increase significantly as the relationships departs from this condition. The effects of stratigraphic and dip variations on thickness maps are considered here. The effect of faults on isopach maps are discussed in Sect. 8.5.

4.3.1
Isopach Maps

An isopach map is used to show thickness trends from measurements at isolated points (Fig. 4.13a). An isopach map can be interpreted as a paleotopographic map if the upper surface of the unit was close to horizontal at the end of deposition. If the paleotopography was controlled by structure, then it can be considered to be a paleostructure map. The thickness variations represent the structure at the base of the unit as it was at the end of deposition of the unit. The trend of increased thickness down the center of the map in Fig. 4.13a could imply a filled paleovalley.

The slope of the base of the paleovalley can be determined from the thickness difference and the spacing between the contours according to the geometry of Fig. 4.13b:

$$\delta = \arctan\left(\Delta t \,/\, h\right) \ , \tag{4.17}$$

where δ = the slope, Δt = the difference in thickness between two contours, and h = the horizontal (map) distance between the contours, measured perpendicular to the contours. For the map in Fig. 4.13a, the slope implied for the western side of the paleovalley is about 0.5° ($\Delta t = 10$, $h \approx 900$). Stratigraphic thickness variations could be caused by growing structures. The dip calculated from an isopach map using Eq. 4.17 could represent the structural dip that developed during deposition. According to this interpretation, Fig. 4.13a could represent a depositional syncline.

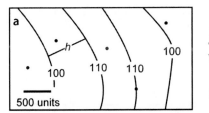 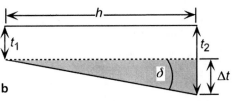

Fig. 4.13. Paleoslope from thickness change. **a** Isopach map. *Dots* are measurement points; h is the location of a cross section. **b** Cross section perpendicular to the trend of the thickness contours, interpreted as if the upper surface of the unit were horizontal

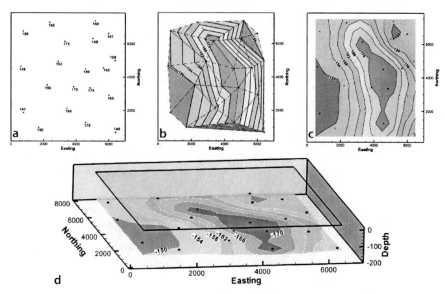

Fig. 4.14. Interpretation of a thickness map. **a** Thickness data. **b** Triangulation contouring. **c** Kriged map on a 10 × 10 grid. **d** Paleostructural interpretation produced by making the thicknesses negative on the kriged map. The plane of zero elevation is shown above the map and would represent the upper surface of the unit at the end of deposition

As an example, an isopach map is constructed from the data in Fig. 4.14a. For the purpose of discussion, the points are contoured by both triangulation (Fig. 4.14b) and kriging (Fig. 4.14c). The paleogeographic implications of the map should be considered before either map is accepted. The triangulated map suggests a stream channel whereas the kriged map suggests an isolated depocenter. Both computer contouring methods close the contours within the map area. Re-examination of the data reveals that if the unit represents a channel, it could be extended off the map to both the north and south and still be consistent with all the control points. Accepting the depocenter interpretation, it can be visualized in 3-D as a paleostructure map by reversing the sign on the contours so that the thickest part plots as the deepest (Fig. 4.14d).

Tthickness trends on isopach maps could alternatively represent unrecognized faults that are too small to be identified directly. A normal fault will cause a thinning of the isopachs and a reverse fault will cause a thickening. Section 8.5 discusses faults on isopach maps. Figure 4.14 could represent a reverse fault that is too small to repeat the top and base of the unit.

4.3.2
Isocore Maps

Isocore maps are particularly valuable for determining the volume of a unit present in the area of interest. The area enclosed by each isocore contour is multiplied by the

Fig. 4.15.
Dip of bed related to vertical apparent thickness, t_v: vertical thickness; t: true thickness, δ: dip of bed

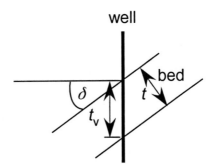

contour interval and then summed to obtain the volume. This is only an approximation because it assumes that the volume consists of a stack of vertical-sided regions. The smaller the contour interval, the better the estimate.

Apparent thickness variations in vertical wells can provide a very sensitive tool for structural analysis if bed thickness is known (Fig. 4.15). True dip is not known in a well unless the interval of interest has been cored or a dipmeter log is available. If the true thickness is known, the dip can be determined by solving Eq. 4.1 to obtain

$$\delta = \arccos(t / t_v) \ . \tag{4.18}$$

If a unit has a true thickness of 100 m, a dip of 10° gives an exaggerated thickness of 102 m, 20° gives 106 m, 30° gives 115 m, 40° gives 131 m, 50° gives 156 m. The importance of this effect will depend on the level of detail being interpreted, but will become significant for nearly any purpose at dips over 20–30°. For example, a measured thickness of 103 ft for a unit having a true thickness of 100 ft may be stratigraphically insignificant, yet implies a dip of 14° which is steeper than the dip producing the closure in many oil fields. If the unit mapped in Fig. 4.13a actually has a constant thickness of 100 units, then the dips in the center of the map where the isocore thickness is 110 units must be 25°. Alternatively, the unit could be horizontal and the wells in which the thicknesses were observed could deviate 25° from the vertical.

As an example of the importance of dip on the variation of apparent thickness, the thickness map in Fig. 4.14a is reinterpreted as representing isocore thicknesses of a folded unit of constant stratigraphic thickness. The thickness variations are converted into dips with Eq. 4.18, assuming a stratigraphic thickness of 150 units, and the values triangulated (Fig. 4.16). What was previously interpreted as a thickness trend is now seen as a dip trend with dips up to 34°. This could represent a significantly folded unit with the steepest dips representing the inflection point on the limb between syncline and anticline. A structure contour map of the unit should be constructed and examined for correspondence between the trends. Note that thicknesses smaller than the assumed constant value yield spurious values when processed with Eq. 4.18. The northeast and southwest corners of the map in Fig. 4.16 would be better interpreted as regions of zero dip because the thicknesses are close to and slightly less than the assumed regional constant value.

Fig. 4.16.
Dip map of data from Fig. 4.14a interpreted as isocore thicknesses measured in a folded, constant-thickness unit. True stratigraphic thickness is 150, dips in degrees

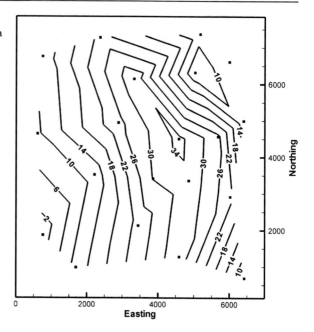

4.4
Derivation: Map-Angle Thickness Equations

The traditional method for determining thickness based on data from a geologic map on a topographic base results in two equations, depending on the relative dip of topography and bedding. The following derivations are after Dennison (1968). If the ground slope and the dip are in the same general direction, Fig. 4.17 shows that t = ah = the true thickness, bc = fe = v = the vertical elevation change, ac = h = the horizontal distance from the upper to the lower contact, angle cae = α = the angle between the measurement direction and the true dip, and angle aej = feg = δ = the true dip. The thickness is

$$t = aj - hj = aj - eg \quad ,$$ (4.19)

where

$$eg = v \cos \delta \quad ,$$ (4.20a)

$$aj = ea \sin \delta \quad ,$$ (4.20b)

$$ea = h \cos \alpha \quad .$$ (4.20c)

Substitute Eqs. 4.20 into 4.19 to obtain

$$t = |h \cos \alpha \sin \delta - v \cos \delta| \quad .$$ (4.21)

Fig. 4.17.
Thickness parameters for a
bed and topographic surface
dipping in the same general
direction

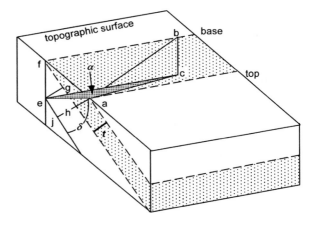

Fig. 4.18.
Thickness parameters for a
bed and topographic surface
dipping in opposite general
directions

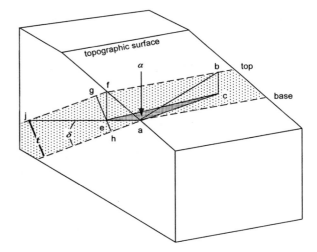

If the dip of bedding is less than the dip of the topography, the second term in Eq. 4.21
(D2.24) is larger than the first, giving the correct, but negative, thickness. Taking the
absolute value corrects this problem.

The thickness of a unit which dips opposite to the slope of topography (Fig. 4.18) is

$$t = eg + eh \quad , \tag{4.22}$$

where

$$eh = ea \sin \delta \quad . \tag{4.23}$$

Substituting Eqs. 4.20a, 4.20c, and 4.23 into 4.22:

$$t = h \cos \alpha \sin \delta + v \cos \delta \quad . \tag{4.24}$$

4.5
Exercises

4.5.1
Interpretation of Thickness in a Well

Based on the data in Table 2.2, what is the isocore thickness of the Smackover? What is the true thickness of the Smackover given its attitude of 12, 056 from the dipmeter log and the orientation of the well from Exercise 2.9.1? Discuss the significance of the difference between the isopach and isocore thickness.

4.5.2
Thickness

Given a bed with dip vector 10, 290, and a measured thickness of 75 m in a vertical well, use the universal-thickness equation to determine its true thickness.

4.5.3
Thickness from Map

Use the map of the Blount Springs area (Fig. 2.27) to answer the following questions. What is the thickness of the Mpm between the structure contours using the map-angle equations and the pole-thickness equation? Are the results the same? If they are different, discuss which answer is better. What is the difference between the true thickness and the vertical thickness of the Mpm? What is the thickness of the Mpm in its northeastern outcrop belt, assuming that the dip is 28° at its northwestern contact and the value determined above occurs at its southeastern contact? Use the concentric fold model and the dip-domain model. Discuss the effect of changing the location of the axial surface on the thickness computed with the dip-domain model. Measure the thickness of the Mpm at 5–10 locations evenly distributed across the map. Measure thicknesses between structure contours where possible. Construct an isopach map from your thickness measurements. Is the unit constant in thickness? What would be the apparent thickness of the Mpm in a north-south, vertical-sided roadcut through the northwestern limb of the anticline?

4.5.4
Isopach Map

Make an isopach map of the sandstone thicknesses on the map of Fig. 4.19. The thickest measurements form a trend that could be a channel or the limb of a monocline. If the thickness anomaly is due to a dip change, what is the amount? If the thickness anomaly is due to paleotopography, what is the maximum topographic slope?

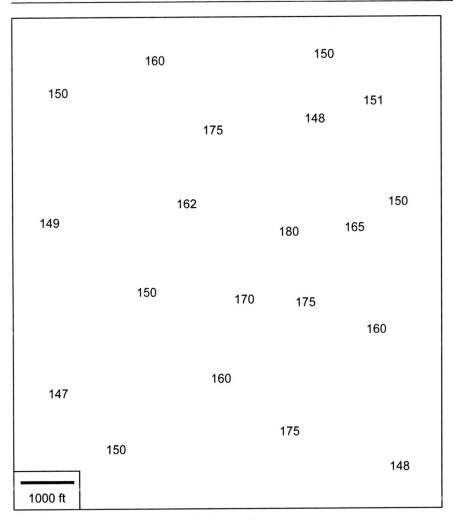

Fig. 4.19. Map of thicknesses (in feet) in the John S sandstone

Chapter 5

Fold Geometry

5.1
Introduction

This chapter describes methods for defining the geometry of folded surfaces in three dimensions and methods for projecting data along and within fold trends. Folds can be divided into domains where the shapes are cylindrical or conical, smoothly curved or planar. The geometries within domains are efficiently described in terms of the orientations and properties of fold axes, plunge lines, crest lines and trough lines. The relationship of these elements to bed attitudes has implications for bed thickness changes and the persistence of the folds along their trend.

5.2
Trend from Bedding Attitudes

The fold trend and plunge is a key element in making and confirming the grain in a map. The change in shape along plunge is given by the fold form, cylindrical or conical. A fold in a cylindrical domain continues unchanged along plunge, whereas a fold in a conical domain will die out along plunge. The trend, plunge, and style of a fold are determined from the bedding attitudes as plotted on stereograms or tangent diagrams. The use of the tangent diagram is emphasized here because of its practical value in separating cylindrical from conical folds and in characterizing the type of conical plunge. The bedding attitude data are collected from outcrop measurements or from dipmeters. Dip-domain style folds may combine both cylindrical and conical elements. If the data show too much scatter for the form to be clear, the size of the domain under consideration can usually be reduced until the domain is homogeneous and has a cylindrical or conical geometry.

5.2.1
Cylindrical Folds

A cylindrical fold is defined by the property that the poles to bedding all lie parallel to the same plane regardless of the specific cross-sectional shape of the fold (Fig. 5.1a). This property is the basis for finding the fold axis. On a stereogram the poles to bedding fall on a great circle (Fig. 5.1b). The pole to this great circle is the fold axis, known as the π axis when determined in this manner. The trend of a cylindrical fold is parallel to its axis. A cylindrical fold maintains constant geometry along its axis as long as the trend and plunge remain constant.

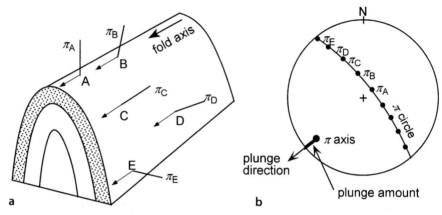

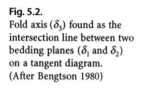

Fig. 5.1. Axis of a cylindrical fold. **a** Fold geometry. *A–E* are measurement points, π_A–π_E are poles to bedding. **b** Axis (π axis) determined from a stereogram, lower-hemisphere projection. (After Ramsay 1967)

Fig. 5.2.
Fold axis (δ_3) found as the intersection line between two bedding planes (δ_1 and δ_2) on a tangent diagram. (After Bengtson 1980)

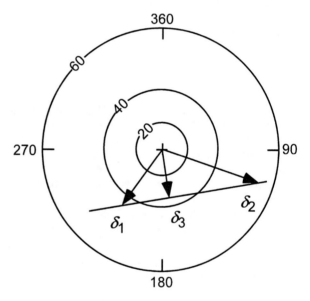

The alternative method for finding the axis is to plot the bedding attitudes on a tangent diagram. The method is based on the principle that intersecting planes have the same apparent dip in a vertical plane containing their line of intersection (Bengtson 1980). Let δ represent the dip vector of a plane. In Fig. 5.2, planes δ_1 and δ_2 are plotted and connected by a straight line. The perpendicular to this line through the origin, δ_3, gives the bearing and plunge of the line of intersection. In a cylindrical fold, all bedding planes intersect in the straight line (δ_3) which is the fold axis.

Each bedding attitude is plotted on the tangent diagram as a point at the appropriate azimuth and dip. If the best fit curve through the dip-vector points is a straight line,

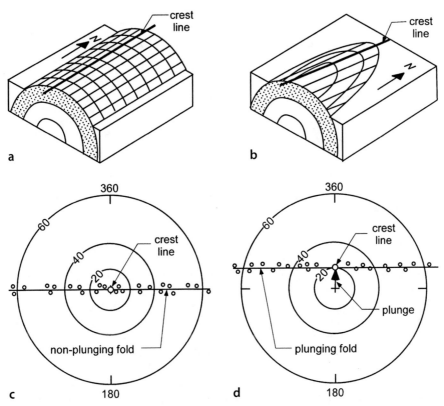

Fig. 5.3. Cylindrical folds showing trend and plunge of the crest line. **a** Non-plunging. **b** Plunging. **c** Tangent diagram of bed attitudes in a non-plunging fold. **d** Tangent diagram of bed attitudes in a plunging fold. (After Bengtson 1980)

the fold is cylindrical. The straight line through the dip vectors goes through the origin for a non-plunging fold (Fig. 5.3a,c) and is a straight line offset from the origin for a plunging fold (Fig. 5.3b,d). A vector from the origin to the line of dip vector points, perpendicular to the dip-vector line, that gives the bearing and plunge of the fold axis.

5.2.2
Conical Folds

A conical fold is defined by the movement of a generatrix line that is fixed at the apex of a cone (Fig. 5.4a); the fold shape is a portion of a cone. A conical fold terminates along its trend. On a stereogram the bedding poles fall on a small circle, the center of which is the cone axis and the radius of which is 90° minus the semi-apical angle (Fig. 5.4b). It is usually difficult to differentiate between cylindrical and slightly conical folds on a stereogram (Cruden and Charlesworth 1972; Stockmal and Spang 1982), yet this is an important distinction because a conical fold terminates along trend whereas a cylindrical fold does not. The tangent diagram is particularly good for making this distinction.

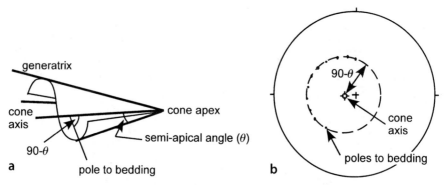

Fig. 5.4. Conical fold geometry. **a** Shape of a conically folded bed (after Stockmal and Spang 1982). **b** Lower-hemisphere stereogram projection of a conical fold (after Becker 1995)

On a tangent diagram, the curve through the dip vectors is a hyperbola (Bengtson 1980), concave toward the vertex of the cone. Type I plunge is defined by a hyperbola concave toward the origin. The fold spreads out and flattens down plunge (Fig. 5.5a,c). Type II plunge is defined by a hyperbola convex toward the origin on the tangent diagram (Fig. 5.5b,d) and the fold comes to a point down plunge. In the strict sense, a bed in a conical fold does not have an axis and therefore does not have a plunge like a cylindrical fold. The orientation of the crest line in an anticline (Fig. 5.5) or trough line in a syncline provides the line that best describes the orientation of a conically folded bed, in a manner analogous to the axis of a cylindrical fold. The plunge of a conical fold as defined on a tangent diagram is the plunge of the crest or trough line, not the plunge of the cone axis. If the crest line is horizontal, the fold is non-plunging but nevertheless terminates along the crest or trough. In general, each horizon in a fold will have its vertex in the same direction but at a different elevation from the vertices of the other horizons in the same fold.

The trend and plunge of the crest or trough of a conical fold is given by the magnitude the vector from the origin in the direction normal to the curve through the bedding dips (Figs. 5.6, 5.7). For points on the fold that are not on the crest or trough, the plunge direction is different. To find the plunge line, draw a line tangent to the curve at the dip representing the point to be projected. A line drawn from the origin, perpendicular to the tangent line gives the plunge amount and direction (Bengtson 1980). In a type I fold (Fig. 5.6), the minimum plunge angle is that of the crest line and all other plunge lines have greater plunge angles. In a type II fold (Fig. 5.7) the plunge line at some limb dip has a plunge of zero. The trend of this plunge line is normal to the tangent line (Fig. 5.7), the same as for all the other plunge lines. At greater limb dips in a type II fold, the down-plunge direction is away from the vertex of the cone (Fig. 5.7).

Because the tangent of 90° is infinity, a practical consideration in using a tangent diagram is that dips over 80° require an unreasonably large piece of paper. The non-linearity of the scale also exaggerates the dispersion at steep dips. A practical solution for folds defined mainly by dips under 80° is to plot dips over 80° on the 80° ring (Bengtson 1981b). This will have no effect on the determination of the axis and plunge of the folds (Bengtson 1981a,b) and will have the desirable effect of reducing the dis-

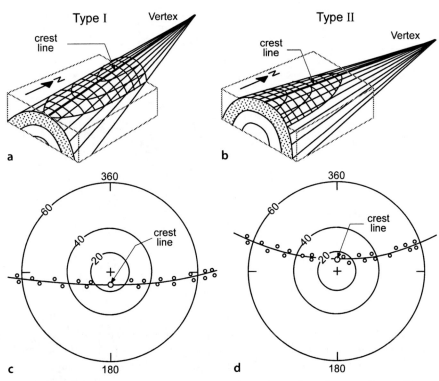

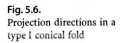

Fig. 5.5. Conical folds showing trend and plunge of crest line. **a** Type I plunge. **b** Type II plunge. **c** Tangent diagram of bed attitudes in a type I plunging fold. **d** Tangent diagram of bed attitudes in a type II plunging fold. (After Bengtson 1980)

Fig. 5.6.
Projection directions in a
type I conical fold

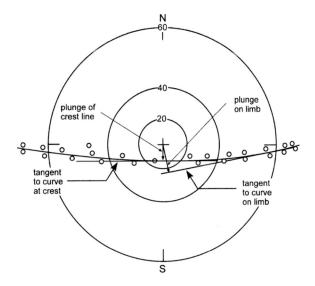

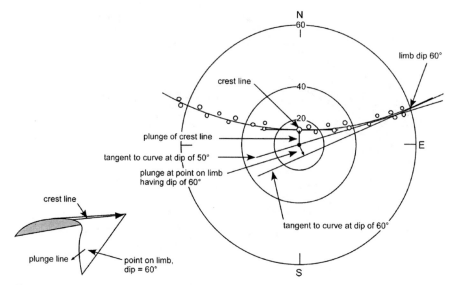

Fig. 5.7. Projection directions in a type II conical fold

person of the steep dips. It is primarily the positions of the more gently dipping points that control the location of the axis and its plunge. The stereogram is the best method for fold axis determination for folds which contain mainly very steep dips.

5.2.3
Tangent Diagram on a Spreadsheet

Plotting dips on a tangent diagram is readily accomplished using a spreadsheet, which then allows the spreadsheet curve-fitting routines to be used to find the best-fit line or curve through the data. The following procedure allows a tangent diagram to be plotted as a simple xy graph on a spreadsheet. To place north (zero azimuth) at the top of the page, shift the origin with

$$Az = Az - 90 \quad , \tag{5.1}$$

where Az = azimuth of the dip. Change the dip magnitude to the tangent of the dip with

$$r = \tan(\delta) \quad , \tag{5.2}$$

where r = radius and δ = dip magnitude. Then change from polar to Cartesian coordinates with

$$x = r \cos Az \quad , \tag{5.3a}$$

$$y = r \sin Az \quad . \tag{5.3b}$$

The new coordinates than can plotted as an *xy* graph and will have the geometry of a tangent diagram. The best-fit curve through the data can be found with standard curve-fitting routines. Experience using simple spreadsheet curve-fitting functions indicates that when a quadratic curve produces a smooth hyperbolic-like shape (eg., Fig. 5.8b), it is an appropriate fit to the data. If a quadratic curve is irregular, then a straight-line best fit is more appropriate.

5.2.4
Example Using a Tangent Diagram

Attitude data from a traverse across the central part of the Sequatchie anticline (Fig. 5.8a) compiled in Table 5.1 provides an example of fold-axis determination using

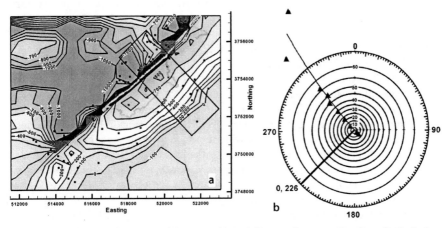

Fig. 5.8. Fold geometry of a portion of the Sequatchie anticline. **a** Index map to location of attitude data (*rectangle*) on structure-contour map. *Squares* represent measurement locations. **b** Tangent diagram for data within rectangle, showing best-fit curve through the data and the axis trend

Table 5.1.
Bedding attitudes across the central Sequatchie anticline from NW to SE

NW				SE	
Dip	Azimuth	Dip	Azimuth	Dip	Azimuth
8	308	56	318	6	144
46	315	75	330	8	145
34	316	83	315	8	144
50	320	70	315	6	127
6	320	0	000	7	136
22	316	5	145	10	136
				9	136

a tangent diagram. A preliminary plot of the data indicated that 83° dip attitude is inconsistent with the rest of the data and exerts too much control on the result, as mentioned above. Removing the 83° point results in a smooth best-fit curve that indicates a nonplunging fold with a slightly conical curvature, opening to the southwest and having a crestal trend of 0, 226 (Fig. 5.8b). This agrees with the geometry of the composite structure contour map (Fig. 5.8a).

5.2.5
Crest and Trough on a Map

A consistent definition of the fold trend, applicable to both cylindrical and conical folds, is the orientation of the crest or trough line (Fig. 5.9). In both cylindrical and conical folds (Figs. 5.3, 5.4) the crest line is a line on a folded surface along the structurally highest points (Dennis 1967). The trough line is the trace of the structurally lowest line. In cross section, the crest and trough traces are the loci of points where the apparent dip changes direction. In cylindrical folds the crest and trough lines are parallel to the fold axis and to each other, but in conical folds the crest and trough lines are not parallel. Crest and trough surfaces connect the crest and trough lines on successive horizons. The trace of a crest or trough surface is the line of intersection of the surface with some other surface such as the ground surface or the plane of a cross section. The crests and troughs of folds are of great practical importance because they are the positions of structural traps. Light fluids like most natural hydrocarbons will migrate toward the crests and heavy liquids will migrate toward the troughs.

The U-shaped trace of bedding made by the intersection of a plunging fold with a gently dipping surface, such as the surface of the earth, is called a fold nose. Originally a nose referred only to a plunging anticline (Dennis 1967; Bates and Jackson 1987) with a chute being the corresponding feature of a plunging syncline (Dennis 1967). Today, common usage refers to both synclinal and anticlinal fold noses. The term nose is also applied to the anticlinal or synclinal bend of structure contours on a single horizon (Fig. 5.9). The dip of bedding at the crest or trough line is the plunge of the line. The

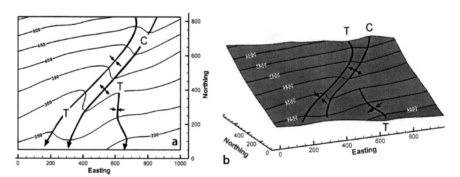

Fig. 5.9. Crest (*C*) and trough (*T*) lines on a structure contour map. **a** Plan view. *Arrows* point in the down-plunge direction. **b** 3-D oblique view to NE

spacing of structure contours along to the trend of the crest or trough gives the plunge of the crest or trough line:

$$\phi = \arctan\left(I / H\right) \quad , \tag{5.4}$$

where ϕ = plunge, I = contour interval, and H = map distance between the structure contours along the plunge.

5.3
Dip Domain Fold Geometry

A dip domain is a region of relatively uniform dip, separated from other domains by hinge lines or faults (Groshong and Usdansky 1988). Dip-domain style folding (Fig. 5.10) is rather common, especially in multilayer folds. A hinge line is a line of locally sharp curvature. On a structure contour map, hinge lines are lines of rapid changes in the strike of the contours. In a fixed-hinge model of fold development, the hinges represent the axes about which the folded layers have rotated.

The orientations of the hinge lines can be found using a stereogram by finding the intersection between the great circles that represent the orientations of the two adjacent dip domains (Fig. 5.11). If the intersection line is horizontal, as for hinge lines 1 and 2 in Fig. 5.10, the hinge line is not plunging. Hinge line 3 between the two dipping domains plunges 11° to the south (Fig. 5.11). The fold geometry between any two adjacent dip domains is cylindrical with the trend and plunge being equal to that of the hinge line. If the bed attitudes from multiple domains intersect at the same point on the stereogram, the fold is cylindrical, and the line is the fold axis, called the β-axis when determined this way (Ramsay 1967).

The tangent diagram (Fig. 5.12) shows the dip-domain fold from Fig. 5.10 to be conical with zero plunge. The concave-to-the-south curvature of the line through the dip vectors means that the fold terminates to the south, as observed (Fig. 5.10). If only the plunging nose of the fold is considered (Fig. 5.12), the hinge line is perpendicular to

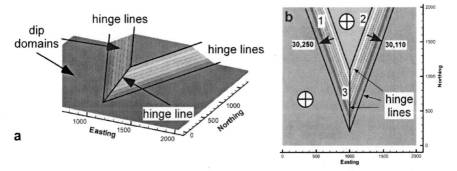

Fig. 5.10. Dip-domain fold (modeled after Faill 1973a). Hinge lines separate domains of constant dip. **a** 3-D oblique view to NW. **b** Bed attitudes and hinge lines (numbered), plan view

Fig. 5.11.
Hinge line orientations for the fold in Fig. 5.10, found from the bed intersections on a lower-hemisphere equal-area stereogram. The primitive *circle* that outlines the stereogram is the trace of a horizontal bed. One of the bedding dip domains has an attitude of 30, 250 and the other of 30, 110. *Numbered arrows* give orientations of hinge lines. Hinge line 1 has the direction 0, 340; hinge line 2 is 0, 020; hinge line 3 is 11, 180

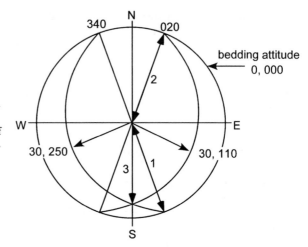

Fig. 5.12.
Tangent diagram of the conical dip-domain fold in Fig. 5.10. *Circles* give attitudes of the three dip domains present in the fold

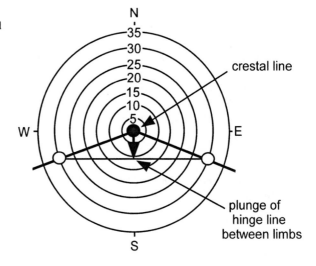

the straight line joining the two dip vectors (Fig. 5.12) and plunges 11° to the south, as observed (Figs. 5.10).

The analytical solution for the azimuth, θ', and the plunge, δ, of a hinge line can be found as the line of intersection between the attitudes of two adjacent dip domains, from Eqs. 12.4, 12.5 and 12.29–12.31:

$$\theta' = \arctan (\cos \alpha / \cos \beta) \ , \tag{5.5}$$

$$\delta = \arcsin (-\cos \gamma) \ , \tag{5.6}$$

where

$$\cos\alpha = (\sin\delta_1\cos\theta_1\cos\delta_2 - \cos\delta_1\sin\delta_2\cos\theta_2)/N \ , \tag{5.7a}$$

$$\cos\beta = (\cos\delta_1\sin\delta_2\sin\theta_2 - \sin\delta_1\sin\theta_1\cos\delta_2)/N \ , \tag{5.7b}$$

$$\cos\gamma = (\sin\delta_1\sin\theta_1\sin\delta_2\cos\theta_2 - \sin\delta_1\cos\theta_1\sin\delta_2\sin\theta_2)/N \ , \tag{5.7c}$$

$$N = [(\sin\delta_1\cos\theta_1\cos\delta_2 - \cos\delta_1\sin\delta_2\cos\theta_2)^2$$
$$+ (\cos\delta_1\sin\delta_2\sin\theta_2 - \sin\delta_1\sin\theta_1\cos\delta_2)^2$$
$$+ (\sin\delta_1\sin\theta_1\sin\delta_2\cos\theta_2 - \sin\delta_1\cos\theta_1\sin\delta_2\sin\theta_2)^2]^{1/2} \ , \tag{5.8}$$

and the azimuth and plunge of the first plane is θ_1, δ_1 and of the second plane is θ_2, δ_2. The value θ' given by Eq. 5.5 will be in the range of $\pm90°$ and must be corrected to give the true azimuth over the range of 0 to 360°. The true azimuth, θ, of the line can be determined from the signs of $\cos\alpha$ and $\cos\beta$ (Table 12.1). The direction cosines give a directed vector. The vector so determined might point upward. If it is necessary to reverse its sense of direction, reverse the sign of all three direction cosines. Note that division by zero in Eq. 5.5 must be prevented.

5.4
Axial Surfaces

The axial surface geometry is important in defining the complete three-dimensional geometry of a fold (Sect. 1.5) and in the construction of predictive cross sections (Sect. 6.4.1).

5.4.1
Characteristics

The axial surface of a fold is defined as the surface that contains the hinge lines of all horizons in the fold (Fig. 5.13; Dennis 1967). The axial trace is the trace of the axial surface on another surface such as on the earth's surface or on the plane of a cross section. The orientation of an axial surface within a fold hinge is related to the layer

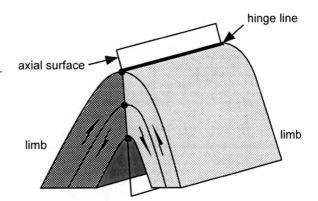

Fig. 5.13.
Characteristics that may be associated with an axial surface. The axial surface contains the hinge lines of successive layers (the defining property). The surface may be the boundary between fold limbs (different *shading*) and may be the plane across which the sense of layer-parallel shear (*double arrows*) reverses direction

thicknesses on the limbs. If, for example, the layers maintain constant thickness, the axial surface bisects the hinge. An axial surface may have one or both of the additional attributes: (1) it may divide the fold into two limbs (Fig. 5.13), and (2) it may be the surface at which the sense of layer-parallel shear reverses direction (Fig. 5.13). In a fold with only one axial surface, the axial surface necessarily divides the fold into two limbs; however, many folds have multiple axial surfaces, none of which bisect the whole fold (Fig. 5.14). The relationship between a fold hinge and the sense of shear on either side of it depends on the movement history of the fold. The movement history of a structure is called its kinematic evolution and a model for the evolution is called a kinematic model. According to a fixed-hinge kinematic model, the hinge lines are fixed to material points within the layer and form the rotation axes where the sense of shear reverses, as shown in Fig. 5.13. Other kinematic models do not require the sense of shear to change at the hinge (Fig. 5.14b).

Dip-domain folds commonly have multiple axial surfaces (Fig. 5.14) which form the boundaries between adjacent dip domains. These axial surfaces are not likely to have the additional attributes described in the previous paragraph. The folds are not split into two limbs by a single axial surface (except in the central part of Fig. 5.14a). In the fixed hinge kinematic model, the sense of shear does not necessarily change across an axial surface although the amount of shear will change (i.e., across axial surface 4, Fig. 5.14b). Horizontal domains may be unslipped and so an axial surface may separate a slipped from an unslipped domain (i.e., axial surfaces 2 and 3, Fig. 5.14b), not a change in the shear direction.

In the strict sense, a round-hinge fold does not have an axial surface because it does not have a precisely located hinge line. In a round-hinge fold it must be decided what aspect of the geometry is most important before the axial surface can be defined

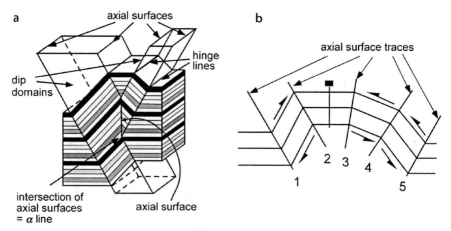

Fig. 5.14. Dip-domain folds. **a** Non-plunging fold showing hinge lines, axial surfaces, and axial surface intersection lines (modified from Faill 1969). Axial surface intersection lines are horizontal and parallel to the fold hinge lines. **b** *Half arrows* indicate the sense of shear in dipping domains. The horizontal domain does not slip, as indicated by the pin line. Axial surfaces are *numbered*

(Stockwell 1950; Stauffer 1973). For the purpose of describing the fold geometry, the virtual axial surface in a rounded fold is defined here as the surface which contains the virtual hinge lines. In relatively open folds, virtual hinge lines can be constructed as the intersection lines of planes extrapolated from the adjacent fold limbs (Fig. 5.15a). In a tight fold the extrapolated hinges are too far from the layers to be of practical use. In a tight fold (Fig. 5.15b) the virtual hinges can be defined as the centers of the circles (or ellipses) that provide the best fit to the hinge shapes, after the method of Stauffer (1973). Defined in this fashion, the axial surface divides the hinge into two limbs, but does not necessarily divide the fold symmetrically nor is it necessarily the surface where the sense of shear reverses. The axial surface, whether defined by actual or virtual fold hinges, does not necessarily coincide with the crest surface (Fig. 5.16) or the trough surface.

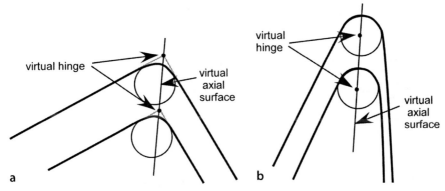

Fig. 5.15. Virtual hinges and virtual axial surfaces in cross sections of round-hinge folds. **a** Virtual hinges in an open fold found by extrapolating fold limbs to their intersection points. **b** Virtual hinges in a tight fold found as centers of circles tangent to the surfaces in the hinge. (After Stauffer 1973)

Fig. 5.16.
A fold in which the crest surface does not coincide with the axial surface

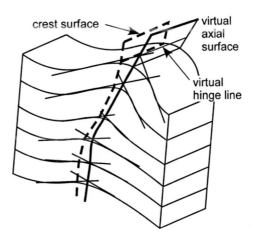

5.4.2
Orientation

Quantitative methods for the determination of the attitude of axial surfaces are given in this section. The orientation of an axial surface is related to the bed thickness change across it. In a cross section perpendicular to the hinge line, if a bed maintains constant thickness across the hinge, the axial surface bisects the hinge. The relationship between bed thickness and axial surface orientation is obtained as follows. Both limbs of the fold meet along the axial surface (Fig. 5.17) and have the common length h and therefore must satisfy the relationship

$$h = t_1 / \sin \gamma_1 = t_2 / \sin \gamma_2 \ . \tag{5.9}$$

If $t_1 = t_2$, then $\gamma_1 = \gamma_2$ and the axial surface bisects the hinge (Fig. 5.17a). If the bed changes thickness across the axial surface, then the axial surface cannot bisect the hinge (Fig. 5.17b). The thickness ratio across the axial surface is

$$t_1 / t_2 = \sin \gamma_1 / \sin \gamma_2 \ . \tag{5.10}$$

If the thickness ratio and the interlimb angle are known, then from the relationship (Fig. 5.17b)

$$\gamma = \gamma_1 + \gamma_2 \ , \tag{5.11}$$

where γ = the interlimb angle and γ_2 = the angle between the axial surface and the dip of limb 2. Substitute $\gamma_1 = \gamma - \gamma_2$ from Eq. 5.11 into 5.10, apply the trigonometric identity for the sine of the difference between two angles, and solve to find γ_2 as

$$\gamma_2 = \arctan \left[t_2 \sin \gamma / (t_1 + t_2 \cos \gamma) \right] \ . \tag{5.12}$$

The axial surface orientation in three dimensions can be found from a stereogram plot of the limbs. If the bed maintains constant thickness, the interlimb angle is bi-

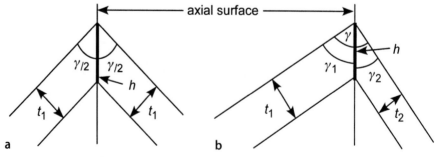

Fig. 5.17. Axial-surface geometry of a dip-domain fold hinge in cross section. **a** Constant bed thickness. **b** Variable bed thickness

sected. The bisector is found by plotting both beds (for example, 2 and *H* in Fig. 5.18), finding their line of intersection (*I*), and the great circle perpendicular to the intersection (*dotted line*). The obtuse angle between the beds is bisected by the double arrows. The great circle plane of the axial surface (shaded) goes through the bisection point and the line of intersection of the two beds. If bed thickness is not constant across the hinge, the partial interlimb angle is found from Eq. 5.12 and the appropriate partial angle is marked off along the great circle perpendicular to the line of intersection of the bedding dips.

Another method to determine the axial surface orientation is to find the plane through two lines that lie on the axial surface, for example, the trace of a hinge line on a map or cross section, and either the fold axis or the crest line (Rowland and Duebendorfer 1994). In the fold in Fig. 5.10, for example, hinge 3 plunges south and the surface trace of the hinge is north-south and horizontal and so the axial surface is vertical and strikes

Fig. 5.18.
Determination of the axial surface orientation using a stereogram. Angle between two constant thickness domains, one horizontal (bed H) and one 30, 110 (bed 2). The axial surface (*shaded great circle*) bisects the angle (*double arrows*) between bed 2 and the horizontal bed. The axial surface attitude is 75, 290. Lower-hemisphere, equal-area projection

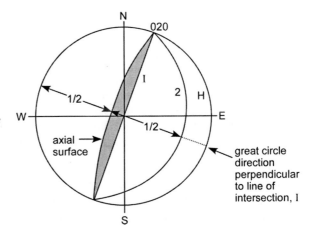

Fig. 5.19.
Plunging dip-domain fold. The axial surface intersection lines (α) plunge in a direction opposite to that of the hinge line

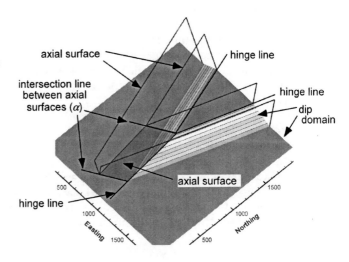

Fig. 5.20.
The intersection line between
axial surfaces. Axial surface 1–2
bisects bed 1 (30,250) and bed 2
(30, 110). Axial surface 2-H
bisects bed 2 and bed H (0, 000).
The line of intersection of axial
surfaces (50, 000) is the α line.
Lower-hemisphere, equal-area
stereogram

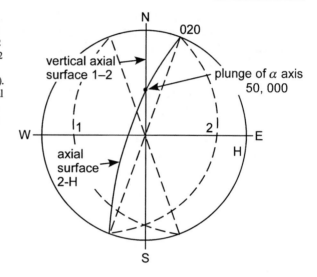

north–south. In the general case, the axial trace and fold axis are plotted as points on
the stereogram and the great circle that passes through both points is the axial surface.
If the axial trace and the hinge line coincide, as is the case for hinges 1 and 2 in Fig. 5.10,
this procedure is not applicable. It is then necessary to have additional information,
either the trace of the axial surface on another horizon or the relative bed thickness
change across the axial surface.

The intersection of two axial surfaces (Fig. 5.19) is an important line for defining
the complete dip-domain geometry, and is here designated as the α line (α for axial
surface). An α line has a direction and, unlike a fold axis, has a specific position in
space. Angular folds may have multiple α lines. In a non-plunging fold an α line is
coincident with a hinge line (Fig. 5.14a), but in a plunging fold (Fig. 5.19) the hinge
lines and the α lines are neither parallel nor coincident.

The orientation of the α line is found by first finding the axial surfaces, and then
their line of intersection. The stereogram solution is to draw the great circles for both
axial surfaces and locate the point where they cross (Fig. 5.20), which is the orientation of the line of intersection. The analytical solution for the α line is that for the
line of intersection between two planes (Eqs. 5.5–5.8), in which the two planes are
axial surfaces.

5.4.3
Location in 3-D

For a 3-D interpretation the axial surfaces and the α lines must be located in space as
well as in orientation. If hinge lines are known from more than one horizon they can
be combined to create a map (Fig. 5.21). The axial surface is the structure contour
map of equal elevations on the hinge lines.

Fig. 5.21.
Structure contour map of an axial surface. *Dashed lines* are hinge lines on three different horizons, *A*, *B*, and *C*. *Solid lines* are contours on the axial surface. Contours are in meters

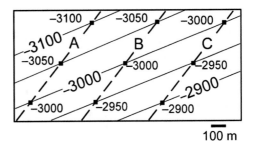

100 m

Fig. 5.22.
3-D axial surface geometry for the fold in Fig. 5.19. The axial surfaces bounding the left limb and the domain-intersection surface (bounded by α lines) are *shaded*, those bounding the right limb are indicated by their structure contours. The *heavy lines* represent a vertical slice through the structure: *ast:* axial surface trace; *bt:* bedding trace

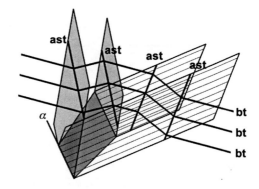

Axial traces can be determined from cross sections, and then mapped to form a surface (Fig. 5.22). The cross sections may cross the structure in any direction and need not be normal to the axial surfaces. In 3-D software, the axial traces can be extracted from successive sections and these lines triangulated to form the axial surface. Having multiple beds on the cross sections makes choosing the trace straightforward. If only one horizon has been mapped, axial traces can still be inferred but then the section must be normal to the hinge line and the effect of thickness changes must be considered.

5.5
Using the Trend in Mapping

It may be necessary to know the fold trend before a reasonable map of the surface can be constructed. Elevations of the upper contact of a sandstone bed are shown in Fig. 5.23a. Preliminary structure contours, found by connecting points of equal elevation, suggest a northwest-southeast trend to the strike. There is nothing in the data in Fig. 5.23a to definitively argue against this interpretation. Given the knowledge that the fold trends north–south with zero plunge, the map of Fig. 5.23b can be constructed. It can now be seen that the original map connected points on opposite sides of the anticline. The correct structure of the map area could not be determined without knowing the fold trend.

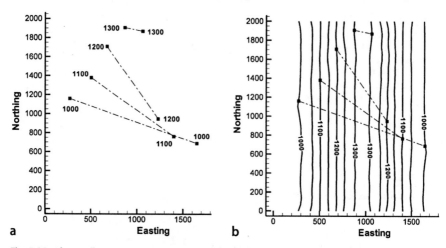

Fig. 5.23. Alternative structure contour maps from the same data. **a** Control points contoured.
b Structure contours based on interpretation that the structure is a north–south trending, non-plunging anticline. Contours from **a** are *dashed*

5.6
Minor Folds

Map-scale folds may contain minor folds that can aid in the interpretation of the map-
scale structure. The size ranking of a structure is the *order* of the structure. The largest
structure is of the first-order and smaller structures have higher orders. Second- and
higher-order folds are particularly common in map-scale compressional and wrench-fault
environments. The bedding attitudes in the higher-order folds may be highly discordant
to the attitudes in the lower-order folds (Fig. 5.24). Minor folds provide valuable information for interpreting the geometry of the first-order structure but, if unrecognized, may
complicate or obscure the interpretation. A pitfall to avoid is interpreting the geometry of
the first-order fold to follow the local bedding attitudes of the minor folds.

If produced in the same deformation event, the lower-order folds are usually coaxial
or nearly coaxial to the first-order structure and are termed parasitic folds (Bates and
Jackson 1987). Plots of the bedding attitudes of the minor folds on a stereogram or
tangent diagram should be the same as the plots for the first-order structures and can
be used to infer the axis direction of the larger folds. Seemingly discordant bedding
attitudes seen on a map can be inferred to belong to minor folds if they plot on the
same trend as the attitudes for the larger structure.

Asymmetric minor folds are commonly termed drag folds (Bates and Jackson 1987).
A drag fold is a higher-order fold, usually one of a series, formed in a unit located
between stiffer beds; the asymmetry is inferred to have been produced by bedding-
parallel slip of the stiffer units. The sense of shear is indicated by the arrows in Fig. 5.24
and produces the asymmetry shown. The asymmetry of the drag folds can be used to
infer the sense of shear in the larger-scale structure. In buckle folds, the sense of shear
is away from the core of the fold and reverses on the opposite limbs of a fold (Fig. 5.24),

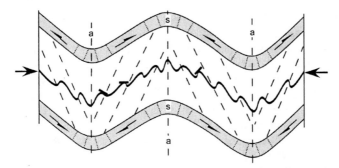

Fig. 5.24. Cross section of second-order parasitic folds in a thin bed showing normal (buckle-fold) sense of shear (*half arrows*) on the limbs of first-order folds. *Short thick lines* are local bedding attitudes. *Dashed lines* labeled *a* are axial-surface traces of first-order folds. *Unlabeled dashed lines* are cleavage, fanning in the stiff units (*s*) and antifanning in the soft units. *Large arrows* show directions of boundary displacements

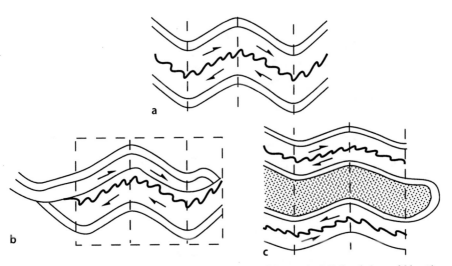

Fig. 5.25. Interpretations of drag folds having the same sense of shear on both limbs of a larger fold. **a** Observed fold with vertical axial traces. **b** Sense of shear interpreted as caused by a folded bedding-parallel thrust fault. **c** Sense of shear interpreted as being the upper limb of a refolded recumbent isoclinal fold

thereby indicating the relative positions of anticlinal and synclinal hinges. The inclination of cleavage planes in the softer units (Fig. 5.24), in the orientation axial planar to the drag folds, gives the same sense of shear information. Cleavage in the stiffer units is usually at a high angle to bedding, resulting in cleavage dips that fan across the fold.

Individual asymmetric higher-order folds are not always drag folds. The asymmetry may be due to some local heterogeneity in the material properties or the stress field, and the implied sense of shear could be of no significance to the larger-scale structure. Three or more folds with the same sense of overturning in the same larger fold limb are more likely to be drag folds than is a single asymmetric fold among a group of symmetric folds.

Folds in which the sense of shear of the drag folds remains constant from one fold limb to the next (Fig. 5.25a) require a different interpretation. One possibility is the presence of a bedding-plane fault with transport from left to right (Fig. 5.25b); the folds would be fault-related drag folds. Alternatively the beds may be recumbently folded and then refolded with a vertical axial surface. In this situation (Fig. 5.25c), the sense of shear would be interpreted as in Fig. 5.24, and Fig. 5.25a then represents the upright limb of a recumbent anticline, the hinge of which must be to the right of the area of Fig. 5.25a, as shown in Fig. 5.25c.

Fold origins other than by buckling may yield other relationships between the sense of shear given by drag folds and position within the structure. Structures caused by differential vertical displacements, for example salt domes or gneiss domes, could result in exactly the opposite sense of shear on the fold limbs from that in buckle folds (Fig. 5.26). The pattern in Fig. 5.26 has been called Christmas-tree drag.

Fig. 5.26.
Cross section of parasitic drag folds in the center bed as related to first-order differential vertical folding. *Large arrows* show directions of the boundary displacements

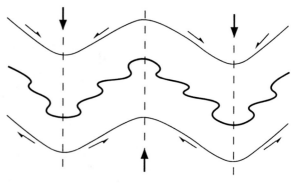

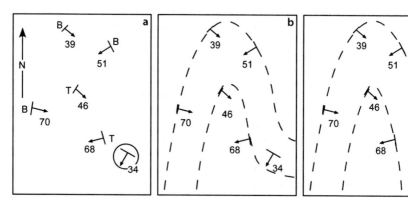

Fig. 5.27. Geological data and different possible geometries based on alternative interpretations of the significance of the *circled* attitude measurement. **a** Data. Bedding attitudes with *B:* observed base of formation; *T:* observed top of formation. **b** Interpretation honoring all contacts and attitudes. **c** Interpretation honoring all contacts but not all attitudes

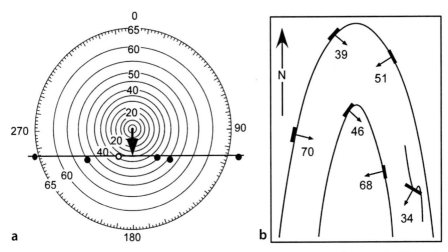

Fig. 5.28. Alternative interpretation of geological data in Fig. 5.27a. **a** Tangent diagram of bedding attitudes. *Open circle* is the 34, 212 point. Plunge of the fold is 30S. **b** Geological map of syncline with a coaxial minor fold on the limb. (After Stockwell 1950)

Potential interpretation problems associated with minor folds are illustrated by the map in Fig. 5.27a. The observed contact locations and bedding attitudes could be explained by the maps in either Fig. 5.27b or 5.27c. The shape of the first-order fold honors the 34SW dip in Fig. 5.27b and ignores it in Fig. 5.27c. The dip oblique to the contact could be justified as being either a cross bed or belonging to a minor fold.

Plotting the data from Fig. 5.27 on a tangent diagram (Fig. 5.28a) shows that all the points, including the questionable point (34, 212), fall on the same line, indicating a cylindrical fold plunging 30° due south. This result leads to rejection of the cross-bed hypothesis and indicates that the oblique bedding attitude is coaxial with the map-scale syncline. If the map of Fig. 5.27c is supported by the contact locations, then the structure has the form given by Fig. 5.28b.

5.7
Growth Folds

A growth fold develops during the deposition of sediments (Fig. 5.29a). The growth history can be quantified using an expansion index diagram, where the expansion index, E, is

$$E = t_d / t_u , \qquad (5.13)$$

with t_d = the downthrown thickness (off structure) and t_u = the upthrown thickness (on the fold crest). The thicknesses should be measured perpendicular to bedding so as not to confuse dip changes with thickness changes (Fig. 5.29a). The expansion index given here is the same as for growth faults (Thorsen 1963; Sect. 7.6). Different but related equations for folds have been given previously by Johnson and Bredeson (1971) and Brewer

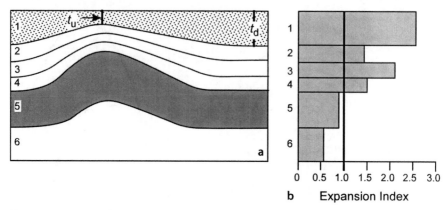

Fig. 5.29. Expansion index for a fold. Units *1–4* are growth units and units *5–6* are pre-growth units. **a** Cross section in the dip direction. **b** Expansion index diagram

and Groshong (1993). Using the same equation for both folds and faults facilitates the comparison of growth histories of both types of structures where they occur together.

The magnitude of the expansion index is plotted against the stratigraphic unit to give the expansion index diagram (Fig. 5.29b). The diagram illustrates the growth history of the fold. An expansion index of 1 means no growth and an index greater than 1 indicates upward growth of the anticlinal crest during deposition. The fold in Fig. 5.29a is a compressional detachment fold (Fig. 11.37) in which tectonic thickening in the pre-growth interval causes the expansion index to be less than 1. The growth intervals show an irregular upward increase in the growth rate. An expansion index diagram is particularly helpful in revealing subtle variations in the growth history of a fold and in comparing the growth histories of different structures. The expansion index is most appropriate for the sequences that are completely depositional. If erosion has occurred across the crest of the fold the expansion index will be misleading.

5.8
Exercises

5.8.1
Geometry of the Sequatchie Anticline

Plot the attitude data from Table 5.1 on a stereogram and on a tangent diagram to find the fold style and plunge. How do the two methods compare?

5.8.2
Geometry of the Greasy Cove Anticline

The bedding attitudes below (Table 5.2) come from the Greasy Cove anticline, a compressional structure in the southern Appalachian fold-thrust belt. What is the π axis of the fold? Use a tangent diagram to find the axis and the style of the fold.

Table 5.2.
Bedding attitudes, Greasy Cove
anticline, northeastern Alabama

Dip	Azimuth	Dip	Azimuth	Dip	Azimuth
46	316	55	316	60	310
55	311	40	295	14	124
12	319	26	281	60	304
20	248	25	275	10	270
10	266	14	294	16	307
12	243	15	173	12	150
24	154	22	154	28	231
20	129	30	128	30	119
35	143	25	114	32	131
25	120	70	151	26	165
20	345	12	255	11	258
15	160	12	190		

5.8.3
Structure of a Selected Map Area

Use the map of a selected structure (for example, Fig. 3.3 or 3.29) to answer the following questions. Measure and list all the bedding attitudes on the map. Plot the attitudes on a stereogram and a tangent diagram. What fold geometry is present? Which diagram gives the clearest result? Explain. Define the locations of the crest and trough traces from the map. Are the directions the same as given by the attitude diagrams? Find the attitudes of the axial planes, and locate the axial-plane traces on the map. What method did you use and why? What are the problems, if any, with the interpretation? Do the axial-surface traces coincide with the crest and trough traces? What are the orientations of the axial-surface intersection lines? Show where these intersection lines pierce the outcrop.

Chapter 6

Cross Sections, Data Projection and Dip-Domain Mapping

6.1
Introduction

A cross section shows the relationships between different horizons and allows the information from multiple map horizons to be incorporated into the interpretation. Cross sections may categorized as illustrative or predictive. The purpose of an illustrative cross section is to illustrate the cross section view of an already-completed map or 3-D interpretation. A slice through a 3-D interpretation is a perfect example. The purpose of a predictive cross section is to assemble scattered information and, utilizing appropriate rules, predict the geometry between control points. A predictive cross section can be used to predict the geometry of a horizon for which little or no information is available.

Data projection is typically part of the cross-section construction process. Relevant data commonly lie a significant distance from the line of section. Rather than ignore this information, it can be projected onto the section plane. The quality of the result depends on selecting the correct projection direction. This chapter describes how to select the projection direction and gives several techniques for making the projection by hand or analytically. Projection within a dip-domain style structure involves defining the 3-D axial-surface network and so becomes a blend of mapping, data projection, and section construction.

6.2
Cross-Section Preliminaries

6.2.1
Choosing the Line of Section

Cross sections constructed for the purpose of structural interpretation are usually oriented perpendicular to the fold axis, perpendicular to a major fault, or parallel to these trends. The structural trend to use in controlling the direction of the cross section is the axis of the largest fold in the map area or the strike of the major fault in the area. Good reasons may exist for other choices of the basic design parameters. For example, the cross section may be required in a specific location and direction for the construction of a road cut or a mine layout. If other choices of the parameters, such as the direction of the section line or the amount of vertical exaggeration, are required, it is recommended that a section normal to strike be constructed and vali-

Fig. 6.1.
Cross sections through a circular cylinder. **a** Normal section. **b** Oblique section. **c** Offset section. **d** Axial section

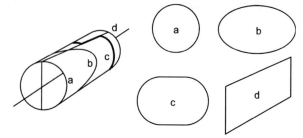

Fig. 6.2.
Structure contour map of an anticline showing the section lines. *A–A'*: normal (transverse) cross section perpendicular to the trend of the anticline. *B–B'*: longitudinal cross section. *C–C'*: well-to-well cross section

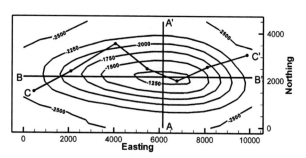

dated (Chap. 10 and 11) first. A grid of cross sections is needed for a complete three-dimensional structural interpretation.

The reason that a structure section should be straight and perpendicular to the major structural trend is that it gives the most representative view of the geometry. The simplest example of this is a cross section through a circular cylinder (Fig. 6.1). If the entire cylinder is visible, then it would readily be described as being a right circular cylinder. The cross section that best illustrates this description is Fig. 6.1a, normal to the axis of the cylinder, referred to as the normal section. Any other planar cross section oblique to the axis is an ellipse (Fig. 6.1b). An elliptical cross section is also correct but does not convey the appropriate impression of the three dimensional shape of the cylinder. A section that is not straight (Fig. 6.1c) also fails to convey accurately the three-dimensional geometry of the cylinder, although, again, the section is accurate. Section c in Fig. 6.1 could be improved for structural interpretation by removing the segment parallel to the axis, producing a section like Fig. 7.1a. A section parallel to the fold (or fault) trend (Fig. 6.1d) is also necessary to completely describe the geometry.

Predictive cross sections are constructed using bed-thickness and fold-curvature relationships that are appropriate for the structural style. In order to use these geometric relationships, or rules, to construct and validate cross sections, it is necessary to choose the cross section to which the rules apply. Such a rule, in the case of the circular cylinder in Fig. 6.1, is that the beds are portions of circular arcs having the same center of curvature. In this simple and easily applicable form, the rule applies only to section a. More complex rules could be developed for the other cross sections, but it is quicker and less confusing to select the plane of the cross section that fits the simplest rule than to change the rule to fit an arbitrary cross-section orientation.

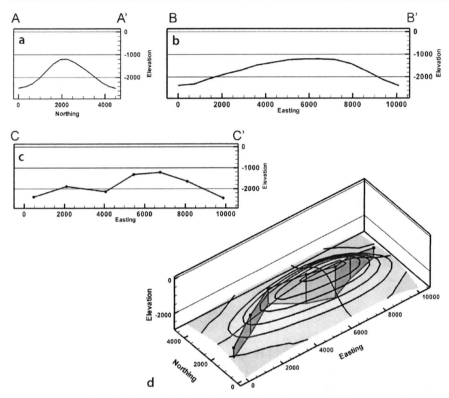

Fig. 6.3. Sections through the map in Fig. 6.2. **a** Normal section perpendicular to the fold crest. **b** Longitudinal section parallel to the fold crest. **c** Zig-zag or well-to-well cross section. **d** Oblique 3-D view of the structure showing the three cross sections

The effect of a curved section (Fig. 6.2) on the implied geometry of an elongate dome is shown in Fig. 6.3. The correct geometry of the structure is shown by the normal section, a straight-line cross section perpendicular to the axial trace of the structure (Fig. 6.3a) and the longitudinal section (Fig. 6.3b) parallel to the crest of the structure. A line of section that is not straight, such as one that runs through an irregular trend of wells or a seismic line that follows an irregular road, produces a false image of the structure. The zig-zag section across the map (Fig. 6.3c) incorrectly shows the anticline to have two local culminations instead of just one. This is a serious problem if the cross section is used to locate hydrocarbon traps or to infer the deep structure using the predictive section drawing techniques described in Sect. 6.4.

The first line of section across a structure chosen for interpretation should avoid local structures, like tear faults, oblique to the main structural trend. Oblique structures introduce complexities into the main structure that are more easily interpreted after the geometry of the rest of the structure has been determined. Returning to the cylinder, now shown offset along a tear fault (Fig. 6.4), cross sections at a and c will

Fig. 6.4.
Cross-section lines across a
cylinder offset along an oblique fault

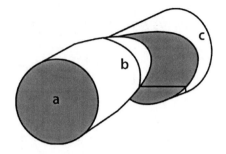

reveal the basic geometry of the cylinder. A cross section at b that crosses the fault will
be very difficult to interpret until after the basic geometry is known from sections a
or c. The simplest method for constructing the structure along section b would be to
project the geometry into it from the unfaulted parts of the cylinder.

6.2.2
Choosing the Section Dip

Only a cross section perpendicular to the plunge (the normal section) shows the true
bed thicknesses. In all other sections the thicknesses are exaggerated. This is important if the section is going to be used for predictive purposes. The plunge of a cylindrical fold is the orientation of its axis, which can be found from the bedding attitudes using the stereonet or tangent diagram techniques given in Sect. 5.2. A conical
fold does not have an axis and so, in the strict sense, there is no normal section. The
orientation of either the crestal line or the cone axis is an approximate plunge direction for a conical fold. On a structure contour map the trend and plunge of the crestal
line is readily identified. The plunge angle is given by the contour spacing in the plunge
direction.

Within a domain of cylindrical folding, changing the dip of the section plane is
equivalent to changing the vertical or horizontal exaggeration. This relationship is the
basis of the map interpretation technique known as down-plunge viewing. The map
pattern in an area of moderate topographic relief represents an oblique, hence exaggerated, section through a plunging structure. Viewed in the direction of plunge, the
map pattern becomes a normal section (Mackin 1950).

The down-plunge view of a fault should give the correct cross-section geometry and
the sense of the stratigraphic separation. The plunge direction of a fault is parallel to
the axis or crest or trough line of ramp-related folds or drag folds. If the fault is listric
or antilistric, the plunge direction should be the axis of the curved surface, just as if it
were a folded surface. If the fault is planar and there are no associated folds, the appropriate plunge direction is parallel to the cutoff line of a displaced marker against the
fault (Threet 1973).

If a vertical cross section is constructed normal to the trend of the plunge, the vertical exaggeration due to the plunge angle can be removed by rotating the section using
the method given in Sect. 6.5. The same approach can be used to convert a map view
into a normal section.

6.2.3
Vertical and Horizontal Exaggeration

Both vertical and horizontal exaggeration are used to help visualize and interpret the structure on cross sections. Vertical exaggeration is a change of the vertical scale (usually an expansion) while maintaining a constant horizontal scale and is a common mode of presentation of geological cross sections. Vertical exaggeration makes the relief on a subtle structure more visible on the cross section (Fig. 6.5a). Horizontal exaggeration is a change of the horizontal scale while maintaining a constant vertical scale

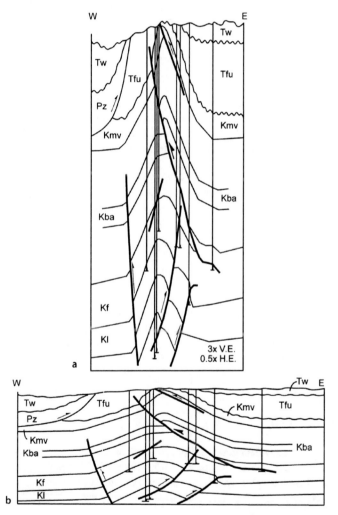

Fig. 6.5. Cross sections across Tip Top field, Wyoming thrust belt. **a** 3:1 vertical exaggeration and a 0.5:1 horizontal exaggeration, as might be seen on a seismic reflection profile. **b** Unexaggerated cross section. (Section modified from Groshong and Epard 1994, after Webel 1987)

Fig. 6.6.
Vertical and horizontal exaggeration. A bed of original
thickness *t* is shown in *white*.
a Unexaggerated cross section.
b Horizontally exaggerated
(squeezed) cross section.
c Vertically exaggerated cross
section

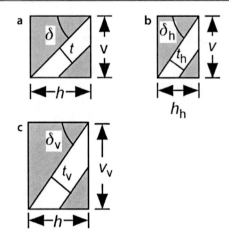

and is common, along with vertical exaggeration, in the presentation of seismic lines
(Stone 1991). Reducing the horizontal scale (squeezing) makes a wide, low amplitude
structure more visible and makes the break in horizon continuity at faults more obvious. Squeezing exaggerates the structure without producing an unmanageably tall cross
section.

Vertical exaggeration (V_e) is equal to the length of one unit on the vertical scale
divided by the length of one unit on the map, and horizontal exaggeration (H_e) is the
length of one unit on the horizontal scale divided by the length of one unit on the map
(Fig. 6.6):

$$V_e = v_v / v \ , \tag{6.1}$$

$$H_e = h_h / h \ , \tag{6.2}$$

where v_v = exaggerated vertical dimension, v = vertical dimension at map scale,
h_h = exaggerated horizontal dimension, and h = horizontal dimension at map scale.
As derived at the end of the chapter (Eqs. 6.21 and 6.22), the true dip is related to the
exaggerated dip by

$$\tan \delta_v = V_e \tan \delta \ , \tag{6.3}$$

$$\tan \delta_h = \tan \delta / H_e \ , \tag{6.4}$$

where δ_v = vertically exaggerated dip, δ_h = horizontally exaggerated dip, and δ = true
dip. Equation 6.3 is plotted in Fig. 6.7. In its effect on the dip, a vertical exaggeration
is equivalent to the reciprocal of a horizontal exaggeration (from Eq. 6.24 at the end of
the chapter):

$$V_e = 1 / H_e \ . \tag{6.5}$$

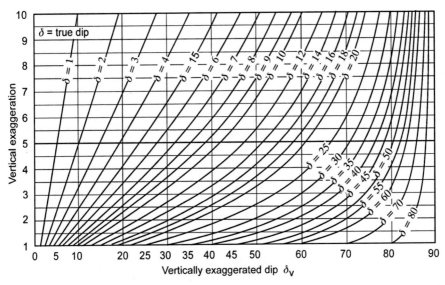

Fig. 6.7. Relationship between true dip and vertically exaggerated dip (Eq. 6.3) for various amounts of vertical exaggeration. (After Langstaff and Morrill 1981)

The effect of exaggeration on the thickness of a unit is given by (derived as Eqs. 6.26 and 6.28)

$$t_h / t = \cos \delta_h / \cos \delta \ , \tag{6.6}$$

$$t_v / t = V_e \, (\cos \delta_v / \cos \delta) \ . \tag{6.7}$$

The symbols are the same as in Eqs. 6.1–6.4. Horizontal exaggeration has no effect on the thickness of a horizontal bed, whereas vertical exaggeration changes the thickness of a horizontal bed by an amount equal to the exaggeration. Horizontal exaggeration changes the thickness of a vertical bed by the full amount of the exaggeration, whereas vertical exaggeration has no effect on the thickness of a vertical bed. For beds dipping between 0 and 90°, both horizontal and vertical exaggeration cause the apparent thickness to increase.

Exaggeration creates several problems in the interpretation of a cross section. The first is that a large vertical exaggeration or horizontal squeeze may so distort the structure that the structural style becomes unrecognizable. This will lead to difficulties in interpretation or to misinterpretations. For example, the exaggerated cross section in Fig. 6.5a looks more like a wrench-fault style than the correct thin-skinned contraction style. Cross-section construction and validation techniques and models for the dip angles and angle relationships do not apply to the exaggerated geometry. Exaggeration also causes thicknesses to be a function of dip (Fig. 6.8). Care must be taken not to interpret exaggerated thicknesses as being caused by tectonic thinning

Fig. 6.8.
Effect of a 3:1 vertical exaggeration on thickness. **a** Profile vertically exaggerated 3:1. Bed thickness increases as the dip decreases. **b** Unexaggerated profile. Bed thickness is constant

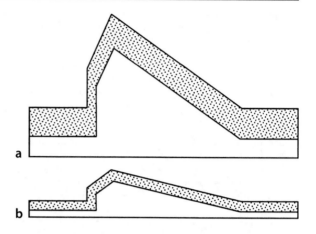

or thickening or by structural growth during deposition. The profile can be easily corrected when the true horizontal and vertical scales are known. The correction factor is the inverse of the horizontal or vertical exaggeration. Create an unexaggerated profile by multiplying the correction factor times the scale of the exaggerated axis. If the cross section is in digital form, this is a simple operation using a computer drafting program.

Seismic time sections are commonly displayed with both horizontal and vertical exaggerations (Stone 1991). Horizontal exaggeration may be applied to obtain a legible horizontal trace spacing. The amount of horizontal exaggeration is most conveniently determined by comparing the distance between shot or vibration points marked on the profile with the scale between the corresponding points on the location map. The vertical scale on a time section is in two-way-travel time, and the determination of the vertical exaggeration requires depth conversion as well as scaling. A few simple techniques can provide the necessary scaling information without geophysical depth migration. If the depth to a particular horizon is known independently, as from a well, then the vertical exaggeration at that well can be determined directly from the definition (Eq. 6.1). If the true dip is known for a unit or a fault, then the vertical exaggeration can be found by solving Eq. 6.3, given the exaggerated dip from the profile.

If there is a unit on a seismic time section that can be expected to have constant depositional thickness and minimal structural thickness changes, then any observed thickness change in the unit is caused by the exaggeration (Fig. 6.9a,b). The vertical exaggeration of the time section can be removed by restoring the bed thickness to constant (Stone 1991). A package of reflectors should be chosen that retains its reflection character and proportional spacing regardless of dip. The reflectors should be parallel to one another and not terminate up or down dip. The thickness changes of such a package are more likely to be caused by exaggeration than by deposition. A simple procedure for removing the vertical exaggeration is to change the vertical scale until the unit maintains constant thickness regardless of dip (Fig. 6.9c). This provides a quick depth migration that applies to the depth interval over which the unit

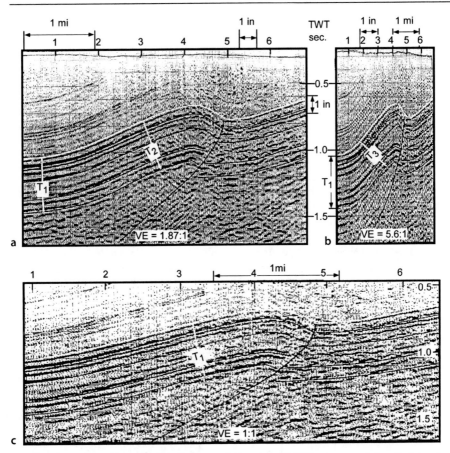

Fig. 6.9. Time migrated seismic profile from central Wyoming. *TWT*: Two-way travel time; T_i: interval thickness. **a** Original profile having a vertical scale of 7.5 in per second and a horizontal scale of 12 traces per in. Vertical exaggeration (*ve*) is 1.87:1. **b** The vertical scale is the same as in **a**, the horizontal scale is reduced by two-thirds. Vertical exaggeration is 5.6:1. **c** Unexaggerated version produced by expanding the horizontal scale. Thickness T_1 is now constant across the profile. (After Stone 1991)

occurs. Because seismic velocity varies with depth, the profile might remain exaggerated at other depths. Normally seismic velocity increases with depth and so the vertical exaggeration decreases with depth. This method does not take into account horizontal velocity variations. A further caution is that the thickness variations seen in Fig. 6.9a are just like those that are caused by deformation. The inferred vertical exaggeration should always be cross checked by other methods whenever possible. If the profile is also horizontally exaggerated, then this method will give the correct exaggeration ratio of 1:1, but both the horizontal and vertical scales could be exaggerated. Eliminate the horizontal exaggeration while maintaining the ratio constant to produce a depth section.

6.3
Illustrative Cross Sections

The purpose of an illustrative cross section is to illustrate the geometry present on a structure contour map. It is comparable to a topographic profile in concept and construction.

6.3.1
Construction by Hand or with Drafting Software

Begin by drawing the line of section on the map (A–A', Fig. 6.10a). The line of section in Fig. 6.10a has been selected to be perpendicular to the crest of the anticline. Clearly show the end points of the section because they will serve as the reference points for all future measurements. The section will be compiled on a graph where the vertical axis represents elevations and the horizontal axis is the distance along the profile (Fig. 6.10b). For an unexaggerated profile, the vertical scale should be the same as the map scale and the lines spaced accordingly. To construct a vertically exaggerated profile, let the vertical scale be some multiple of the map scale.

Cross sections are usually drawn either to be vertical or to be inclined such that the plane of the section is perpendicular to the direction of plunge of the structure of

Fig. 6.10.
Initial stage of cross-section construction from a map. The line of section is A–A'. **a** Structure contour map and a dip measurement (at elevation 820). **b** Graph for cross-section construction. For no vertical exaggeration both the horizontal and vertical scales are the same as the map scale

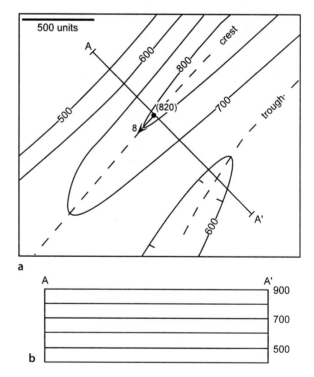

Fig. 6.11.
Direct projection of data from structure contour map to cross section. *Dotted lines* are right-angle projection lines. *Circles* show projected points: *filled circles* are projected from known elevations; the *open circle* is an interpolated elevation. The drafting tools are *shaded*

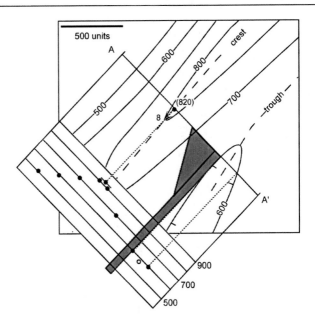

interest. For many purposes, it is most convenient to construct a vertical profile. For a cylindrical structure, a vertical profile is easily transformed into a normal section (Sect. 6.5). The techniques of section construction will begin with vertical profiles. Construction of an initially tilted profile is considered in Sect. 6.6.1.

The next step in constructing the cross section is to transfer the data from the map to the profile. One convenient projection method is to align the cross-section graph parallel to the line of section, tape the map and section together so that they cannot slip, and project data points at right angles onto the cross section with a straight edge and a right triangle (Fig. 6.11). Any type of map information can be transferred to the cross section by this method. In computer drafting it is usually more convenient to draw straight lines vertically or horizontally; therefore the map should be rotated so that the projection direction is either horizontal or vertical. Data points are located at their correct distances from the ends of the section and at their proper elevations. The probable locations of turning points of the structure contour map are also marked as points (Fig. 6.11), for example, between the two adjacent 600 contours. The location of a turning point is constrained to be between the next higher and lower elevations.

An alternative method is to mark the locations of the data points on a strip of paper for working by hand (Fig. 6.12a) or on a line drawn on top of the section line in a drafting program. After marking, the line of data locations is rotated to be parallel to the cross-section horizontal (Fig. 6.12b). In a drafting program, group the points before rotating. Then project the data points from the line onto the cross section (Fig. 6.12c). The points can be projected with a right triangle as in the previous method, or the overlay can be moved to the correct elevation on the section and each point marked at the appropriate distance from the end of the section. After compiling the

Fig. 6.12.
Transferring data from map to cross section using an overlay (*dashed line*). **a** Data points are marked on the overlay. **b** The overlay is aligned with the section. **c** Points are projected onto the section (*dotted lines*). *Filled circles* are projected from known elevations; the *open circle* is an interpolated elevation

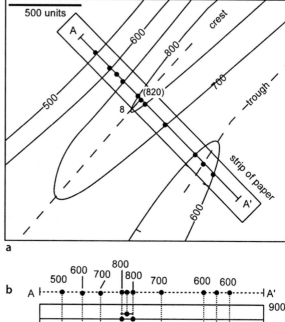

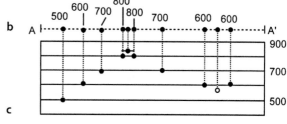

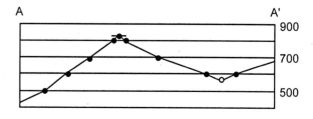

Fig. 6.13.
Cross section *A–A'* from the structure contour map of Fig. 6.12. Vertical section, no vertical exaggeration. The *short horizontal line* is the apparent dip from the bedding attitude on the map

data onto the profile, it should be checked. Then the profile is constructed by connecting the dots (Fig. 6.13). If the correct shape of the profile is not clear, points can be added by interpolation between contours on the map.

6.3.2
Slicing

With 3-D software a cross section can be constructed by slicing the 3-D model (Fig. 6.14). The slice automatically shows the apparent dips of beds and faults, contact locations, and the apparent thicknesses of the beds.

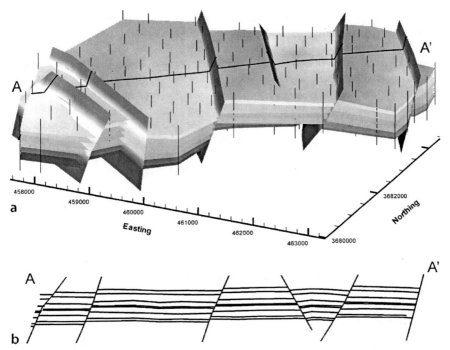

Fig. 6.14. 3-D map of coal-cycle tops and normal faults in the SE Deerlick Creek coalbed methane field, Black Warrior Basin, Alabama. *Vertical black lines* are wells (data from Groshong et al. 2003b). **a** Oblique view to the NW. *Wide line* crossing the region is the slice line. **b** Vertical slice from SW (*left*) to NE (*right*) across the area

6.4
Predictive Cross-Section Construction

A predictive cross section uses a set of geometric rules to predict the geometry between scattered control points. This process requires interpolation between data points and perhaps extrapolation beyond the data points. The best method depends on the nature of the bed curvature and on the nature of the bed thickness variations, or the lack of bed thickness variations. Where closely spaced control points are available, any method of interpolation will produce similar results. Where the data are sparse, the model that best fits the structural style will provide the basis for drawing the best cross section.

Two predictive methods will be presented here. The first method, here termed the dip-domain technique, is based on the assumption that the beds occur in planar segments separated by narrow hinges or faults (Coates 1945; Gill 1953). Easily done by hand, this method is also widely used in computer-aided structural design programs and in structural models. The second method is the method of circular arcs, which is based on the assumption that the beds maintain constant thickness and form segments of circular arcs (Hewett 1920; Busk 1929). Most cross sections published before the 1950s use this method.

Begin either technique by compiling the hard data onto the line of section. The cross section that shows just the original data will be called the *data* section. The cross-section interpretation should be done as an overlay on the data section. The interpolation between the control points may change dramatically during the interpretation process, but the locations of the control points and bedding attitudes should remain the same. Including the data section along with the final interpretation separates the data from the interpretation, a fundamental distinction that should always be made. The data may be subject to revision, of course. For example, inconsistencies in the cross section may indicate that a geologic contact has been mislocated or that a fault is required. This is one of the important reasons for constructing a predictive cross section. In the best scientific procedure, the original data and the interpreted result are both presented in the final report.

Data to be compiled will include dips and contact locations. If the section is not perpendicular to the fold axis, all dips shown on the cross section must be the apparent dips in the plane of the section. Measured attitudes must be converted using Eq. 2.18 or with a tangent diagram or stereogram. On an exaggerated profile (not recommended), the dip must be the exaggerated dip from Eq. 6.3 or 6.4. When the data have been transferred to the cross section, the section should again be checked against the information seen along the line of section on the map. A common mistake is to produce a cross section that fails to match the map along the line of section. All elevations, geological contacts, attitudes (apparent dips), and attitude locations must match exactly along the line of section. Projection of data to the line of section is commonly required and is discussed in Sect. 6.6.

It is useful to summarize the stratigraphic thicknesses in a "stratigraphic ruler" which will greatly speed up the drawing of an unexaggerated cross section. Draw the stratigraphic section at the scale of the cross section on a narrow piece of paper or, in computer drafting, make it a group of its own. This can be used as a ruler to mark off the stratigraphic units on the cross section and provides a quick check to see if the thickness of a unit is consistent with its dip. This only works on unexaggerated cross sections and is one of the important reasons for not using vertical or horizontal exaggeration.

Many structural interpretations are based on seismic reflection profiles which are already displayed in the form of a cross section. A seismic line that is to be interpreted structurally should satisfy the same criteria with respect to the choice of the plane of section and vertical exaggeration as a geological cross section. If geological data are available, transferring the data from maps to the seismic line will provide constraints that will help in the construction or validation of the depth interpretation. A successfully depth-converted seismic line must follow the same geometric rules as a geologic cross section.

6.4.1
Dip-Domain Style

The dip-domain method is based on the assumption that the beds occur as planar segments separated by narrow hinges. This method was originally proposed as section construction technique by Gill (1953) who called it the method of tangents. The basis for the technique lies in the relationship between the bedding thickness and the symmetry of the hinge (discussed previously in Sect. 5.4.2). For constant thickness

Fig. 6.15.
Dip-domain fold hinges in a
constant thickness layer. *t* Bed
thickness; γ_i-half-angles of the
interlimb angle

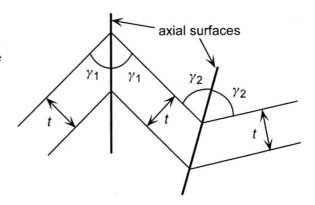

beds, the axial surface bisects the interlimb angle between adjacent dip domains
(Fig. 6.15). This maintains constant bed thickness. If the beds change thickness across
the axial surface, then the axial surface cannot bisect the hinge. The technique is described here in the context of constant thickness beds. The technique is the same for
beds that change thickness except that the axial surfaces do not bisect the hinges. See
Eq. 5.13 for a method to calculate the axial surface orientation in folds that do not
maintain constant bed thickness (see also Gill 1953).

6.4.1.1
Method

The following steps outline the dip-domain construction technique.

1. On the map or cross section, define the dip domains and locate the boundaries
   between domains as accurately as possible (Fig. 6.16a). A certain amount of variability from constant dip is expected in each domain (perhaps a 2–5° range).
2. Define the axial surfaces between domains (Fig. 6.16b). If bed thickness is constant,
   the axial surfaces bisect the hinges, but if bed thickness changes are known, use
   Eq. 5.13 to find the dips of the axial surfaces. Where axial surfaces intersect, a dip
   domain disappears and a new axial surface is drawn between the newly juxtaposed
   domains (for example, locations X in Fig. 6.16b). Note that a single fold is likely to
   have multiple hinges, as illustrated in Fig. 6.16.
3. Draw a key bed or group of beds through the structure, honoring the domain dips and
   the stratigraphic tops (Fig. 6.16c). Sometimes the data do not allow a single key bed to
   be completed across the whole structure. Shifting up or down a few beds to a new key
   bed will usually allow the section to be continued. Note that axial-surface intersections
   do not necessarily coincide with named stratigraphic boundaries. It is usually helpful
   to draw an horizon through the axial-surface intersection points (Fig. 6.16c).
4. Complete the section by drawing all the remaining beds with their appropriate
   thicknesses (Fig. 6.16c).
5. If desirable on the basis of the structural style, round the hinges an appropriate
   amount using a circular arc with center on the axial surface, or a spline curve.

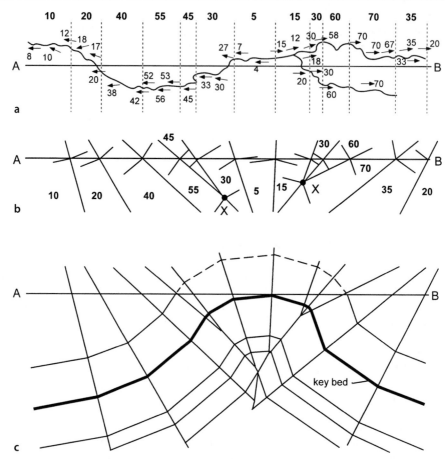

Fig. 6.16. Dip-domain cross-section construction technique. **a** Map of dips measured along a stream traverse and the boundaries (*dotted lines*) between interpreted dip domains. **b** Initial stage of cross-section construction showing domain dips and hinge locations with axial surfaces that bisect the hinges. *X:* axial-surface intersection points. **c** Completed cross section. (After Gill 1953)

6.4.1.2
Cylindrical Fold Example

The steps in building a cross section and interpolating the geometry using the constant bed thickness dip-domain method is illustrated with the Sequatchie anticline (Fig. 6.17). The map is characterized by domains of approximately constant dip, making it a good candidate for a dip-domain style cross section. The fold is nearly cylindrical within the map area and so the geometry of the structure should be constant along the axis. The crestal line is horizontal (Fig. 5.8b), making a vertical section the most appropriate. Prior to drawing the section, the stratigraphic thicknesses are determined and summarized in a stratigraphic ruler at the same scale as the map (Fig. 6.18).

Fig. 6.17.
Geologic map of a portion
of the Sequatchie anticline
at Blount Springs, Alabama,
showing the line of cross section. Geologic contacts: *wide
lines,* topographic contours
(ft): *thin lines,* measured bedding attitudes are shown by
arrows. c: Attitude computed
from three points

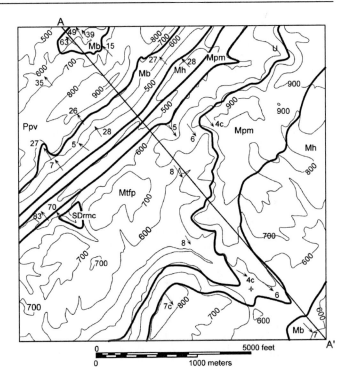

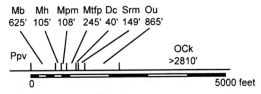

Fig. 6.18. Stratigraphic column for the Sequatchie anticline map area at the same scale as the map,
to be used as a stratigraphic ruler. Thicknesses are in feet. Thicknesses of Ppv through Mpm are
from outcrop measurements. The top of the Ppv is not present in the map area. Thicknesses of Mtfp
through OCk are from the Shell Drennen 1 well (Alabama permit No. 688) interpreted by McGlamery
(1956), and corrected for a 4° dip. The well bottomed in the OCk and so the drilled thickness is less
than the total for this unit

The first step is to transfer the data from the map to the cross section. The line of
section is drawn on the map (Fig. 6.17), at right angles to the fold axis. The topography
is drawn using the method of Fig. 6.12, with the vertical scale equal to the map scale
(Fig. 6.19). The geologic contacts are shown by arrows and the dips close to the line of
section are shown as short line segments. The stratigraphic ruler is shown intersecting
the topography at the projected surface location of the well that provided the thicknesses of the subsurface units. The geological data in solid lines on Fig. 6.19 form the
data section which should not be subject to significant revision.

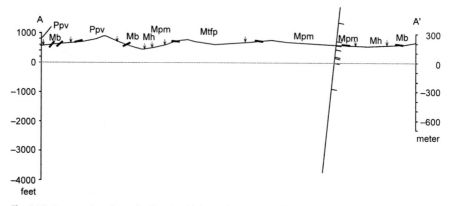

Fig. 6.19. Data section along the line A–A' (Fig. 6.17). No vertical exaggeration. The stratigraphic column is shown where the trace of the well projects onto the line of section. *Short arrows* at the topographic surface are the geological contact locations. *Wide short lines* are bedding dips

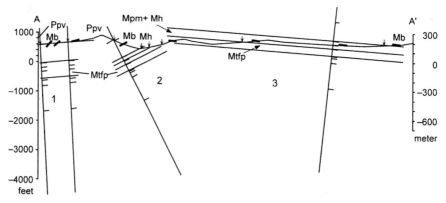

Fig. 6.20. Comparison between domain dips, stratigraphic thicknesses, and contact locations. *Short arrows* at the topographic surface are the geological contact locations. *Wide short lines* are bedding dips. Dip domains are *numbered*

The next step is to establish the domain dips and see how well the domains fit the locations of the formation boundaries (Fig. 6.20). As a first approximation, the fit to one backlimb and two forelimb domains is tested. (The forelimb is the steeper limb.) The dip of domain 1 (3NW) is given by the dip of the line connecting the base of the Ppv on opposite sides of the Mb inlier. The steeper dips of Mb within the inlier are caused by second-order structures and do not apply at the scale of the cross section. The domain 2 dip is the 27NW dip seen at the surface. The domain 3 backlimb dip of 6SE is seen in outcrop but is selected primarily because with this dip the unit thicknesses match the contact locations. Portions of the beds are drawn in with constant bed thickness to compare with the contact locations. The domain 2 dip fits both contacts of the Mh, even though this information was not used to define the dip.

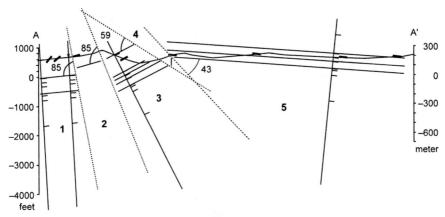

Fig. 6.21. Axial surface traces (*dotted lines*) that bisect the interlimb angles. Exact locations of the axial surfaces are not yet fixed in this step. Dip domains are *numbered*

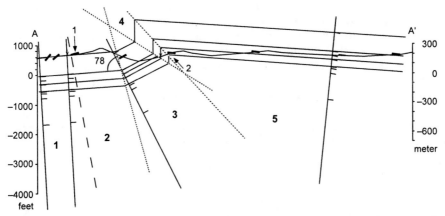

Fig. 6.22. Dip-domain cross section with axial surfaces (*dotted lines*) moved so that the dip domains match the stratigraphic contact locations. The *dashed* axial surface will be deleted and domains *1* and *2* combined

The axial surface orientations are determined next (Fig. 6.21). Following the relationship in Fig. 6.15 for constant bed thickness, the axial surfaces bisect the hinges. The interlimb angles are measured, bisected and the axial surfaces drawn between each domain. Two dip domains (2 and 4) are added to those shown in Fig. 6.20 so that the dips can be honored at the ground surface. It is tempting to insert a fault at the location of domain 4, but the map (Fig. 6.17) shows a vertical to near-vertical domain to the southwest in the same position as on the vertical dip on the cross section. Not far to the southwest of the map area, the units are directly connected across the two limbs (Cherry 1990) with no fault present. The positions of the axial surfaces in Fig. 6.21 are only approximate; the next step is to determine their exact locations.

The locations of the axial surfaces are now adjusted until the dip domains match the stratigraphic contacts (Fig. 6.22). The dip change of the Ppv at location 1 must be ignored and the corresponding axial surface between domains 1 and 2 removed in order to match the locations of the stratigraphic contacts. A new axial surface dip is determined as the boundary between the two domains in contact (1 + 2 and 3) after the incorrect axial surface is removed. The vertical dip selected for the forelimb provides a good match to all the contacts except for the top of the Dc at location 2. A slight rounding of the contact at this location will provide a match to the map geometry. The internal consistency of the section based on constant thicknesses, planar domain dips and the mapped contact locations and depths in the well is strong support for the interpretation.

Axial surfaces are shown as crossing in Fig. 6.22, an impossibility. Where two axial surfaces intersect, the dip domain between them disappears and a new axial surface is defined between the two remaining dip domains (Fig. 6.23). The final cross section (Fig. 6.23) is an excellent overall fit to the dips and contact locations. Locations 1 and 2 are the only misfits. The misfits are quite small. At location 1, the base of the Mh does not match the mapped outcrop location which could be caused by a second-order fold at that point or by the mislocation of a poorly exposed contact. A very small domain of thickened bedding is required at location 2 in order to keep the top of the Dc below the surface of the ground and so that the contacts of the Dc and the Sm meet across the axial surface. It is no surprise that bed thickness is not perfectly constant in such a tight hinge. The surprise is that such a small region of thickening is required in the hinge. The effect of the thickening of the Mtfp is to round the hinge, a feature that might continue upward along the axial surface as well, but is shown as ending within the Mtfp. Both the vertical domain and the thickened domain disappear at point 3 where a new axial surface bisects the angle between the remaining two domains (2 and 4). The match of the top of the OCk across this axial surface is an additional confirmation of the cross-section geometry because the location of the axial surface is defined by intersection point 3, not by projection of the OCk contact.

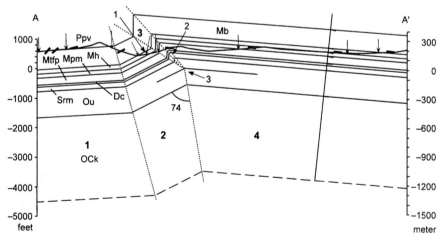

Fig. 6.23. Final constant-thickness, dip-domain cross section across the Sequatchie anticline. No vertical exaggeration. The *numbered arrows* are explained in the text. *Small arrows* mark the contact locations. The *dashed line* is the level of the deepest horizon drilled

This example illustrates the importance of the cross section to structural interpretation. The rule of constant bed thickness allows a few dips and the formation contact locations to tightly constrain the geometry of the cross section. The rule works well even though there is a small amount of thickening in the tightest hinge. The cross section can, in turn, be used to revise the geologic map and the composite structure contour map. The cross section provides the needed control for mapping the deeper geometry. Extrapolation to depth using the composite-surface technique breaks down if vertical lines through the control points pass through axial surfaces, as happens in the forelimb of the Sequatchie anticline (Fig. 6.23). Composite surface maps (Sect. 3.6.2) provide a good first approximation, but the final interpretation should be controlled directly by cross sections based on multiple horizons.

6.4.2
Circular Arcs

The method of circular arcs is based on the assumptions that bed segments are portions of circular arcs and that the arcs are tangent at their end points (Hewett 1920; Busk 1929). This type of curve can be drawn by hand using a ruler and compass. The resulting cross section will have smoothly curved beds. The method of circular arcs produces a highly constrained geometry in which both the shape of the structure and the exact position of each bed within the structure are predicted. When these predictions fit all the available data, the cross section is very likely to be correct. If the stratigraphic and dip data cannot be matched by the basic construction technique, as often happens, dips can be interpolated that will produce a match. The basic method is given first, then two techniques for dip interpolation.

6.4.2.1
Method

If the dips are known at the top and bottom of the bed (Fig. 6.24), the geometry of a circular bed segment is constructed by drawing perpendiculars through the bed dips (at A and B), extending the perpendiculars until they intersect (at O) which defines the center of curvature. Circular arcs are drawn through A and B to define the top and bottom of the bed.

This process is repeated for multiple data points to draw a complete cross section (Fig. 6.25). The first center of curvature (O) is defined as the intersection of the normals to the first two dips (A and B). The marker horizon located at point A is extended to the bedding normal through B along a circular arc around point O. The next center of

Fig. 6.24.
Cross section of a bed that is a portion of a circular arc. (After Busk 1929)

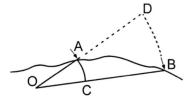

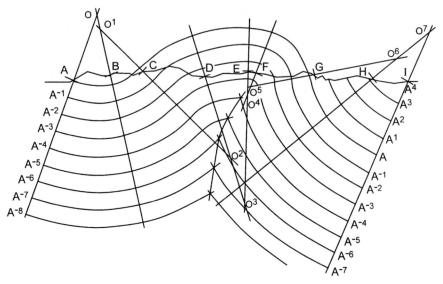

Fig. 6.25. Cross section produced by the method of circular arcs. *A–I:* outcrop dip locations; A^i: marker horizons; O^i: centers of curvature. (After Busk 1929)

Fig. 6.26.
Sensitivity of the crest location on a circular-arc cross section to the dips in the adjacent syncline. *A:* Dip on surface anticline used for linear projection of the fold limb; *B–D:* dips in adjacent syncline used for circular-arc construction of the limb; O^i: centers of curvature. Wells attempting to drill the lowest unit at the crest are shown. (After Busk 1929)

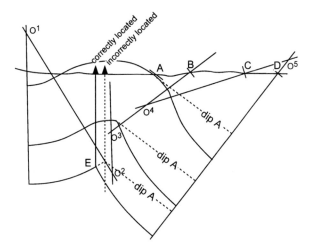

curvature is located at O^1. The marker horizon is extended to the normal through C as a circular arc with center O^1. The same procedure is followed across the section to complete the key horizon A–A (Fig. 6.25). The remaining stratigraphic horizons are drawn as segments of circular arcs around the appropriate centers. Constructed in this fashion, the beds have constant thickness. To maintain constant bed thickness, the beds form cusps in the core of the fold.

To properly control the geometry of a cross section at depth, data may be needed at a long distance laterally from the area of interest (Fig. 6.26). For example, in order to correctly locate the crest of an anticline at depth, dips are needed from the adjacent synclines. If the last dip in the anticline (Fig. 6.26) was collected at A, then the steep limb of the structure would be drawn with the long dashed lines and the crest on the lowest horizon would be at the location of the incorrect well. Using the dips at B, C, and D, the structure is drawn with the solid lines, and the crest is found to be at E (Fig. 6.26). This is a general property of cross-section geometry and also applies to dip-domain constructions.

6.4.2.2
Dip Interpolation

Frequently the predicted geometry and the bed locations do not agree. The predicted location of horizon A (Fig. 6.27) on the opposite limb of the anticline is at B, but that horizon may actually crop out at B' or B". This result means that insufficient data are available to force a correct solution. It is necessary to modify the data or to interpolate intermediate dip values between A and B in order to make the horizon intersect the section at B' or B". Two methods of dip interpolation will be given; the first is to interpolate a planar dip segment and the second is to interpolate an intermediate dip.

The simplest method is to insert a straight line segment (AY, Fig. 6.28) between the two arc segments that produce the disagreement. This method is usually successful and provides an end-member solution. The procedure is from Higgins (1962):

Fig. 6.27.
Cross section showing the mismatch between the predicted location of the key bed at *A* and its mapped location (*B'* or *B"*) at *B*. (After Busk 1929)

Fig. 6.28.
Interpolation using a straight line with a circular arc. (After Higgins 1962)

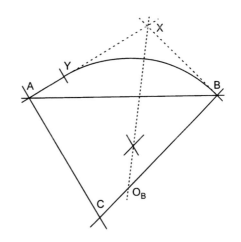

1. Extend the dips at A and B so that they intersect at X.
2. On AX locate point Y such that YX = XB.
3. Bisect angle YXB. The bisector will intersect BC, the normal to B, at O_b.
4. With center O_b and radius BO_b, draw the arc from B to Y. This arc is tangent to AY, the straight-line extension of the dip from A.

The second method is to insert a dip such that the two data points are joined by two circular arcs that are tangent at the data points and at the interpolated dip. The result is a cross section with continuously curving beds. This method is given by Busk (1929) and Higgins (1962). Beginning with the two dips A and B (Fig. 6.29):

1. Draw AA' perpendicular to the lesser dip at A; draw BB' perpendicular to the greater dip at B.
2. Draw the chord AB. Angle CAB must be greater than angle CBA; if not, switch the labels on points A and B.
3. Erect the perpendicular bisector of AB. This line intersects AA' at Z.

Fig. 6.29.
Interpolation using circular-arc segments. (Modified from Higgins 1962)

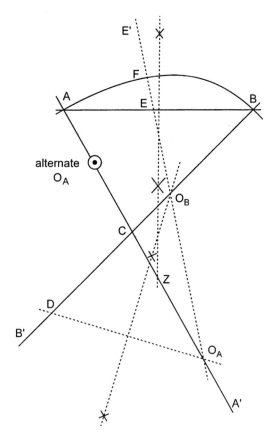

4. Choose point O_A anywhere on line AA' on the opposite side of Z from A. (If the length O_A–Z is very large, it is equivalent to drawing a straight line through A.)
5. On BB' locate point D such that BD = AO_A.
6. Draw DO_A connecting D and O_A.
7. Erect the perpendicular bisector of DO_A. This line intersects BB' at O_B.
8. Draw O_AE' through O_A and O_B, intersecting AB at E.
9. With center O_A and radius AO_A, draw an arc from A, intersecting O_AE' at F.
10. With center O_B and radius BO_B, draw an arc from B intersecting O_AE' at F. This completes the interpolation.

If a correct solution is not obtained, it may be because the sense of curvature changes across an inflection point, causing the centers of curvature to be on opposite sides of the key bed. Modify step 4 above by using the alternate position of O_A (Fig. 6.29: alternate O_A), located between C and A.

6.4.2.3
Other Smooth Curves

Interactive computer drafting programs provide several different tools for drawing smooth curves through or close to a specified set of points. Typically they are parametric cubic curves for which the first derivatives, that is the tangents, are continuous where they join (Foley and Van Dam 1983). In this respect the curves are like the method of circular arcs, for which the tangents are equal where the curve segments join, but cubics are able to fit more complex curves than just segments of circular arcs. Two different smooth curve types are widely available in interactive computer drafting packages, Bézier and spline curves. The two curve types differ in how they fit their control points and in how they are edited. Both types are useful in producing smoothly curved lines and surfaces (Foley and Van Dam 1983; De Paor 1996).

A Bézier curve consists of segments that are defined by four control points, two anchor points on the curve (P_1 and P_4, Fig. 6.30a) and two direction points (P_2 and P_3) that determine the shape of the curve. The curve always goes through the anchor points. The shape is controlled in interactive computer graphics applications by moving the direction points. In a computer program the direction points may be connected to the anchors by lines to form handles (Fig. 6.30b) that are visible in the edit mode. At the join between two Bézier segments, the handles of the shared anchor point are colinear, ensuring that the slopes of the curve segments match at the intersection.

A spline curve only approximates the positions of its control points (Fig. 6.31) but is continuous in both the slope and the curvature at the segment boundaries, and so the curve is even smoother than the Bézier curve (Foley and Van Dam 1983). The shape is controlled in interactive computer graphics applications by moving the control points that are visible in the edit mode. This curve type should be drawn separately from the actual data points because editing the curves changes the locations of the points that define the curve. The control points can be manipulated until the match between the curve and the data points is acceptable.

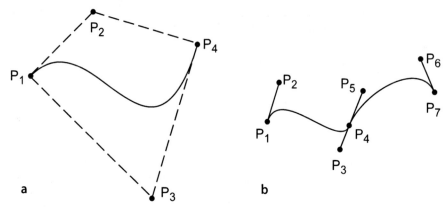

Fig. 6.30. Bézier curves. **a** The four control points that define the curve. **b** Two Bézier cubics joined at point P_4. Points P_3, P_4, and P_5 are colinear. (After Foley and Van Dam 1983)

Fig. 6.31.
Spline curve and its control points

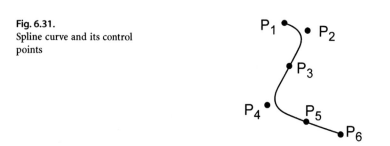

Drawing a cross section (or a map) using the smooth curves just described requires care to maintain the correct geometry. Constant bed thickness, for example, is not likely to be maintained if the section is drawn from sparse data. The appropriate bed thickness relationships can be obtained by editing the curves after a preliminary section has been drawn. The cross section of the Sequatchie anticline illustrates the problems. The original section (Fig. 6.23) was redrawn by changing the lines from polygons to spline curves in a computer drafting program. The resulting cross section (Fig. 6.32) may be more pleasing to the eye than the dip-domain cross section, but it is less accurate. The unedited spline-curve version (Fig. 6.32a) is much too smooth. Each bedding surface is defined by 4 to 6 points, a data density that might be expected with control based entirely on wells. Bedding thicknesses are not constant as in the dip-domain version, and the amplitude of the structure is reduced. These are the typical results of analytical smoothing procedures, including the smoothing inherent in gridding as used for map construction. Editing the spline curves produces a better fit to the true dips (Fig. 6.32b). A more accurate spline section can be produced by introducing many more control points, which is the appropriate procedure for producing a final drawing of a known geometry. The addition of control points to improve an interpretation based on a sparse data set requires additional information, such as the bedding dips, or the requirement of constant bed thickness.

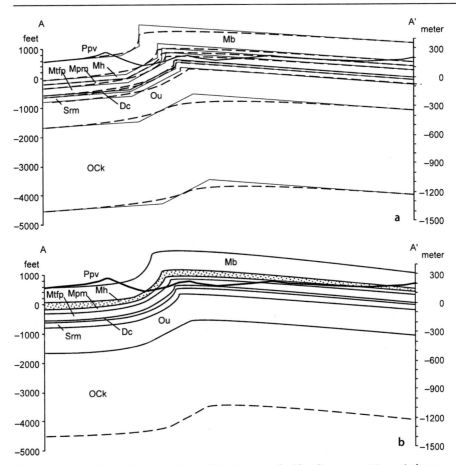

Fig. 6.32. Cross section of the Sequatchie anticline interpreted with spline curves. No vertical exaggeration. **a** Dip-domain cross section (*thin solid lines* from Fig. 6.23) and computer-smoothed spline interpretation (*thick dashed curves*). **b** Spline curve section edited to more closely resemble the dip-domain section

6.5
Changing the Dip of the Section Plane

If a cross section is not perpendicular to the fold axis, it is helpful for structural interpretation to rotate the section plane until it is a normal section. Alternatively, it might be necessary to rotate a normal section to vertical. The necessary relationships are equivalent to removing (or adding) a vertical exaggeration, as done in the visual method of down-plunge viewing.

From the geometry of Fig. 6.33, the exaggeration in a vertical section across a plunging fold (Eq. 6.1) is

$$V_e = t_v / t = 1 / \cos \phi \ . \tag{6.8}$$

Fig. 6.33.
Vertical exaggeration in cross section parallel to the plunge direction, caused by a plunge angle of ϕ. The true thickness is t; the exaggerated thickness is t_h in the horizontal plane and t_v in the vertical plane

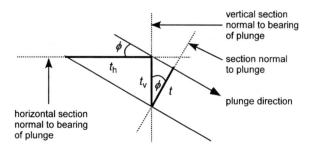

The section is changed from a normal section to a vertical section by exaggerating the vertical scale with Eq. 6.8. The vertical exaggeration on a vertical section due to the plunge is removed by multiplying the vertical scale of the section by the reciprocal of the vertical exaggeration, $\cos \phi$.

The exaggeration on a horizontal section (map view), t_h/t, (Fig. 6.33) is

$$V_e = t_h / t = 1 / \sin \phi \ . \tag{6.9}$$

The exaggeration on a horizontal section due to the plunge is removed by multiplying the vertical scale of the section by the reciprocal of the vertical exaggeration, $\sin \phi$.

The same procedure can be used to rotate the plane of a cross section around a vertical axis. Treat Fig. 6.33 as being the map view and the vertical exaggeration as being a horizontal exaggeration. Equation 6.8 then gives the horizontal exaggeration of the profile, with ϕ = the angle between the normal to the line of section and the desired direction of the section normal. Rotate the section by multiplying the horizontal scale by the reciprocal of the horizontal exaggeration, $\cos \phi$.

6.6
Data Projection

In order to make maximum use of the available information, it is usually necessary to project data onto the plane of the cross section from elsewhere in the map area. Data from a zig-zag cross section or seismic line should be projected onto a straight line to correctly interpret the structure. Wells should be projected onto seismic lines for best stratigraphic correlation and to confirm the proper depth migration of the seismic data. The additional data that are obtained by projection from the map to the line of section help constrain the interpretation of the cross section and help ensure that the interpretation is compatible with the structure off the line of section. Projection of data to the line of section is an important step in the geological interpretation, not a simple mechanical process.

Incorrect projection places the data in the wrong relative positions on the cross section and renders the interpretation incorrect or impossible. The effect of the projection technique is illustrated with an example (Fig. 6.34) originally presented by Brown (1984). A cross section of the structure in Fig. 6.34a has been constructed by projecting the wells onto the line of section along the strike of the structure contours (Fig. 6.34b). The resulting profile is poor in terms of structural style. The cross section shows multiple small faults instead of a single smooth fault. Note that no well shows

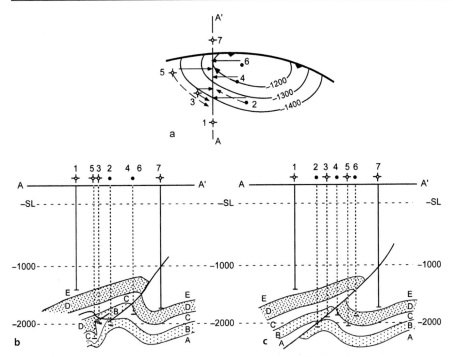

Fig. 6.34. Different cross sections obtained by different methods of data projection. **a** Structure contour map of horizon *E*, showing the alternative projection directions. *Solid lines* are parallel to the fold axis; *dashed lines* are parallel to structure contours. **b** Cross section produced by projecting wells along structure contours. **c** Cross section produced by projecting wells along the plunge of the fold axis. *SL* sea level. (After Brown 1984)

more than one fault, yet the cross section shows locations where a vertical well should cut two faults. Projection along plunge (Fig. 6.34c) significantly improves the cross section. The west half of the structure has a uniform cylindrical plunge to the west and so projection along plunge produces a reasonable cross section. Only one fault is present and it is relatively planar, as expected.

Three general approaches to projection will be presented: projection along plunge, projection with a structure contour map, and projection within dip domains. For structures where the plunge can be defined from bedding attitude data, projection along plunge is effective. Where formation tops are relatively abundant, but attitudes are not available, projection by structure contouring is straightforward and accurate. Computer mapping programs usually take this approach. It must be recognized that the structure contours themselves are interpretive and may not be correct in detail until after they have been checked on the cross section. Iterating between maps and the cross section in order to maintain the appropriate bed thicknesses is a powerful technique for improving the interpretation of both the maps and the cross section. For dip-domain style structures, defining the dip-domain (axial-surface) network is an efficient method for projecting the geometry in three dimensions.

Not all features in the same area will necessarily have the same projection direction. For example, stratigraphic thickness changes may be oblique to the structural trends and should therefore be projected along a trend different from the structural trend. Folds and cross-cutting faults may have different projection directions. The respective trends and plunges should be determined from structure contour maps (Chap. 3), isopach maps (Sect. 4.3.1) and dip-sequence analysis (Chap. 9).

6.6.1
Projection Along Plunge

Projection of information along plunge is most appropriate where the local data are too sparse to generate a structure contour map, but where the trend and plunge can be determined from bedding attitudes, for example, from a dipmeter.

6.6.1.1
Projecting a Point or a Well

The projection of a point, such as a formation top in a well, along plunge to a new location, such as a vertical cross section or a seismic line (Fig. 6.35) is done using

$$v = h \tan \phi \ , \tag{6.10}$$

where v = vertical elevation change, h = horizontal distance in the direction of plunge from projection point to the cross section, and ϕ = plunge. For example, if $h = 1$ km and the plunge is 15° (Fig. 6.35), the elevation of the projected point is 268 m lower on the cross section than in the well.

6.6.1.2
Plunge Lines

Projection along plunge is conveniently done using plunge lines, which are lines in the plunge direction, inclined at the plunge amount (Wilson 1967; De Paor 1988). Plunge lines

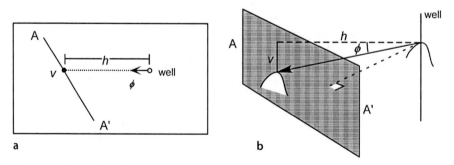

Fig. 6.35. Projection of a well along plunge to a cross section or seismic profile. **a** Map view of the projection of the well to cross section A–A'. The *arrow* gives the plunge direction; ϕ plunge amount; h horizontal distance in direction of plunge from projection point to cross section; v vertical elevation change. **b** 3-D view

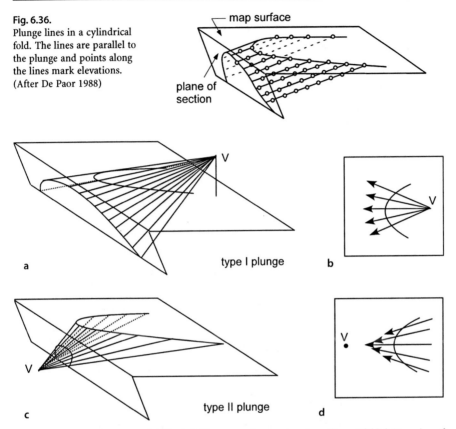

Fig. 6.36.
Plunge lines in a cylindrical
fold. The lines are parallel to
the plunge and points along
the lines mark elevations.
(After De Paor 1988)

Fig. 6.37. Plunge lines in conical folds. *V* Fold vertex. **a** Perspective view of type I fold. **b** Map view of plunge lines in type I fold. *Arrows* point down the plunge direction. **c** Perspective view of type II fold. **d** Map view of plunge lines in type II fold. *Arrows* point down the plunge direction

provide an effective means for quantitatively describing and projecting the 3-D geometry of a fold. A series of plunge lines defines the shape of the structure (Fig. 6.36). The plunge line for a cylindrical fold is parallel to the fold axis and is the same direction for every point within the fold (Fig. 6.36). Each plunge line in a conical fold has its own bearing and plunge (Fig. 6.37). The plunge lines fan outward from the vertex. In a type I conical fold (Fig. 6.37a,b) the plunge is away from the vertex and in a type II fold the plunge is toward the vertex (Fig. 6.37c,d). Plunge line directions in conical folds are best determined from the tangent diagram as described previously (Sect. 5.2.2).

A plunge line lies in the surface of the bed. Begin the projection by drawing a line on the map parallel to the plunge through the control point to be projected. Starting from the known elevation of the control point, mark spot heights (Fig. 6.38) spaced according to

$$H = I / \tan \phi \ , \tag{6.11}$$

Fig. 6.38.
Projection along plunge in a
vertical cross section. The
projection is parallel to plunge
along the plunge line from
point *1* to point *2*. *Open circles*
are spot heights along the
plunge line. For explanation
of symbols, see text

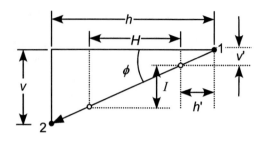

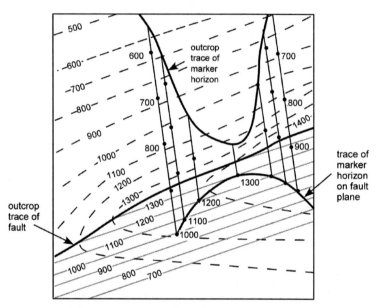

Fig. 6.39. Projection of a marker horizon to a fault plane along plunge lines. *Dashed lines* are topographic contours above sea level. *Dotted lines* are subsurface structure contours on the fault. Plunge lines are *solid* and marked by spot elevations. (After De Paor 1988)

where H = horizontal spacing of points, I = contour interval, and ϕ = plunge. If the control point is not at a spot height, the distance from the control point to the first spot height is

$$h' = H\,v'/I \;,\tag{6.12}$$

where h' = the horizontal distance from the control point to the first spot height and v' = the elevation difference between the control point and the first spot height.

Projection along plunge lines is particularly suited to projecting data from an irregular surface, such as a map, onto a surface, such as a cross section or fault plane, that itself can be represented as a structure-contour map. Figure 6.39 shows plunge lines derived from a map of a folded marker horizon on a topographic base. The fold

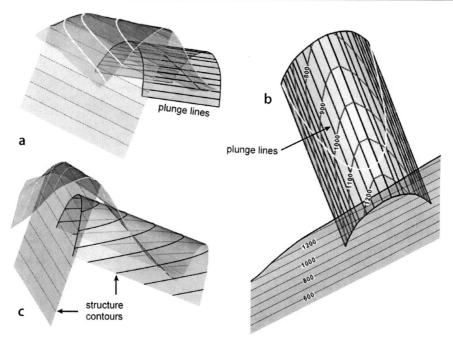

Fig. 6.40. 3-D views of the map in Fig. 6.39. **a** Oblique view to NW. Topographic surface with *white contours*, fault with *thin black contours*, fold with *thick black plunge lines*. **b** Vertical view, N up. Structure contours (with elevations) and plunge lines on the fold, structure contours (with elevations) only on the fault. *White line* is outcrop trace of fold. **c** Oblique view to NW. Same as **a** except fold shape indicated by structure contours

is projected south, up plunge, along the plunge lines onto the structure contour map of a fault. The intersection points where the plunge lines have the same elevation as the fault contours are marked and then connected by a line that represents the trace of the marker horizon of the fault plane (Fig. 6.39). In 3-D (Fig. 6.40a), the outcrop trace is projected up and down plunge from the outcrop trace to more completely illustrate the fold.

A structure contour map can be constructed from the plunge lines by joining the points of equal elevation (Fig. 6.40b). Figure 6.40b demonstrates that the plunge lines are not parallel to the structure contours and that projections should be made parallel to the plunge lines, not parallel to the structure contours. The structure contours provide an additional cross check on the geometry of the structure and on the internal consistency of the data. Once the fold geometry is constructed the plunge lines can be deleted and the shape shown by structure contours alone (Fig. 6.40c).

This technique can be performed analytically using the method of De Paor (1988). An individual point P (Fig. 6.41), given by its *xyz* map coordinate position, can be projected along plunge to its new position P' $(x', 0, z')$ on the cross-section plane (defined by $y' = 0$). Select the map coordinate system such that x is parallel to the line of cross section and y is perpendicular to the line of section. Choose $y = z = 0$ to lie in the plane

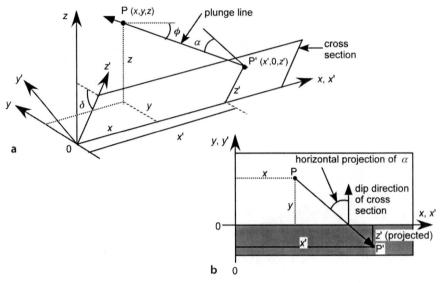

Fig. 6.41. Projection along plunge into the plane of the cross section. **a** Perspective diagram. **b** Horizontal map projection. The projection of the section plane onto the map is *shaded*

of the cross section. The sign convention requires that the positive (down) plunge direction be in the negative *y* direction. The elevation of a point is *z*. The dip of the plunge line = ϕ, the angle between the plunge line and the normal to the cross section = α, and the dip of the cross section = δ. The plunge line is constant in direction in a cylindrical fold but may be different for every location in a conical fold.

The general equations for the projected position of a point P', derived at the end of the chapter (Eqs. 6.36 and 6.41), are:

$$x' = x + y \tan \alpha + \tan \alpha \ (z \cos \alpha - y \tan \phi)/(\tan \phi + \tan \delta \cos \alpha) \ , \tag{6.13}$$

$$z' = (z \cos \alpha - y \tan \phi)/(\tan \phi \ \cos \delta + \sin \delta \ \cos \alpha) \ . \tag{6.14}$$

For a vertical cross section, from Eqs. 6.42 and 6.43,

$$x' = x + y \tan \alpha \ , \tag{6.15}$$

$$z' = z - (y \tan \phi)/\cos \alpha \ . \tag{6.16}$$

A cross section perpendicular to the fold axis is possible only for a cylindrical fold. The equations for projection to the normal section are (from Eqs. 6.44 and 6.45)

$$x' = x \ , \tag{6.17}$$

$$z' = z \cos \phi - y \sin \phi \ . \tag{6.18}$$

In a conical fold the plunge amount and direction changes with location. A cross section perpendicular to the crestal line will be closely equivalent to a normal section in slightly conical structures. Substitute the plunge of the crestal line in Eqs. 6.17 and 6.18 to approximate a normal section. The shorter the projection distance, the better the approximation.

6.6.1.3
Graphical Projection

The concept of projection along plunge is the basis of the graphical cross-section construction technique of Stockwell (1950). This method makes it possible to project data onto cross sections that have steep dips, such as vertical sections or sections normal to gently dipping fold axes. The graphical method is given by the following steps. Refer to Fig. 6.42 for the geometry.

1. Create the graph on which the cross section will be constructed at the same scale as the map and align it perpendicular to the plunge direction. Draw the line AC parallel to the plunge direction at the edge of the map. AC represents a horizontal line on the plunge projection. The plunge projection will be constructed from this line.
2. Begin the projection with the point that is to be projected the farthest (P_1). Project this point along the dotted line P_1A, perpendicular to the line AC to point A. From point A, draw the line AB at the plunge angle, ϕ, from AC.
3. Draw the orientation of the plane of the cross section (line CB) at the desired orientation to the vertical (angle ACB). In Fig. 6.42 the plane of the cross section has been chosen to be perpendicular to the plunge (angle ABC = 90°). The plane of ABC represents a vertical cross section in the plunge direction through point P_1.
4. The length CB is the distance in the plane of the east-west section from the map elevation of P_1 to its location P_1' on the cross section. Draw a vertical construction line (dotted line P_1C') through P_1 onto the line of section and measure the length C'B' (solid line) = CB down from the map elevation of the point to find P_1'.
5. Repeat step 4 to project all other points, for example P_2 is projected to P_2'.

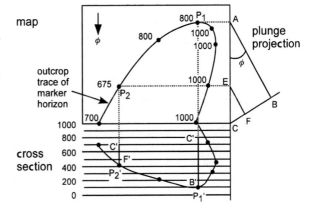

Fig. 6.42.
Projection of map data onto a cross section normal to plunge. The plunge is ϕ to the south. *Numbers* on the map (*square box*) are the topographic elevations of the points to be projected

Fig. 6.43.
Projection along plunge onto a vertical cross section. *Numbers* on the map (*square box*) are the topographic elevations of the points to be projected

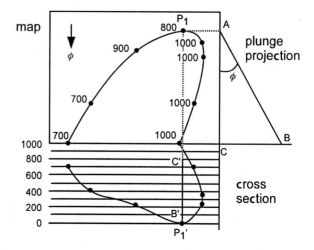

The ratio of the vertical projection length AB to the cross-section projection length CB is constant for all points, CB : CA = CF : CE = sin ϕ. Projection by hand is very rapid if a proportional divider drafting tool is used. Set the divider to the ratio CB / AB; then as the projection length AC is set, the divider gives the required length CB.

The method can be modified for other cross-section orientations by changing the orientation of the line of section on the plunge projection (Fig. 6.43). The orientation of the plane of section on the plunge projection is CB. In Fig. 6.43, CB is at 90° to AC, making the section plane vertical. The ratio CB : CA is constant for all points. Follow steps 1 to 5 above, changing the orientation of the line CB.

The plunge of a fold typically changes along the axis. Cylindrical fold axes may be curved along the plunge and cylindrical folds will change into conical folds at their terminations. Projection along straight plunge lines should be done only within domains for which the geometry of the structure is constant. Variable plunge can be recognized from undulations of the crest line on a structure-contour map or as excessive dispersion of the bedding dips around the best-fit curves on a stereogram or tangent diagram. If the plunge is variable, then the geographic size of the region being utilized should be reduced until the plunge is constant and all the bedding points fit the appropriate line on the stereogram or tangent diagram. If the sequence of plunge angle changes along the fold axis direction can be determined, then the straight line AB (Figs. 6.42, 6.43) could be replaced by a curved plunge line.

6.6.2
Projection by Structure Contouring

Structure contours represent the position of a marker horizon or a fault between the control points. Contouring provides a very general method for projecting data and can be used where the plunge cannot be defined from attitude measurements. This is a convenient method in three-dimensional interpretation. The projection technique is to map the marker surfaces between control points and draw cross illustrative sections through the maps.

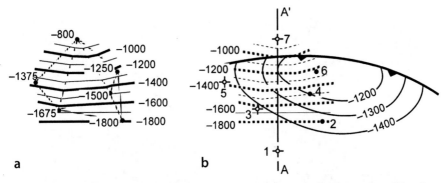

Fig. 6.44. Fault-surface map based on the wells in Fig. 6.34 that cut the fault. **a** *Points* give well locations and depths of fault cuts; *dotted lines* connect the nearest neighbors for contouring. Structure contours (*solid lines*) are derived by triangulation. **b** Structure contours on the fault (*dotted lines*) from a, superimposed on the structure contours of horizon E (*solid lines*) to show their parallelism to the fault trace

Where sufficient data are available, projection by contouring is equivalent to along-plunge projection. For example, the wells in Fig. 6.34 can be projected to the line of section by contouring without knowing the plunge amount or direction. To demonstrate this, the fault in Fig. 6.34 is contoured. The wells are inspected individually to be sure that each one shows only one fault cut, indicating that each well may cut the same fault. The fault contours generated from the elevations of the fault cuts (Fig. 6.44a) are smooth, as expected for a single fault. The fault contours superimposed on the original map (Fig. 6.44b) give the projected elevations of the fault along the line of section. The contours trend almost east-west, parallel to the plunge direction of the fold. A cross section of the fault along A–A' in Fig. 6.44b would be nearly identical to that in Fig. 6.34c. Each map horizon could be mapped to give a 3-D reconstruction of the structure which could then be sliced to create a cross section. When surfaces are mapped separately from sparse data, the spacings between them (thicknesses) are likely to show irregular variations. These variations should be corrected in the final cross section to maintain constant thicknesses (assuming it is geologically appropriate).

6.7
Dip-Domain Mapping from Cross Sections

The construction of maps from cross sections is a valuable technique where the structure is complex and/or the folds are conical. For cylindrical folds and faults the dip domains can be constructed is sections normal to the axis and linked together. Conical folds pose a special problem because no cross section can be drawn that will preserve bed thickness everywhere. The most general method is to define and map the dip domains in three dimensions. The approach will be illustrated here with the simple conical fold introduced in Sect. 5.3.

The horizontal domains in the conical fold (Figs. 6.45, 6.46) show unexaggerated thickness only in a vertical section. Only sections normal to the hinge lines show the unexaggerated thicknesses on the limbs (Fig. 6.46), requiring two different cross-sec-

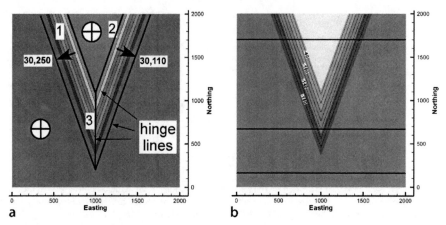

Fig. 6.45. Dip domains in conical fold. **a** Dip-domain map of middle horizon. **b** Structure contour map of middle horizon showing lines of cross section (*heavy EW lines*)

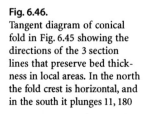

Fig. 6.46.
Tangent diagram of conical fold in Fig. 6.45 showing the directions of the 3 section lines that preserve bed thickness in local areas. In the north the fold crest is horizontal, and in the south it plunges 11, 180

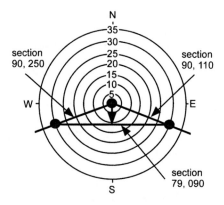

tion trends (110° and 250°). In the region of the 11° south plunge, a section normal to plunge is normal to bedding (Fig. 6.46) but will give an exaggerated thickness where beds are horizontal.

The simplest procedure is to map axial surfaces on straight, vertical cross sections (Fig. 6.47), or from multiple map horizons. An axial surface is by definition the surface through successive hinge lines. This relationship applies regardless of whether or not the profile is perpendicular to the bedding or to the hinge lines. Axial surfaces on successive cross sections are correlated and then mapped in 3-D (Fig. 6.48a). Axial surfaces in conical folds will intersect in three dimensions, and the intersection lines must be located (Fig. 6.48b). Once the axial surfaces and their intersections are constructed, the marker horizons can be mapped across the region (Fig. 6.49). Because a dip domain is a region of uniform dip, once the domain boundaries have been located, bedding attitudes may be projected anywhere within a single domain. Bedding surfaces can be projected throughout the entire domain from a single observation point.

Fig. 6.47.
Axial surface traces defined
on a vertical slice oblique to
hinge lines (northern section
across Fig. 6.45b); *ast:* axial
surface trace; *hp:* hinge point

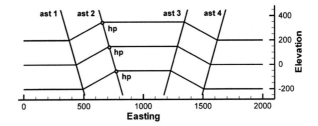

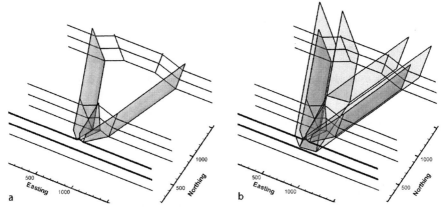

Fig. 6.48. Axial surfaces in the fold of Fig. 6.45. **a** Constructed by linking traces on profiles. **b** Completed axial surface network superimposed on previous construction

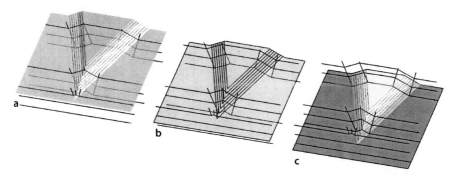

Fig. 6.49. Completed marker surfaces for map in Fig. 6.45. **a** Top. **b** Middle. **c** Bottom

Some types of information, for example fracture density, may be related to the proximity of the observation point to the fold hinge and so should be projected parallel to the closest hinge line. 3-D axial surface maps provide essential information for predicting the deep structure using kinematic models of cross-section geometry (especially using flexural-slip models Sect. 11.6).

6.8
Derivations

6.8.1
Vertical and Horizontal Exaggeration

From Fig. 6.50a, the dip of a marker on an unexaggerated profile is

$$\tan \delta = v / h \ ,\tag{6.19}$$

and the thickness of a unit in terms of its vertical dimension is

$$t = L \sin (90 - \delta) = L \cos \delta \ ,\tag{6.20}$$

where δ = unexaggerated dip, t = unexaggerated thickness, and L = unexaggerated vertical thickness. Let the vertical exaggeration be $V_e = v_v / v$ and the horizontal exaggeration be $H_e = h_h / h$, where v and h are the original horizontal and vertical scales and the subscripts h and v indicate the exaggerated scale. The equations for the exaggerated dips (Fig. 6.50b) have the same form as Eq. 6.19:

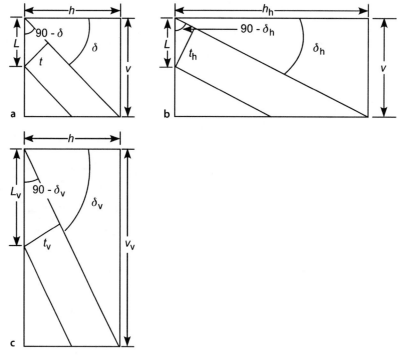

Fig. 6.50. Horizontal and vertical exaggeration. **a** Unexaggerated cross section. **b** Exaggerated horizontal scale; horizontal exaggeration (H_e) = 2:1. **c** Exaggerated vertical scale; vertical exaggeration (V_e) = 2:1

$$\tan \delta_v = v_v / h \ , \tag{6.21a}$$

$$\tan \delta_h = v / h_h \ . \tag{6.21b}$$

Replace h in Eq. 6.21a and v in 6.21b with the values from Eq. 6.19 and use the definition of the exaggeration to obtain the relationship between original and exaggerated dips:

$$\tan \delta_v = V_e \tan \delta \ , \tag{6.22a}$$

$$\tan \delta_h = \tan \delta / H_e \ . \tag{6.22b}$$

To relate the horizontal to the vertical exaggeration, substitute the value of $\tan \delta$ from Eq. 6.22a into 6.22b to obtain

$$V_e \, H_e = \tan \delta_v / \tan \delta_h \ . \tag{6.23}$$

To obtain the same exaggerated angle by either horizontal or vertical exaggeration, set $\delta_v = \delta_h$ in Eq. 6.23:

$$V_e = 1 / H_e \ . \tag{6.24}$$

The thickness of a unit on a horizontally exaggerated profile (Fig. 6.50b), t_h, is

$$\sin (90 - \delta_h) = \cos \delta_h = t_h / L \ . \tag{6.25}$$

Eliminate L by dividing Eq. 6.25 by 6.20:

$$t_h / t = \cos \delta_h / \cos \delta \ . \tag{6.26}$$

The thickness of a unit on a vertically exaggerated profile (Fig. 6.50c), t_v, is

$$\cos \delta_v = t_v / (V_e \, L) \ . \tag{6.27}$$

Eliminate L by dividing Eq. 6.27 by Eq. 6.20:

$$t_v / t = V_e \, (\cos \delta_v / \cos \delta) \ . \tag{6.28}$$

6.8.2
Analytical Projection along Plunge Lines

The point P is to be projected parallel to plunge to point P' on the cross section (Fig. 6.51a). The plunge direction makes an angle of α to the direction of the perpendicular to the cross section (Fig. 6.51d) and the plunge is ϕ. Following the method of De Paor (1988), the x coordinate axis is taken parallel to the line of the section and the plane of section intersects the x axis at zero elevation. In the plane of the cross section,

Fig. 6.51.
Projection along plunge. **a** Perspective diagram. **b** Vertical plane through plunge line *PP'*. **c** Vertical plane normal to the cross section through line *OP'*. **d** Plan view. Projection of the cross section is *shaded*. Point *Q* is vertically above *A* at *A–Q* and point *P'* is vertically above *O* at *O–P'*

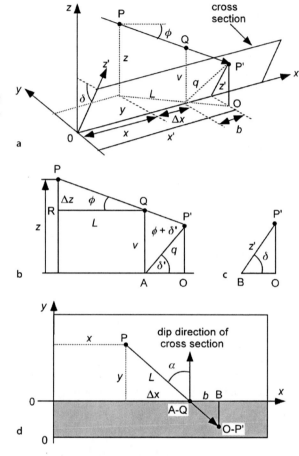

the position of point P (x, y, z) is P' (x', z'). The apparent dip of the intersection line, q, of the vertical plane through PP' with the cross section is δ'. The relationship between the apparent dip and the true dip, δ, of the z' line, is given by Eq. 2.18 as

$$\tan \delta' = \tan \delta \cos \alpha \ . \tag{6.29}$$

Begin by finding z'. In the plane normal to the cross section (Fig. 6.51a,c),

$$z' = OP' / \sin \delta \ . \tag{6.30}$$

In the plane of the plunge (Fig. 6.51b), using triangles AQP' and PQR,

$$OP' = q \sin \delta' \ , \tag{6.31}$$

$$\Delta z = L \tan \phi \ , \tag{6.32}$$

$$v = z - L \tan \phi \ , \tag{6.33}$$

and by using the law of sines with angles AQP' and QP'A in triangle QAP', along with $\cos \phi = \sin (90 - \phi)$,

$$q = v / (\tan \phi \cos \delta' + \sin \delta') \ . \tag{6.34}$$

In the plane of the map (Fig. 6.51d)

$$L = y / \cos \alpha \ . \tag{6.35}$$

Substitute Eqs. 6.29, 6.31, 6.33, 6.34 and 6.35 into 6.30 to obtain

$$z' = (z \cos \alpha - y \tan \phi) / (\tan \phi \cos \delta + \sin \delta \cos \alpha) \ . \tag{6.36}$$

The x' coordinate is found from (Fig. 6.51a)

$$x' = x + \Delta x + b \ . \tag{6.37}$$

In the plane of the map (Fig. 6.51d)

$$\Delta x = y \tan \alpha \ , \tag{6.38}$$

$$b = OB \tan \alpha \ . \tag{6.39}$$

In the plane normal to the cross section (Fig. 6.51c)

$$OB = OP' / \tan \delta \ . \tag{6.40}$$

Substitute Eqs. 6.29, 6.31–6.35 and 6.38–6.40, into 6.37 to obtain

$$x' = x + y \tan \alpha + \tan \alpha \ (z \cos \alpha - y \tan \phi) / (\tan \phi + \tan \delta \cos \alpha) \ . \tag{6.41}$$

For a vertical cross section, $\delta = 90°$ and Eqs. 6.36 and 6.41 reduce to

$$x' = x + y \tan \alpha \ , \tag{6.42}$$

$$z' = z - y \tan \phi / \cos \alpha \ . \tag{6.43}$$

For a cross section normal to the plunge line, possible only for a cylindrical fold, $\alpha = 0$, $\delta = (90 - \phi)$ and Eqs. 6.36 and 6.41 reduce to

$$x' = x \ , \tag{6.44}$$

$$z' = z \cos \phi - y \sin \phi \ . \tag{6.45}$$

6.9
Exercises

6.9.1
Vertical and Horizontal Exaggeration

Draw the cross section in Fig. 6.52 vertically exaggerated by a factor of 5:1. Draw the cross section in Fig. 6.52 horizontally squeezed by a factor of 1:2.

6.9.2
Cross Section and Map Trace of a Fault

Draw an east-west cross section across the northern part of the structure contour map in Fig. 6.53. Suppose a fault that dips 40° south cuts the structure in the blank area between the arrows. What would its trace be on the structure contour map? Is the fault normal or reverse? Draw a north-south cross section showing the fault.

6.9.3
Illustrative Cross Section from a Structure Contour Map 1

Draw a cross section perpendicular to the major structural trend in Fig. 6.54. Discuss any assumptions required. What are the dips of the faults? Are the faults normal or reverse?

Fig. 6.52.
Cross section of a fold having constant bed thickness in shaded unit

Fig. 6.53.
Unfinished structure contour map. *Arrows* indicate the general position of the fault trace

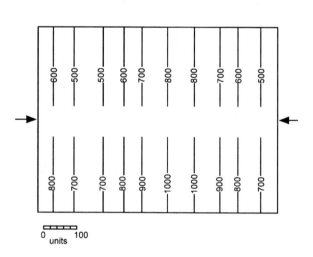

Fig. 6.54.
Structure contour map of the
top of the Gwin coal cycle.
Elevations of the top Gwin are
posted next to the wells. Units
are in feet, negative below sea
level

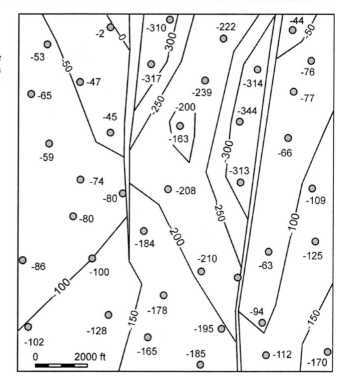

6.9.4
Illustrative Cross Section from a Structure Contour Map 2

Draw cross sections along the three lines indicated on Fig. 6.55. Using the fault dip
determined from the map, extend the faults above and below the marker horizon until
they intersect. Which fault(s) formed last?

6.9.5
Illustrative Cross Section from a Structure Contour Map 3

Draw cross sections along the three lines indicated in Fig. 6.56. Determine the dips of
the faults from the map and then extend the faults above and below the marker horizon until they intersect. Which fault is youngest?

6.9.6
Predictive Dip-Domain Section

Complete the cross section in Fig. 6.57 by extending it into the air and deeper into the
subsurface. How far can the section be realistically extended?

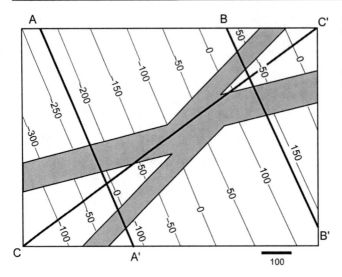

Fig. 6.55.
Structure contour map of a normal-faulted surface. The horizon surface is missing in the *shaded* fault zones. Locations of lines of section *A–A'*, *B–B'*, and *C–C'* are shown

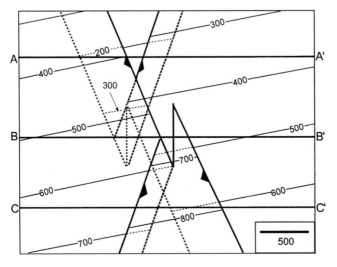

Fig. 6.56.
Structure contour map of a reverse-faulted surface. The horizon surface is repeated by the fault zones. The fault cutoffs are *wider lines*. Hidden contours are *dashed*. The locations of lines of section *A–A'*, *B–B'*, and *C–C'* are shown

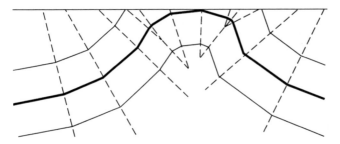

Fig. 6.57.
Partially complete dip-domain cross section. *Dashed lines* are axial-surface traces

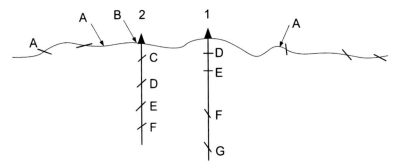

Fig. 6.58. Cross section through the Burma No. 1 and 2 wells. *Short lines* are surface dips. Letters *A–G* are marker horizons seen at the locations of dip measurements that can be correlated. *Arrows* point to locations where markers can be identified in outcrop but the dip cannot be measured. The dips in the wells are from oriented cores

6.9.7
Predictive Cross Sections from Bedding Attitudes and Tops

Complete the cross section in Fig. 6.58, keeping bed thicknesses constant. Use both the dip-domain technique and the method of circular arcs. Scan the section into a computer and complete using the smooth curves provided by a drafting program. Compare the results of the different techniques.

6.9.8
Fold and Thrust Fault Interpretation

Construct illustrative cross section A–A' from the map in Fig. 3.29 using the structure contour map constructed in Exercise 3.7.4. Use the dip-domain technique to construct the same cross section using only the surface geology along the profile. Use the circular arc technique to construct the same cross section using only the surface geology along the profile. What is the plunge of the central portion of the structure from a stereogram or tangent diagram? Project the northern part of the structure onto section B–B' using the method of along-plunge projection. Compare and contrast the cross sections. The wells to the Fairholme were drilled to find a hydrocarbon trap but were not successful. Use the map and cross sections to determine a structural reason for drilling the wells and a structural reason that they were unsuccessful.

6.9.9
Projection

Project the top reservoir onto the seismic line assuming the structure is normal to the seismic line (Fig. 6.59). Project the top reservoir onto the seismic line assuming the structure plunges 10° in the direction 225°.

Fig. 6.59.
Map showing trace of a seismic
line and the location of a well

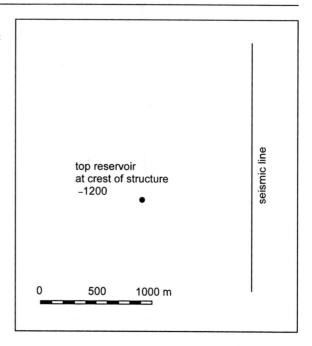

Chapter 7

Properties of Faults

7.1
Introduction

This chapter focuses on the recognition of faults, their geometrical properties, displacement distributions, and the correlation of multiple fault cuts into continuous fault surfaces. Unconformities are treated here because, by truncating units at the map scale, they share characteristics with faults and need to be distinguished from faults.

7.2
Recognition of Faults

Faults are recognized where they cause discontinuities in the traces of marker horizons on maps and cross sections, discontinuities in the stratigraphic sequence, and anomalies in the thicknesses. Faults may also be recognized from the diagnostic shape of drag folds adjacent to the fault and by the distinctive rock types caused by faulting.

7.2.1
Discontinuities in Geological Map Pattern

At the map scale, a fault is inferred where it causes a break in the continuity of the units on a geologic map. A time slice through a 3-D seismic-reflection volume is similar to a geologic map in a region of low topographic relief and will also show faults as discontinuities. As an example, the coal seams in the north half of Fig. 7.1 are abruptly truncated along strike where they intersect faults (points A). Fault dips and hence fault separations (Sect. 7.4.2) cannot be determined from the near-horizontal map surface, but these are known to be normal-separation faults. The truncation of units by a contact (B, Fig. 7.1) indicates that the contact is a fault or that the beds are cut by an unconformity. The base of the continuous bed that crosses the truncated beds (C, Fig. 7.1) is either a fault contact or an unconformity. The contact indicated by B and C in Fig. 7.1 is a fault because the truncated beds to the north (bc, ml) are younger than the cross-cutting unit to the south (by) and so should be above the unconformity, not below it. At D, E, and F, the contacts are parallel and so provide no direct evidence of faulting, although the absence of stratigraphic units at the contacts suggests the presence of faults. The contact at D can be traced into a location (C) where it is faulted; hence it is probably a fault. The contact at F is a reverse fault because the dips of bedding on both sides of the contact show that the older units to the south lie on top of the younger units to the north. The contact indicated by E places upright lower Cambrian rocks adjacent to overturned younger units to the north, suggesting re-

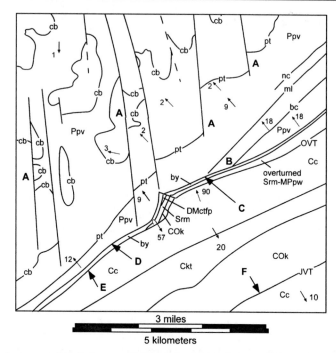

Fig. 7.1. Geologic map of the Ensley area, Alabama. The topography is nearly flat, making the map close to a horizontal section. Units (oldest to youngest): Cambrian: *Cc:* Conasauga Ls., *Ckt:* Ketona dolomite; Cambro-Ordovician: *COk:* Knox dolomite; Silurian: *Srm:* Red Mountain Formation; Devonian-Mississippian: *DMctfp:* Chattanooga Shale, Tuscumbia Limestone, Fort Payne Chert; Mississippian-Pennsylvanian: *MPpw:* Parkwood Formation; Pennsylvanian: *Ppv:* Pottsville, containing the following marker units (oldest to youngest): *by:* Boyles sandstone, *bc:* Black Creek coal, *ml:* Mary Lee coal, *nc:* Newcastle (upper Mary Lee) coal, *pt:* Pratt (American) coal, *cb:* Cobb coal. *OVT:* Opossum Valley thrust, *JVT:* Jones Valley thrust. Contact relationships: *A:* offset of strike traces, *B:* truncation of units at contact, *C:* unit crosses contacts, *D:* missing section, *E:* missing section, *F:* older over younger. (After Butts 1910 and Kidd 1979)

verse fault drag (Sect. 7.2.7). The contact indicated by C and D is a reverse fault. Once a contact has been identified as a fault, it is usually shown as a heavier line on the map.

The juxtaposition of different units across a fault favors some form of topographic expression for the fault because of the likelihood that the two units will not weather and erode identically. The boundary between the juxtaposed units is likely to form a topographic lineament. The fault-zone material (Sect. 7.2.6) may erode differently from any of the surrounding country rocks and thus cause a topographic valley or ridge along the fault.

7.2.2
Discontinuities on Reflection Profile

The primary criterion for recognizing a fault on a seismic- or radar-reflection profile is as a break in the lateral continuity of a reflector or a group of reflectors (Fig. 7.2a, arrows A). Picking faults can be difficult because sometimes the reflectors hang over the actual fault

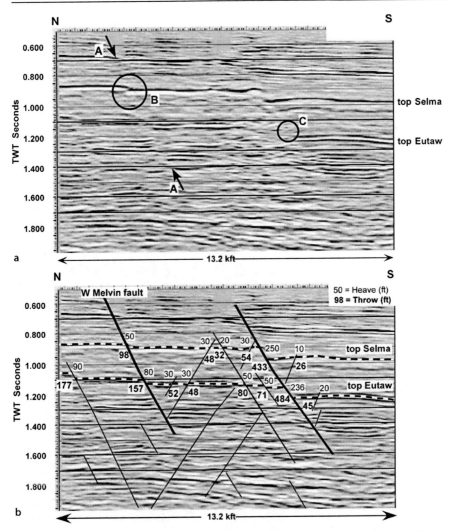

Fig. 7.2. Normal faults on a vertical profile from a time-migrated 3-D seismic reflection volume. V.E. about 1:1. The profile is from the Gilbertown graben system, southern Alabama (modified from Groshong et al. 2003a). **a** Uninterpreted. *A:* fault trace between arrows; *B:* reflectors hang over fault trace; *C:* disturbed zone along fault trace. **b** Interpreted. The faults indicated with *heavier lines* have been identified in nearby wells. *Numbers* next to the faults are heave (regular type) and throw (bold). Throws are determined from the heaves using Throw = Heave times tan (fault dip). Only the most obvious faults are interpreted below the top of the Eutaw

trace (Fig. 7.2a, location B), obscuring the exact location of the fault and its separation. At location B (Fig. 7.2a) the overhang makes the fault look like it is reverse, although it is actually a normal fault (c.f., Fig. 7.2b). Local disruption of reflectors or lack of reflectors in the fault zone may also make the location uncertain (C, Fig. 7.2a). The termination of multiple reflectors along a straight line or a smooth curve, together with a consistent sense of separation, provide the most convincing evidence of a fault.

In some areas the faults produce reflections as, for example, along the large normal fault in the Gulf of Mexico (Fig. 7.3). Fault reflections are favored along low-angle faults and where the impedance contrast (difference in physical properties) is large across the fault. Steeply dipping reflectors, whether originating from beds or faults, are difficult to image on conventional seismic lines because the reflections return to the surface beyond the farthest receiver.

Fig. 7.3.
Corsair fault, a thin skinned growth fault on the Texas continental shelf. 48-fold, depth-converted seismic line. *F:* fault reflectors. (After Christensen 1983)

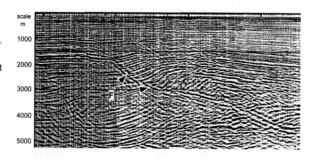

Fig. 7.4.
Faults in a portion of the Ruhr coal district, Germany.
a Seismic reflection profile: dynamite source, 12-fold common-depth-point stack, time migrated. *A:* reflector discontinuity at a large thrust fault; *B:* reflector discontinuity at a small, complex, reverse fault zone; *C:* crossing reflectors in a fault zone; *D:* reflector discontinuity at an angular unconformity; *E:* zone of disturbed reflectors on the steep limb of a fold.
b Geological cross section based on the seismic profile and the wells shown. *Letters* designate the same features as in **a**. (Adapted from Drozdzewski 1983)

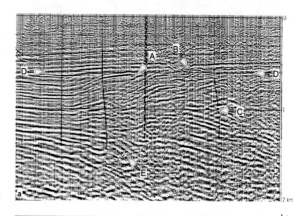

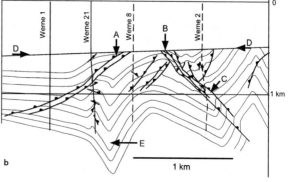

The abrupt termination of a reflecting horizon at a fault provides a point source of diffractions, arcuate reflectors that emanate from the fault and cross other reflectors. Time migration of the seismic profile is designed to restore the reflectors to their correct relative locations and dips and to remove the diffractions. Nevertheless, some diffractions may remain in time-migrated profiles, as seen in the region between B and C in Fig. 7.4a.

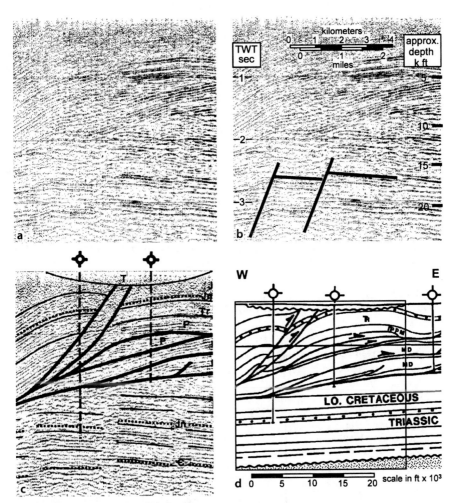

Fig. 7.5. Velocity discontinuities create features that look like faults on seismic profiles. **a** Segment of a seismic line across Wyoming thrust belt (dynamite source, eight-fold common-depth-point stack, migrated time section, approximate vertical exaggeration 1.3 at 2.7 s; Williams and Dixon 1985). **b** Discontinuities in seismic reflectors that might be normal faults. **c** Interpretation by Williams and Dixon (1985). **d** Geological cross section using well control and the seismic line; no vertical exaggeration (Williams and Dixon 1985). The *box* outlines the area of the seismic line. No normal faults are present. *TWY*: two-way traveltime (s); *C*: Cambrian; *MD*: Mississippian-Devonian; *IPPM*: Pennsylvanian, Permian, Mississippian undifferentiated; *P*: Permian; *Tr, TR*: Triassic; *J*: Jurassic undifferentiated; *Jn*: Jurassic Nugget sandstone; *T*: Tertiary

Features other than faults may resemble faults. Reflectors are also truncated at an unconformity (Fig. 7.4a, between the arrows labeled D). The presence of parallel reflectors above the unconformity supports the interpretation that the truncation is indeed an unconformity, not a fault. Reflections from steeply dipping beds may fail to be recorded or may not be correctly migrated, leading to zones of disturbed reflectors that can easily be mistaken for fault zones. The lack of reflector continuity around and above location E (Fig. 7.4a) is due to the steep limb of a fold (Fig. 7.4b), not to a fault.

Regions of complex structure are commonly associated with faults and may have steep dips and large lateral velocity changes, either of which can create discontinuities at deeper levels on the seismic profile (Fig. 7.5). Discontinuities in the otherwise continuous reflectors below 2-s two-way travel time (Fig. 7.5a) might easily be interpreted as being normal faults (Fig. 7.5b), although the interpreters of the line correctly did not do so (Fig. 7.5c). This region is below a major thrust fault that places rocks as old as Devonian on top of Cretaceous units. The older rocks have significantly higher seismic velocities than the Cretaceous rocks. The geological interpretation (Fig. 7.5d), based on well control and the seismic profile, indicates no offset of the Cretaceous and older units below the thrust, but rather a continuous westward regional dip of the subthrust units. The apparent eastward dip of the subthrust units is a velocity pull up caused by the high velocities of the rocks in the thrust sheet. The apparent normal faults in the subthrust sequence are probably caused by the rapid lateral velocity gradients associated with fault slices within the thrust sheet.

Reflection profiles can be one of the most powerful tools for structural interpretation. Because the profile itself is an interpretation based on the inferred velocity structure of the region, the geological interpretations must be tested with all of the same techniques that are applied to geological data.

7.2.3
Discontinuities on Structure Contour Map

A linear trend of closely spaced structure contours that form a monoclinal fold may represent an unrecognized fault (F, Fig. 7.6). The monocline could be replaced by a fault. Linear fold trends are, of course, perfectly reasonable and so independent evi-

Fig. 7.6.
Structure contour map showing the possible locations of faults (*F*) between the *arrows*. Contours are in feet

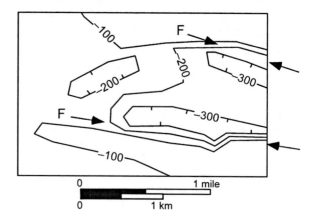

dence for faulting is desirable before the fold is reinterpreted as a fault as, for example, a direct observation of the fault somewhere along the trend. In some areas, especially in extensional terrains, folds with steep limb dips are rare or do not occur at all and so the closely spaced contours could be replaced by a fault on the basis of consistency with the local structural style.

7.2.4
Stratigraphic Thickness Anomaly

A linear thickness anomaly in a generally uniform stratigraphic unit may be caused by a fault. The anomaly can be caused by the stratigraphic section missing or repeated by an unrecognized fault. A fault appearing as a thickness anomaly is a very likely occurrence where the stratigraphic separation on the fault is less than the stratigraphic resolution. A unit penetrated by three wells (Fig. 7.7a) might be interpreted as having a stratigraphic thin spot in the middle well. The thin spot might also represent a normal fault cut (Fig. 7.7b). Where a thickness anomaly is recognized as a possible fault, the data should be re-examined for other types of evidence of a fault cut. A cross section involving multiple units, rather than just one unit as in Fig. 7.7, should be constructed. Thickness anomalies that line up on a cross section with a dip appropriate for a fault probably represents a fault.

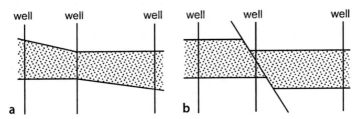

Fig. 7.7. Cross sections showing alternative interpretations of a stratigraphic thickness anomaly in the *shaded* unit of the center well. **a** Thinning interpreted as due to stratigraphic change. **b** Thinning interpreted as due to a normal fault

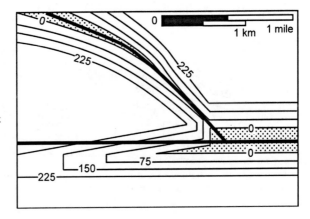

Fig. 7.8.
Isopach map showing the effect of two intersecting normal faults on the thickness of a formation that is regionally 225 or more units thick. Contours are of thickness, the formation is *absent* in the shaded areas. *Heavy lines* mark the fault trends; because the faults are dipping planes, the lines do not represent the exact map traces of the faults

On an isopach map a thickness anomaly will follow the trace of the fault (Fig. 7.8). A fault with normal separation will cause the unit to be thinner and a reverse separation will cause the unit to be thicker. The greater the stratigraphic separation of the fault, the greater the thickness change from the regional value. The maximum thinning is to zero thickness but the maximum thickening can be any amount, depending on the number of repetitions that occur within the boundaries of the formation. The example shows an east–west trending fault that is losing displacement to the west while an intersecting fault to the north increases in displacement to the west. See Sect. 8.5 for the quantitative interpretation of fault separation from isopach maps.

7.2.5
Discontinuity in Stratigraphic Sequence

The recognition of a fault at a point such as in an outcrop or a well log is commonly based on a break in the continuity of the normal stratigraphic sequence (Fig. 7.9). Except for the special case of fault slip exactly parallel to the stratigraphic boundaries (Redmond 1972), some amount of the stratigraphic section is always missing or repeated at the fault contact. The measure of the fault magnitude that can be obtained from the information available at a point on a fault is the stratigraphic separation, equal to the thickness of the missing or repeated section (Bates and Jackson 1987). Only the stratigraphic separation can be determined at a point on the fault, not the slip. Determination of the slip requires additional information. The true slip on the faults in Fig. 7.9 could be oblique or even strike slip if the dip of bedding is oblique to the dip of the fault.

The method for determining the position of the fault cut and the amount of the stratigraphic separation is the same in the field as in the subsurface. A well log is the same as a section measured across a fault. The faulted section must be correlated to a reference section. Correlate upward in the footwall (Fig. 7.10) until the sections no longer match. Then correlate downward in the hangingwall until the sections no longer match. The mismatch will occur at the same place in the faulted section if a single fault is present. A mismatch in upward and downward correlation that fails to occur at the same place in the faulted section could be caused by the presence of more than one fault or by a fault zone of finite thickness. In the former situation, the process should be repeated on a finer scale until all the faults are located.

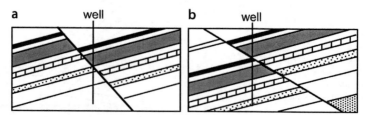

Fig. 7.9. Cross sections of faults recognized from a stratigraphic discontinuity where the well cuts a fault. **a** Normal separation. Part of the normal stratigraphic sequence is missing in a vertical well. **b** Reverse separation. Part of the normal stratigraphic sequence is repeated in a vertical well

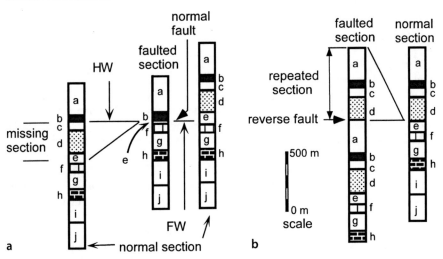

Fig. 7.10. Location and amount of stratigraphic separation determined from the missing or repeated section across the fault. **a** About 300 m of section is missing across a normal fault. **b** About 600 m of section is repeated across a reverse fault

An overturned fold limb can easily be mistaken for a fault zone. The upward and downward correlations will break down at the axial surfaces that bound the overturned limb. The overturned sequence will appear to be an unfamiliar unit and so is easily interpreted as being a fault zone. Try correlating an upside-down stratigraphic column to the possible fault zone to see if an overturned section is present. Beds near the fault may be steeply dipping, causing a great exaggeration of the bed thicknesses in a vertical well, representing another factor to consider when correlating to the type section.

The precision of the determination of stratigraphic separation depends on the level of detail to which the stratigraphy is known. For example, if the fault cut in Fig. 7.10a was at the top of unit e rather than near the base, the missing section would be significantly (50–75 m) greater. The error in the amount of missing section is the sum of the stratigraphic uncertainties in the units on both sides of the fault. Lack of stratigraphic resolution means that small faults may not be detectable by correlation of sections.

The *fault cut* refers to the position of the fault in a well and the to the amount of the missing or repeated section compared to the reference section (Tearpock and Bischke 2003). The location of the fault cut, the amount of the fault cut and the reference section should be recorded for each fault cut. It is not unusual to find that after preliminary mapping has been completed, the original stratigraphic thicknesses were inappropriate. For example, the units in a reference well may be shown by mapping to have a significant dip and their thicknesses therefore will have been exaggerated. Mapping may also reveal that the reference section is faulted. Any change in thickness in the reference section requires changes in the magnitudes of all the fault cuts determined from it.

7.2.6
Rock Type

Rock types diagnostic of faulting (Fig. 7.11) include the cataclasite suite, produced primarily by fragmentation, and the mylonite suite, produced by large crystal-plastic strains and/or recrystallization (Sibson 1977; Ramsay and Huber 1987). A cataclastic fault surface, or slickenside, is usually grooved, scratched, or streaked by mineral fibers or may be highly polished. The parallel striations or mineral elongation directions are slickenlines which indicate the slip direction at some stage in the movement history. A mylonite typically has a thinly laminated compositional foliation and a strong penetrative mineral lineation which is elongated in the displacement direction. Soft-sediment fault zones typically have macroscopic textures similar to those of mylonite zones, although no metamorphism has occurred. At the microscopic scale, the grains in a soft-sediment fault zone are usually undeformed.

A fault zone may be recognizable in well logs. An uncemented cataclastic fault zone is mechanically weak and very friable. This is the reason why an outcropping fault zone may erode to a valley. Where a well encounters a mechanically weak unit, such as an uncemented fault zone, an enlargement of the well-bore diameter occurs. A caliper log is produced by a tool that measures the size of the well bore and a fault may be recorded as an expansion in the size of the borehole, generally short in length. Well enlargement may occur in any mechanically weak unit such as coal, uncemented clay or sand, salt, etc. If the normal stratigraphic reasons for well enlargement can be eliminated, however, then the presence of a fault zone is a good possibility. A high-resolution dipmeter provides a log that resembles a photograph of the wall of the borehole. A fault zone may be inferred from the same observational criteria that would apply to

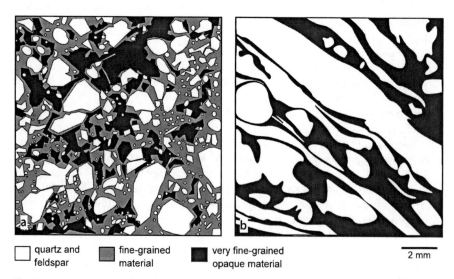

| ☐ quartz and feldspar | ▦ fine-grained material | ■ very fine-grained opaque material | 2 mm |

Fig. 7.11. Representative fault-rock textures. The scale is approximate. **a** Cataclasite. **b** Mylonite. Drawn from thin section photographs in Ramsay and Huber (1987)

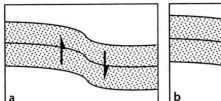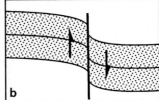

Fig. 7.12. Drag folds caused by permanent bending strain before faulting. **a** Fold before the units reach their ductile limit. **b** A fault forms after the ductile limit is exceeded, and the permanent strain remains as a drag fold

an outcrop, in particular, a high concentration of fractures or an abrupt change in lithology not part of the normal stratigraphic sequence.

7.2.7
Fault Drag

The systematic variation of the dip of bedding adjacent to the fault shown in Fig. 7.12b is known as "normal" drag, or simply *drag*, as the term will be used here. A drag fold gives the sense of slip on the fault. This geometry is probably best interpreted as being the result of the permanent strain that occurs prior to faulting (Fig. 7.12a), a sequence of formation well known from experimental rock mechanics. The term "drag" is thus a misnomer because the fold forms prior to faulting, not as the result of frictional drag along a pre-existing fault. In fact, the strain associated with slip on a pre-existing fault tends to produce bending in the opposite direction relative to the sense of slip (Reches and Eidelman 1995) called "reverse" drag. The reverse drag described by Reches and Eidelman (1995) is significantly smaller in magnitude than the normal drag produced before faulting and might not be noticeable at the map scale. Map-scale reverse drag is also well known as a consequence of slip on a downward-flattening normal fault (Hamblin 1965). Because the term drag is long established for the fold close to a fault and because the curvature of drag folds gives the correct interpretation of the sense of shear, use of the term is continued here. The drag-fold geometry usually extends no more than a few to tens of meters away from the fault zone. Although the drag geometry is common, not all faults have associated drag folds.

7.3
Unconformities

Unconformities are important surfaces for structural and stratigraphic interpretation. They provide evidence for structural and sea-level events, and mineral deposits, including hydrocarbons, are commonly trapped below unconformities. An unconformity truncates older geological features and so shares a geometric characteristic with a fault which also truncates older features (Figs. 7.4 and 7.13). Sometimes the two can be difficult to distinguish. The flat portion of a fault may cut bedding at a very low angle, just like the typical unconformity.

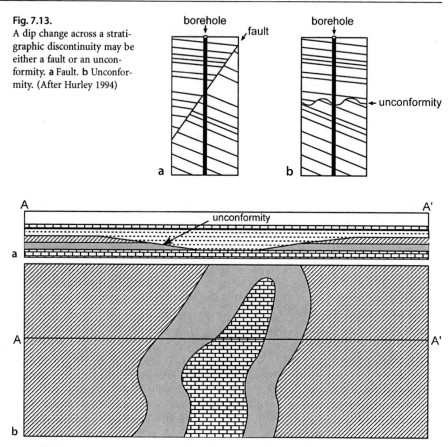

Fig. 7.13.
A dip change across a stratigraphic discontinuity may be
either a fault or an unconformity. **a** Fault. **b** Unconformity. (After Hurley 1994)

Fig. 7.14. The geometry of a low-angle unconformity. **a** Cross section showing a filled paleo-valley. Vertical exaggeration is about two to three times. **b** Subcrop map of units below the unconformity. (After Calvert 1974)

An unconformity can be mapped like a stratigraphic horizon, but it is significantly different because the units both above and below the unconformity can change over the map area (Fig. 7.14). An unconformity surface should be mapped separately from the units that it cuts or that terminate against it. The stratigraphic separation across an unconformity can be mapped and the separation can be expected to die out later-ally into a continuous stratigraphic sequence (the correlative conformity) or at an-other unconformity. A subcrop map shows the units present immediately below an unconformity (Fig. 7.14b). This type of map has several important uses. It provides a guide to the paleogeology, can indicate the trends of important units, for example hydrocarbon reservoirs, and can suggest the paleostructure. An analogous map type is a subcrop map of the units below a thrust fault.

The dip change across an unconformity can be very small, a few degrees or less. A plot of the cumulative dip versus depth (Hurley 1994) is very sensitive to such small

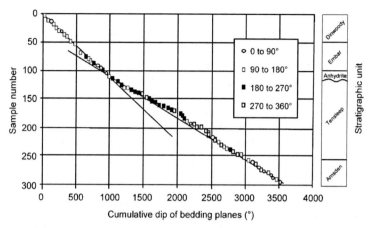

Fig. 7.15. Cumulative dip versus sample number from a well in the Bighorn basin, western U.S. Samples are *numbered* from the top of the well. Bedding planes are derived from a high-resolution dipmeter log. Points falling into different quadrants of the compass are coded with different *symbols*. (After Hurley 1994)

changes (Fig. 7.15). The vertical axis can be either depth or the sample number. Equal spacing of points on the vertical axis gives a more interpretable result and so if the bedding planes are sampled at irregular intervals, it is better to plot sample number rather than true depth or distance (Hurley 1994). The dips of the bedding planes are added together in the direction of the traverse to obtain the cumulative dip. In the example (Fig. 7.15), the change in average slope of the line through the data points indicates the position of an unconformity at the top of the Tensleep sandstone. The dip change in Fig. 7.15 is only 1.3° across the unconformity, yet it is clearly visible.

An additional analysis technique for the cumulative dip data is a first derivative plot (Hurley 1994). The first derivative (slope) between each two data points is plotted as the vertical axis and the cumulative dip of bedding planes as the horizontal axis. The slope between two successive dip points is the difference in depth or the sample-number increment (1) divided by the difference in cumulative dip between adjacent samples. Abrupt dip changes, even small ones, appear as large departures from the overall trend on this type of plot.

7.4
Displacement

Displacement is the general term for the relative movement of the two sides of a fault, measured in any direction (Bates and Jackson 1987). The components of displacement depend on five measurable attributes: the attitude of the fault, the magnitude of the stratigraphic separation, the attitudes of the beds on both sides of the fault, and direction of the slip vector in the fault plane. Different sets of these attributes are used to define the displacement components of slip or separation as will be discussed next. Throw and heave are the separation components most useful in structure-contour map construction because they appear on the map.

7.4.1
Slip

Slip (or net slip) is the relative displacement of formerly adjacent points on opposite sides of the fault, measured along the fault surface (Bates and Jackson 1987). Slip is commonly described in terms of its components in the plane of the fault, dip slip and strike slip (Fig. 7.16a). Dip slip components are normal or reverse and strike slip components are right lateral or left lateral. The offset of a marker horizon, as recorded on a map, can be produced by slip in a variety of directions. The geometry of the fault and the offset horizon are identical in Figs. 7.16a,b but the fault in 7.16a is oblique slip and in 7.16b is pure dip slip.

There are two traditional approaches for finding the net slip on a fault. The most direct is to find the offset of a geological line that can be correlated across the fault. A geological line may be a geographic or a paleogeographic feature such as a stream channel or a facies boundary. The alternative is based on the distance in the slip direction between correlative planar surfaces. In this method the slip direction must be known, for example, from the slickenline direction. The orientation of the slip vector may be recorded as the bearing and plunge of the lineation, or as the rake, which is the angle from horizontal in the plane of the fault.

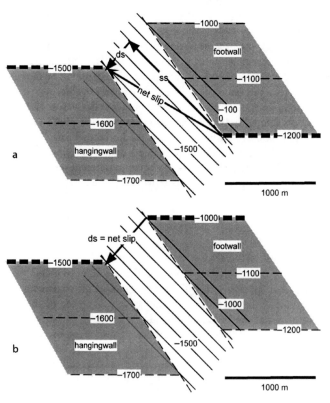

Fig. 7.16.
Displacement of a marker horizon (*shaded*, with *dashed contours*) across a fault (*unshaded*, with *solid contours*). The final geometries of **a** and **b** are identical although the slip is different. *Wide contour lines* represent correlative linear features displaced by the fault. *ds:* dip slip; *ss:* strike slip. **a** Right-lateral, normal oblique slip. **b** Normal dip slip

The best features for determining true displacements are the offsets of pre-existing linear stratigraphic trends or the intersection lines produced by cross-cutting features, for example, a dike-bedding plane intersection line. Offset structural lines may be used in some circumstances. Fold hinge lines on identical marker horizons are potentially correlative marker lines. The offsets of the hinge lines in Fig. 7.17 give the true displacement for both the dip slip and strike slip examples. Offset hinge lines are good markers of displacement if the folds developed prior to faulting. Tear faults present at the beginning of folding may divide a region into separate blocks in which the folds can develop independently. The folds in adjacent blocks may have never been connected and so the separation of their hinge lines is not a measure of the displacement. If more than one fold hinge line can be matched across the fault and more than one hinge line gives approximately the same displacement, the probability is higher that the fault displaces pre-existing folds and the displacement determination is valid. The most complete method for the determination of the slip from the offset of geological features that intersect the fault is based on Allan maps of cutoff lines in the fault surface (see Sect. 8.4.3).

Fault slip can be obtained if marker beds can be matched across the fault in the slip direction. The slip direction may be given by the direction of grooves or slickenlines observed on the fault surface or by the trend of the lineation in the fault-zone rock.

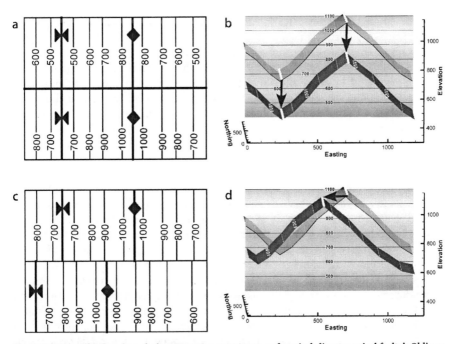

Fig. 7.17. Folds offset by a later fault. **a** Structure contour map of vertical slip on vertical fault. **b** Oblique view to N of 3-D version of **a**. **c** Structure contour map of strike-slip on a vertical fault. **d** Oblique view to N of 3-D version of **c**. North-south lines on **a** and **c** are structure contours labeled with elevations. *White lines* on **b** and **d** are fold hinge lines, *black arrows* are slip vectors, fault is shaded with contours labeled

The slip direction is usually perpendicular to the axis of the drag fold at the border of the fault zone, which can be observed in outcrop or be inferred by dip sequence (SCAT) analysis (Chap. 9, see also Becker 1995). The axes of folds contained within a fault zone are approximately normal to the slip direction, but may be arcuate, in which case the slip direction bisects the arc (Hansen 1971). Slip vector determination contains inherent ambiguities that must be considered, however. Different slip directions may be overprinted with only the last increment being observed or the slip trajectories may be curved.

To determine the fault slip, draw the slip vector on the structure contour map of the fault so that it connects the correlated points on the hangingwall and the footwall as in Fig. 7.16. If the end points of the slip vector are specified by their xyz coordinates, the length of the slip vector in the plane of the fault is (from Eq. 2.4)

$$L = [(x_2 - x_1)^2 + (y_2 - y_1)^2 + (z_2 - z_1)^2]^{1/2} \; , \tag{7.1}$$

where L = the slip and the subscripts "1" and "2" = the coordinates of the opposite ends of the slip vector. If the end points are specified by the horizontal and vertical distance between them (from Eq. 4.7)

$$L = (v^2 + h^2)^{1/2} \; , \tag{7.2}$$

where L = the slip, v = vertical distance between end points, and h = the horizontal distance between end points.

Without the information provided by the correlation of formerly adjacent points across the fault or knowledge of the slip vector, it is impossible to determine whether a fault is caused by strike slip, dip slip, or oblique slip. Map interpretation must frequently be done without knowledge of the slip amount or direction, a situation for which the fault separation, given next, provides the framework for the interpretation.

7.4.2
Separation

Separation is the distance between any two index planes disrupted by a fault (Dennis 1967; Bates and Jackson 1987). Several different separation components are commonly used to describe the magnitude of the fault offset. Strike and dip separations are, respectively, the separations measured between correlative surfaces along the strike and directly down the dip of the fault (Billings 1972). The stratigraphic separation (Fig. 7.18) is the stratigraphic thickness of the beds missing or repeated across a fault (after Bates and Jackson 1987). Stratigraphic separation is a thickness, and therefore represents a measurement direction perpendicular to bedding. The stratigraphic separation is equal to the fault cut, that is, the amount of section missing or repeated across a fault at a point. It is possible for a fault to have a large displacement and yet show no stratigraphic separation. For example, strike-slip displacement of horizontal beds produces no stratigraphic separation. Other combinations of slip direction and dip of marker beds can also result in zero separation (Redmond 1972).

Fig. 7.18.
Vertical and stratigraphic fault
separation. **a** Reverse fault.
b Normal fault: the marker
horizon must be projected
across the fault in order to
measure the vertical separation

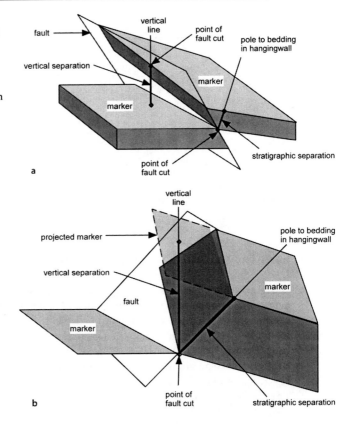

Vertical separation is the distance, measured vertically, between the two parts of a displaced surface (Fig. 7.18; Dennis 1967). In the case of a normal-separation fault, one of the surfaces must be projected across the fault to make the measurement. The vertical separation of an offset marker horizon is shown in a vertical cross section in the direction of dip of the bedding across the fault from the fault cut (Fig. 7.19). The separation is measured from the marker horizon at the point of the fault cut to the position of the same marker horizon across the fault (reverse separation; Fig. 7.19a) or from the marker horizon at the fault cut to the *projection* of the marker horizon from its location across the fault (normal separation, Fig. 7.19b). The amount of the vertical separation is

$$v = t / \cos \delta \ , \tag{7.3}$$

where v = vertical separation, t = stratigraphic separation = amount of the fault cut, and δ = cross-fault bedding dip. The cross-fault bedding dip is the dip of bedding across the fault from the marker horizon at the fault cut. The vertical separation of a vertical bed is undefined.

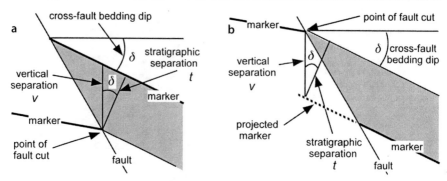

Fig. 7.19. Vertical separation calculated in a vertical cross section in the direction of the cross-fault bedding dip. **a** Reverse separation fault. **b** Normal separation fault

7.4.3
Heave and Throw from Stratigraphic Separation

Heave and throw have a special significance because they are the separation components visible on the structure contour map of a faulted horizon. Throw is the vertical component of the dip separation and heave is the horizontal component of the dip separation, both being measured in a vertical cross section in the dip direction of the fault (Dennis 1967; Billings 1972). Throw and heave can be found directly from the stratigraphic separation. The following discussion refers to the geometry shown in Fig. 7.20. Let point P_1 be the location of the fault cut in a well or exposed in outcrop. The marker horizon is shaded. P_2 is the location of the marker horizon in a vertical plane oriented in the direction of fault dip. The calculation of throw and heave is a projection across the fault from the control location (P_1) to a predicted location (P_2). The stratigraphic separation at the point of the fault cut is the thickness of the missing or repeated section, t. The dip separation, S_d, is equal to the apparent thickness of the missing or repeated section in the direction of the fault dip (dip vector) between points P_1 and P_2. The dip separation can be found from Eq. 4.1 as

$$S_d = t / \cos \rho \quad , \tag{7.4}$$

where t = the stratigraphic separation and ρ = the angle between the pole to the cross-fault bedding attitude and the dip vector of the fault. The dip of bedding used is always that belonging the side of the fault to which the projection is being made (P_2), that is, across the fault from the fault cut on the marker horizon. The value of ρ can be found with a stereogram (Sect. 4.1.1.1) or from any of the analytical methods in Sect. 4.1.1.2.

The heave and throw are determined by taking the horizontal (H) and vertical (T) components (Fig. 7.20) of S_d from Eq. 7.4:

$$H = t \cos \phi / \cos \rho \quad , \tag{7.5}$$

$$T = t \sin \phi / \cos \rho \quad , \tag{7.6}$$

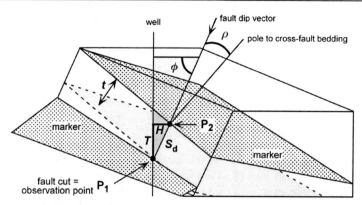

Fig. 7.20. Fault heave (H) and throw (T) at a fault cut (point P_1). S_d = dip separation; $H = L \cos \phi$; $T = L \sin \phi$; t stratigraphic separation = fault cut; ϕ dip of the fault; ρ angle between the fault dip and the pole to the cross-fault bedding attitude at P_2. The marker horizon is *shaded*

where H = heave, T = throw, t = stratigraphic separation, ϕ = the dip of the fault, and ρ = the angle between the cross-fault bedding attitude and the dip vector of the fault. If bedding is horizontal, its pole is vertical, causing ρ to be equal to $90 - \phi$, and Eqs. 7.5 and 7.6 reduce to

$$H = t / \tan \phi \ , \tag{7.7}$$

$$T = t \ . \tag{7.8}$$

In the case of zero dip of bedding, and only in the case of zero dip of bedding, the stratigraphic separation is equal to the fault throw and equal to the vertical separation.

As an example of the calculation, consider a fault having a stratigraphic separation of 406 m, the cross-fault bedding attitude is δ = 10, 180, and the fault dip is ϕ = 37, 220. The pole to bedding is found to be 80, 360 and the angle between the pole to bedding and the fault dip is ρ = 60°. From Eq. 7.5, H = 648 m, and from Eq. 7.6, T = 489 m. These values are derived for the fault in Fig. 7.16 and could have been caused by any number of different combinations of net slip magnitudes and directions.

Neither the stratigraphic thickness nor the attitude of bedding are necessarily the same on both sides of a fault, which is important when the throw and heave are calculated from Eqs. 7.5 and 7.6. The appropriate dip and thickness values are always the cross-fault magnitudes found in the block across the fault from the marker horizon at the fault cut. In subsurface mapping based only on formation tops and fault cuts, it is likely that none of the required dips will be known at the early stage of mapping. This means that throw and heave cannot be calculated accurately at this stage. Throw and heave can be estimated for preliminary mapping purposes by estimating the dip of the fault and the dip of cross-fault bedding. If no other information is available, normal-separation faults can be estimated to dip 60°, reverse separation faults to dip 30°, and bedding can be assumed to be horizontal. As the required dips are determined by mapping, the throw and heave magnitudes can be corrected.

7.5
Geometric Properties of Faults

In this section the typical characteristics of an individual fault surface and its displacement distribution are described.

7.5.1
Surface Shape

A fault surface is usually planar to smoothly curved or gently undulating (Fig. 7.21). The structure contour map of a fault surface should usually be smooth. Viewed over their entire surface, many faults are smoothly curved into a spoon-like shape. Primary surface undulations, if present, are typically aligned in the slip direction and provide a good criterion for the slip direction (Thibaut et al. 1996).

7.5.2
Displacement Distribution

The displacement on a fault dies out at a tip line which is the trace in space of the terminations of a fault (Boyer and Elliott 1982). If the displacement dies out in all directions (Barnett et al. 1987), the fault is surrounded by a tip line (Fig. 7.22) like a dislocation in the theory of crystal plasticity (Nicolas and Poirier 1976). Faults are typically significantly longer in the strike direction than in the dip direction. Faults always

Fig. 7.21.
Perspective block diagram of a
curved fault surface

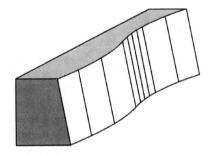

Fig. 7.22.
Dislocation-like fault for which
the displacement (*arrow*) dies
out to zero in all directions at
the tip line. The trace of the
displaced marker horizon is
dashed on the footwall of the
fault and *solid* on the hangingwall

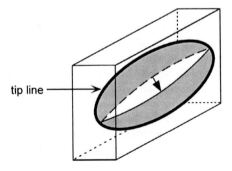

tip line

die out along strike (Figs. 7.22, 7.23a) or transfer displacement to some other fault or to a fold. Displacement in the dip direction may or may not go to zero. A fault may flatten into a detachment without losing displacement (Fig. 7.23b; Gibbs 1989). The variation in displacement in the dip direction is a function of the structural style and a variety of relationships are possible.

An accurately mapped example from the Westphalian coal measures in the United Kingdom (Fig. 7.24a) has a displacement distribution nearly as simple as that in Fig. 7.22. The fault-displacement maps are derived from mapping at five different levels in underground coal mines and so require little inference. Another example from the same area (Fig. 7.24b) is more complex but the displacement is still seen to die out along strike. For this area, a representative aspect ratio is 2.15 (strike length/dip length) for a group of isolated normal faults that die out in all directions for over a length range of 10 m to 10 km, and is in the range of 0.5 to 8.4 for normal faults that intersect other faults or reach the surface (Nicol et al. 1996).

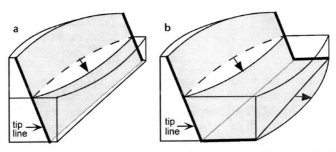

Fig. 7.23. Faults for which the displacement does not die out in the dip direction. The displacement (*arrow*) dies out along strike at the tip lines marked by *heavy lines*. The trace of the displaced marker horizon is *dashed* on the footwall of the fault and *solid* on the hangingwall. **a** Fault of unspecified extent in the dip direction. **b** Fault that bends into a planar lower detachment without losing displacement

Fig. 7.24.
Two examples of contours of fault throw on normal faults dipping about 65°. Contours are projected onto a vertical plane that has the strike of the fault. Scales in meters.
a Elliptical heave distribution. The fault ends to the right against the boundary fault.
b Complex heave distribution. (After Rippon 1985)

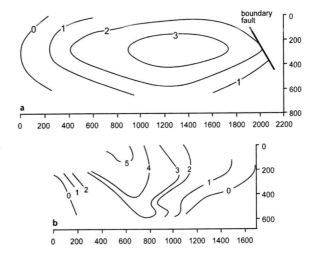

Fig. 7.25.
Displacement versus distance plots for completely exposed small normal faults. **a** Approximately circular-arc distribution. **b** Approximately sinusoidal distribution. **c** Two overlapping faults with D-shaped distributions. *VE:* vertical exaggeration. (Schlische et al. 1996)

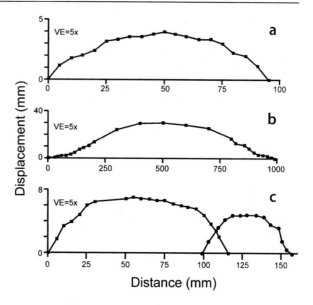

The change in displacement along the strike of a fault can be illustrated with a displacement-distance graph. The displacement-distance curve typically ranges from a circular arc (Fig. 7.25a) to sinusoidal (Fig. 7.25b) to D-shaped (Fig. 7.25c). All three distributions in Fig. 7.25 come from one small area in a quarry and are completely exposed. More complex displacement distributions (Fig. 7.24b) may be the result of the coalescence of multiple faults, like the two faults in Fig. 7.25c.

The smooth variation of displacement along a fault in the style of Figs. 7.24a and 7.25 has been described as the bow and arrow rule (Elliott 1976). The trace of a faulted horizon on one side of the fault (fault cutoff line) forms the bow and a straight line joining the tips of the fault is the bow string. A line perpendicular to the bow string at its center is the arrow. The distance along the arrow between the bow and the bow string is an estimate of the displacement amount and the direction of the arrow is an estimate of the displacement direction. These are reasonable first approximations for dip slip faults although they should be used with caution. If the displacement distribution on the fault resembles that in Fig. 7.22, then the displacement with respect to the correlative horizon across the fault is twice that given by the bow and arrow rule because both hangingwall and footwall are displaced from their original positions (the position of the bow string). The rule does not apply to strike-slip faults.

The length of a fault in the strike direction is usually much greater than its maximum displacement. A comparison between the maximum displacement on a fault and its length, down dip or along strike, shows that ratios of 1 : 8 to 1 : 33 are common, regardless of location, size, or fault type (Table 7.1). The maximum displacement is expected to occur near the center of the fault as in Fig. 7.24a and 7.25. Fault displacements measured from maps or cross sections may not go through the point of maxi-

Table 7.1. Ratio of maximum displacement to length for a variety of faults that die out in tip lines in the direction of the measurement. The first two measurements from Rippon (1985) are from Fig. 7.24a, the third is from Fig. 7.24b

Max. displ./ length	Direction measured	Size range	Fault type	Location	Reference
1/10 to 1/20	Dip	40 – 450 cm	Normal	Japan	Muraoka and Kamata (1983)
1/10 to 1/20	Strike	10 – 400 km	Thrust	Canadian Rocky Mts.	Elliott (1976)
1/8	Strike	1.5 – 21 km	Strike slip	Iran	Freund (1970)
1/82	Strike	2.5 km	Normal	Alabama	Chap. 7, Fig. 7.46
1/700	Strike	2 km	Normal	Derbyshire, UK	Rippon (1985)
1/450	Dip	1 km	Normal	Derbyshire, UK	Rippon (1985)
1/340	Strike	1.5 km	Normal	Derbyshire, UK	Rippon (1985)
1/90	Strike	10 – 200 m	Normal	Western USA	Dawers et al. (1993)
1/125	Strike	0.2 – 10 km	Normal	Western USA	Dawers et al. (1993)
1/30 to 1/50	Strike	0.2 – 10 km	Normal	Timor Sea	Nicol et al. (1996)
1/33	Strike	1 – 123 cm	Normal	Eastern USA	Schlische et al. (1996)

mum displacement and so the displacement/length ratios could be smaller than the values recorded in Table 7.1. The examples of Rippon (1985), however, are well exposed in three dimensions and the ratios are very large. In summary, long faults usually have large displacements but may have small displacements; short faults do not have large displacements.

The relationship between the maximum displacement and the fault length is generally considered to have the form

$$D = \gamma L^C , \tag{7.9}$$

where D = maximum displacement, γ = a constant of proportionality, L = fault length, and C is between 1 and 2 (Watterson 1986; Marrett and Allmendinger 1991; Dawers et al. 1993). The exact relationship appears to depend on many factors including the mechanical stratigraphy and the nature of interactions between overlapping faults (for example, Cowie and Scholz 1992). For the practical estimation of the displacement-length relationship in map interpretation, it appears satisfactory to let $C = 1$ and recognize that the value of γ may change when the size of the fault changes by an order of magnitude and will be different in different locations (Cowie and Scholz 1992; Dawers et al. 1993; Schlische et al. 1996). With $C = 1$, the value of γ is the D/L ratio given in Table 7.1. This relationship can be used to estimate the length of a fault from its maximum displacement or estimate the maximum displacement from the length.

7.6
Growth Faults

A growth fault moves during deposition and controls the thickness of the deposits on both sides if the fault. The classic Gulf of Mexico thin-skinned extensional growth fault, from the region where the concept was developed (Fig. 7.26), is depositional on both sides and the sediments thicken across the fault. A growth fault in a rifted environment may show footwall uplift and erosion concurrent with deposition on the hangingwall. Growth faults can be normal (Fig. 7.26) or reverse (Fig. 7.27); in fact, any type of fault can record growth, including strike-slip faults.

7.6.1
Effect on Heave and Throw

Because the stratigraphic thicknesses are different on opposite sides of a growth fault, the heave and throw determined from the fault cut depend on whether the marker being mapped is in the hangingwall or footwall of the fault. The appropriate thickness is that belonging to the section across the fault from the fault cut in the marker horizon. For a marker on the hangingwall of a normal fault (Fig. 7.26a), the appropriate fault-cut thickness is that of the footwall stratigraphic section. To map a footwall marker in the same well (Fig. 7.26b), the appropriate thickness is that of the hangingwall strati-

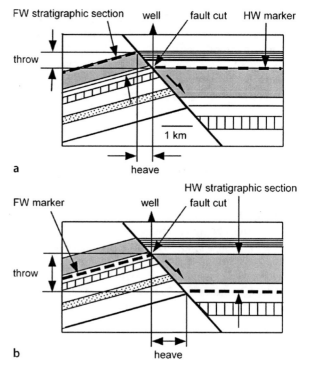

Fig. 7.26. Identical cross sections across a growth normal fault showing effect of growth on throw and heave. The section is vertical and in the direction of fault dip. The fault cut is at the same point in both sections. **a** The fault cutoff of a hangingwall marker (*dashed*) is extrapolated to the footwall cutoff of the same marker using dip and thickness values from the footwall. **b** The fault cutoff of a footwall marker (*dashed*) is extrapolated to the hangingwall cutoff of the same marker using dip and thickness values from the hangingwall

Fig. 7.27.
Identical cross sections of a
growth reverse fault showing
effect of growth on throw and
heave. The section is vertical
and in the direction of fault
dip. The fault cut is at the same
point in both sections. **a** The
fault cutoff of a hangingwall
marker (*dashed*) is extrapolated to the footwall cutoff of
the same marker using dip
and thickness values from the
footwall. **b** The fault cutoff of
a footwall marker (*dashed*) is
extrapolated to the hangingwall cutoff of the same marker
using dip and thickness values
from the hangingwall

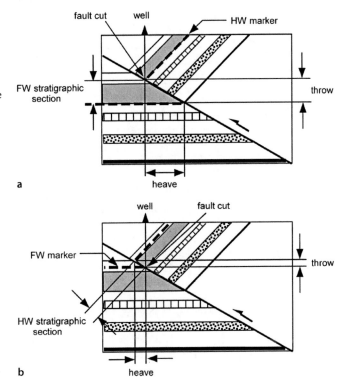

graphic section. The same considerations apply to the correct choice of thickness for
growth reverse faults (Fig. 7.27).

7.6.2
Expansion Index

The growth history of a fault can be illustrated quantitatively with the expansion
index (*E*) of Thorsen (Fig. 7.28):

$$E = t_d / t_u \quad , \tag{7.10}$$

where t_d = the downthrown thickness (hangingwall) and t_u = the upthrown thickness
(footwall). The thicknesses should be measured perpendicular to bedding so as not to
confuse dip changes with thickness changes, and as close to the fault as possible because that is where the maximum thickness changes occur.

The simplest possible growth fault is one that starts, increases in growth rate, then
slows and stops (Fig. 7.28). In the pre-growth interval, *E* = 1.0 (unit j); as the growth
rate increases, so does the expansion index, to a maximum of 2.1 in Fig. 7.28. The growth
of the fault slows and eventually stops and *E* returns to 1.0 (unit a). The plot of
stratigraphic interval versus expansion index gives a visual picture of the growth his-

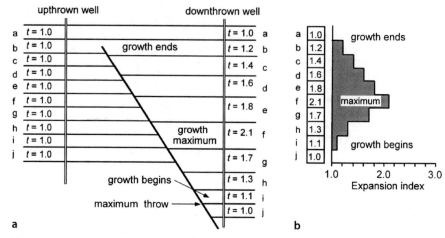

Fig. 7.28. Expansion index for a fault that begins to move, reaches a growth maximum and then stops. **a** Cross section. **b** Expansion index vs. stratigraphic interval graph. (After Thorsen 1963)

tory. If the relative offset is large and if shale units are being compared at less than 3 000 ft of burial, it might be important to correct E for compaction. At greater depths, the relative change in thickness will be small and have little effect on E.

The expansion index plot characterizes the growth history of a fault and might be correlated to other time-dependent features such as the sand/shale ratio or the time of hydrocarbon migration. The use of E eliminates the effect of absolute interval thickness, allowing the growth rates of a generally thin interval to be directly compared to that of a generally thick interval. The use of the similarity between expansion index plots at different fault cuts as an aid to fault correlation will be illustrated in Sect. 7.7.4.

7.7
Fault-Cut Correlation Criteria

Faults are commonly mapped on the basis of observations made at a number of separate locations, called fault cuts. If more than one fault may be present, a very significant problem is to establish which observations belong to the same fault (Fig. 7.29). This problem arises in surface mapping where the outcrop is discontinuous, in sub-

Fig. 7.29.
Map showing three locations where faults have been observed. Which, if any, points are on the same fault? *Numbered circles* are observation points, *dashed lines* are some possible fault correlations

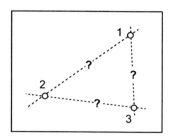

surface mapping based on wells, and in constructing maps from two-dimensional seismic data. Correlating faults in an area of multiple faults is one of the most challenging problems in structural interpretation. A number of fault properties can be used to establish the correlation between fault cuts. Correlation criteria include the fault trend, the sense of throw across the fault, the smoothness of the fault surface, the amount of separation, and the growth history. The rules for correlation that are presented below may have exceptions and should be used in combination to obtain the best result.

7.7.1
Trend and Sense of Throw

The consistency of the fault trend and the consistency of the sense of throw are usually the first two criteria applied in the correlation of fault cuts. The fault trend can sometimes be established at the observation points by direct measurement in outcrop, by physically tracing the fault, or by SCAT analysis of outcrop traverses or dipmeter logs (Chap. 9). If, for example, the fault trend in the map area of Fig. 7.29 can be shown to be east–west, then observation points 2 and 3 are more likely to be on the same fault than are points 1 and 3 or 1 and 2 (Fig. 7.30a), and the fault through point 1 is also likely to trend east–west. A preliminary structure contour map on a horizon for which there is good control may give a clear indication of the fault trends (Fig. 7.30b). Faults that are directly related to the formation of the map-scale folds are commonly parallel to the strike of bedding, especially in regions where the contours are unusually closely spaced. Such zones may represent unrecognized faults.

The high and low areas on a structure contour map of a marker horizon should be explained by the proposed faults. The upthrown and downthrown sides of a fault will commonly be the same along strike. For example, the proposed fault linking observation points 2 and 3 in Fig. 7.30b is downthrown on the north at both locations. A constant sense of vertical separation is a common and reasonable pattern, but not the only possibility on a correctly interpreted fault.

Faults that offset folds may produce complex patterns of horizontal and vertical separations. Vertical displacement on a fault that strikes at a high angle to fold axes

Fig. 7.30.
Maps showing possible correlations between fault cuts at
points 1–3. **a** Correlations
(*dashed lines*) along an east-
west trend. **b** Preliminary
structure contour map on a
marker horizon. Fault correlations (*heavy solid lines*
marked F) along structure contour trends are supported by
the consistent sense of throw
across the faults

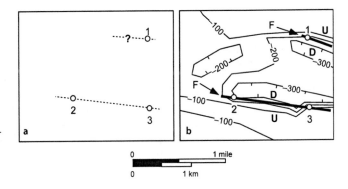

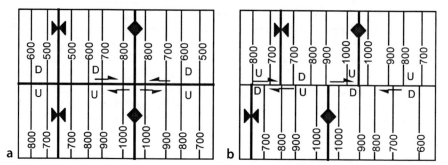

Fig. 7.31. Displacement along faults oblique to the fold trend. North-south lines are structure contours labeled with elevations. **a** Pure vertical slip. The implied strike slip on the fold limbs (*arrows*) is incorrect. **b** Pure strike-slip displacement. The implied vertical slip (*U, D*) is incorrect. For 3-D views of these maps see Fig. 7.17

produces apparent strike-slip displacements of the fold limbs for which the sense of slip reverses at the fold hinge line (Fig. 7.31a) although the sense of throw is constant across the fault. Strike-slip displacement of the folds produces throw that changes sense along the fault (Fig. 7.31b) but the sense of strike separation is constant. Faults for which the sense of throw reverses along the trend of the fault are called scissors faults. Such faults may form by rotation around the points of zero separation or, as in Fig. 7.31b, may be caused by the strike-slip displacement of folds.

7.7.2
Shape

The inferred shape of a fault is an important criterion in correlating fault cuts and in validating fault interpretations. A valid fault surface is usually planar or smoothly curved and has an attitude that is reasonable for the local structural style. Contours on fault surfaces follow the same rules as contours on bed surfaces (Sect. 3.2) with the exception that the fault contours can end in the map area where displacement on the fault ends (Bishop 1960). The dip of the fault is obtained from a structure contour map (Sect. 3.6.1).

Smooth contours on the fault surface demonstrate that the correlations between the fault cuts are acceptable (Fig. 7.32). Because faults are typically planar or gently curved, it is reasonable to assume that both the strike and dip of the fault are approximately constant. Surface undulations, if present, are most likely to be aligned in the slip direction. Unfaulted points or unfaulted wells near an inferred fault provide additional constraints on the geometry because they must lie entirely within the hangingwall or footwall. Negative evidence, e.g., the absence of a fault cut in a well the penetrates an inferred fault plane, requires remapping the fault or faults. The dips implied by the contours must be reasonable for the structural style. Very irregular contours on a fault surface suggest an incorrect correlation of the control points. Folded faults are possible, especially thrusts, but must be consistent with the folds in the hangingwall and footwall.

The process of correlation of fault cuts and bed offsets into individual fault planes based on shape can be made easier and faster with a fault template. A fault template is

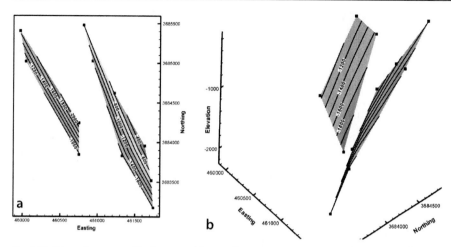

Fig. 7.32. Maps of two normal faults mapped from wells in Deerlick Creek coalbed methane field (modified from Groshong et al. 2003b). *Squares* are the positions of fault cuts, *contours* are on the fault surfaces. **a** Structure contour map. **b** 3-D view to the NW of the two faults

a structure contour map of the expected fault plane at the scale of the map. If mapping on paper, construct a template by picking a suitable contour interval for the fault along with a reasonable dip, then use Eq. 2.21 to find the horizontal spacing between contours on the template. 3-D software will probably allow a plane of a given orientation to be specified. If not, use the equations of a plane (Sect. 2.4.2) to place a plane in an appropriate location. The template can be moved around the map and rotated until a set of fault cut elevations and/or bed offsets are reasonably correlated. Then the real fault plane is constructed directly from the data with the template as a guide. If a representative fault dip for the area is known, then it should be used for the template. If a representative dip is unknown, it is reasonable to begin with 60° for normal faults, 10–30° for low-angle reverse faults, 50–80° for high-angle reverse faults, and 70–90° for strike-slip faults, based on stress theory (Sect. 1.6.4) and common experience.

7.7.3
Stratigraphic Separation

The stratigraphic separation along a fault surface is normally a smoothly varying function, with a maximum near the center of the fault and decreasing to zero at the tip line (Sect. 7.5.2). This simple pattern can be perturbed if the fault grew by linking separate faults (Sect. 8.6), but the variation is nevertheless likely to remain smooth. The stratigraphic separation on the fault is the component usually known, and can be used as a proxy for the displacement as long as the bed geometry adjacent to the fault is not too complex. The distribution of the separation can be displayed by posting the separations on a map of the fault surface (Fig. 7.33a). A point on a fault for which the location of the fault cut and the stratigraphic separation do not agree with the points around it (Fig. 7.33b) is likely to be miscorrelated with the fault.

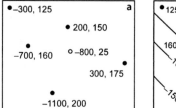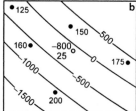

Fig. 7.33. Fault-cut maps with stratigraphic separations. **a** Data: elevation of fault cut, amount of stratigraphic separation (fault cut). **b** Structure contour map on the fault surface; amount of fault cut is posted. The contour interval is 500 units, and negative contours are below sea level. The fault cut (*open circle*) does not correlate with the contoured fault

Fig. 7.34.
Stratigraphic separation diagram representative of relationships seen along the strike direction of a thrust fault. *FW curve:* Stratigraphic position of the thrust in the footwall; *HW curve:* stratigraphic position of the fault in the hangingwall. **a** Single unbroken thrust sheet. **b** Thrust sheet broken by a lateral ramp

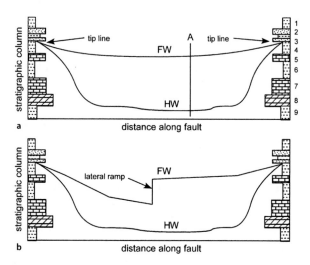

Another method for displaying the variation of the stratigraphic separation along a fault is by means of a stratigraphic separation diagram (Fig. 7.35; Elliott and Johnson 1980; Woodward 1987). The vertical axis of the diagram is the scaled stratigraphic column and the horizontal axis is the distance along the fault. The curves on the diagram show the stratigraphic unit that the fault is in at any particular point. A thrust fault usually places older over younger units and so the footwall block should be the upper curve (Fig. 7.35). The opposite is true for a normal fault. For example, line A (Fig. 7.35a) represents a particular geographic point on the fault, one at which the footwall is at the base of unit 4 and the hangingwall is in unit 9. If the complete fault has been observed, the hangingwall and footwall curves will join at the fault tips. Both hangingwall and footwall curves will be smooth if they are not broken by oblique faults. An oblique structure will produce a rapid change in the stratigraphic level of the curve of the block that contains the feature (Fig. 7.35b). If the rapid change is in the footwall, it is likely to have been caused by a lateral ramp. If the oblique structure is in the hangingwall, it is likely to be a tear fault that subdivides only the hangingwall. If the feature is present

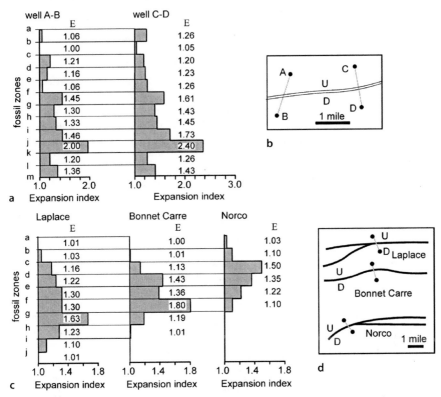

Fig. 7.35. Expansion index (*E*) diagrams across Louisiana growth faults. **a** Expansion indices from two crossings of the same fault. **b** Locations of wells used to determine the expansion indices in **a**. **c** Expansion indices across three adjacent faults. **d** Locations of the wells used to determine the expansion indices in **c** (after Thorsen 1963)

to about the same degree in both the hangingwall and the footwall curves, it may have been caused by a later fault that cuts and displaces the fault in question.

Stratigraphic separation diagrams are most widely used for data derived from the outcrop traces of thrust faults. If the distance coordinate follows the sinuous erosional trace of a low-angle thrust, some of the stratigraphic variation in thrust levels will be caused by the changing depths of erosion and reentrants on the fault. To eliminate this source of variability, it is better to make the diagram follow a straight line or smooth curve that is either parallel to or perpendicular to the transport direction of the fault (Woodward 1987).

7.7.4
Growth History

The stratigraphic evolution of a growth fault provides another criterion for correlating fault cuts. A single fault can be expected to have a similar stratigraphic growth history along its trend. Expansion index diagrams (Sect. 7.6.2) clearly illustrate the details of the growth history and are valuable in fault correlation. In an example from Louisiana, the expansion index diagrams have the same form across the same fault at an interval about 2.5 miles apart (Fig. 7.35a,b). The values of E for the same units are not the same, but the forms of the expansion index curves are the same. Both crossings of the same fault show the maximum growth interval to be of the same age. Three different faults in the same area give three different expansion index diagrams over similar distances (Fig. 7.35c,d). The maximum growth interval becomes younger to the south. Thus, similar growth-history relationships can indicate that the same fault has been crossed and different relationships can indicate that different faults have been crossed.

7.8
Exercises

7.8.1
Fault Recognition on a Map

Use the partially complete geologic map of Fig. 7.36, to do the following: Mark all the faults. Explain the reason for each fault. Indicate the sense of displacement on each fault.

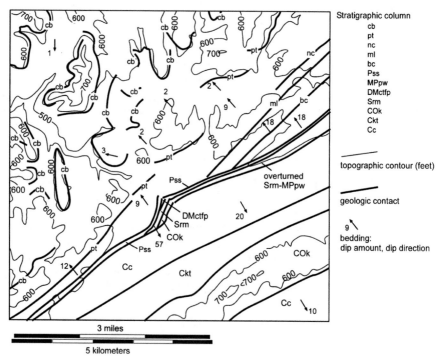

Fig. 7.36. Geologic map of the Ensley area, Alabama, with fault contacts not marked. (After Butts 1910; Kidd 1979)

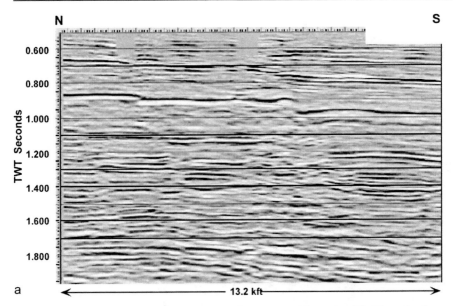

N   S

Fig. 7.37. Seismic line across part of Gilbertown graben system, Alabama (modified from Groshong et al. 2003a). V.E. approximately 1:1

7.8.2
Fault Recognition on a Seismic Line 1

Interpret the faults in Fig. 7.37.

7.8.3
Fault Recognition on a Seismic Line 2

Interpret the faults and unconformity in Fig. 7.38.

7.8.4
Finding Fault Cuts

Locate the position of the faults and the amounts of the fault cuts on the logs from the northern Gulf of Mexico (Fig. 7.40) based on the type log in Fig. 7.39. Example courtesy of Jack Pashin (Pashin et al. 2000).

7.8.5
Correlating Fault Cuts

Contour a fault (Fig. 7.41) that includes all four wells that cut faults. Is the fault surface obtained reasonable? What is the attitude of the fault plane given by the three

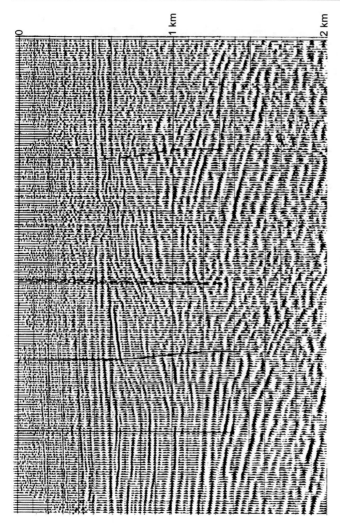

Fig. 7.38.
Seismic reflection profile from
the Ruhr district, Germany.
(Drozdzewski 1983)

points 1, 4, 2? What is the attitude of the fault plane given by the three points 2, 3, 4? What is the attitude of a fault plane through wells 1, 2, and 3? Is this a possible fault plane? Why or why not?

7.8.6
Estimating Fault Offset

Use Table 7.1 to answer the following questions. A fault in the Black Warrior basin of Alabama has a maximum displacement of 100 m. How long is it? What is the length range if the fault is a thrust in the Canadian Rocky Mountains? A fault in the Black

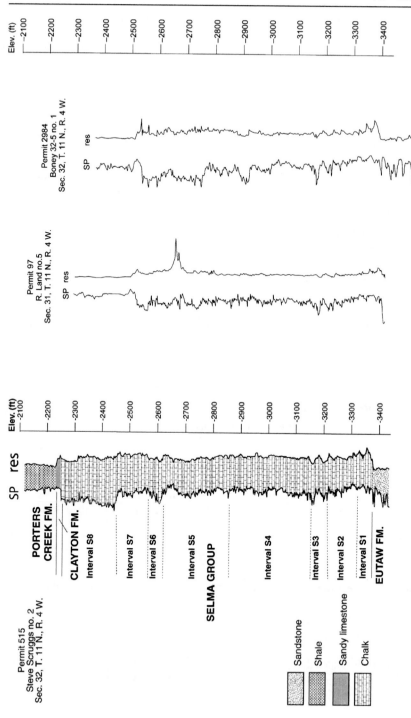

Fig. 7.40. Logs of two faulted wells in Selma Chalk, Gilbertown oil field

Fig. 7.39. Unfaulted type log of Selma Chalk, Gilbertown oil field

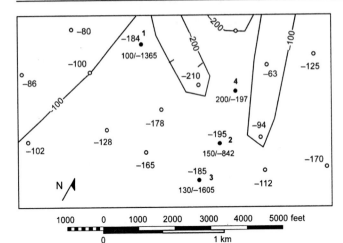

Fig. 7.41.
Map of the southern portion of the Deerlick Creek coalbed methane field. *Open circles* are wells with the elevation of the top of the Gwin coal cycle (ft). *Solid circles* are wells with fault cuts; single number is top of Gwin if present, number pair is stratigraphic separation/elevation of fault cut. Structure contours shown on the top of the Gwin are tentative. Elevations below sea level are negative

Table 7.2. Fault attitude and separation data

Fault attitude	Stratigraphic separation	HW attitude	FW attitude	Fault cut in	ρ	Heave	Throw	Vertical separation
60, 270	100	0	0					
60, 270	100	20, 070	30, 200	HW				
				FW				
30, 200	100	0	0					
30, 200	100	20, 070	30, 070	HW				
				FW				

Warrior basin is 5 kft long. What is its probable maximum displacement? What is the displacement range if the fault is a normal fault in the western United States?

7.8.7
Fault Offset

Find the heave, throw, and vertical separation for the faults listed in Table 7.2.

7.8.8
Growth Faults

Determine the expansion indices across both faults in Fig. 7.42. Discuss the growth of the faults and the relationship of growth to the hydrocarbon occurrence.

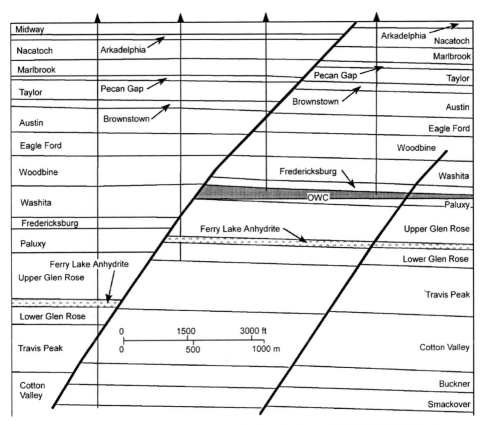

Fig. 7.42. Cross section of the Talco field, northern Gulf of Mexico, Texas. Oil field *shaded; OWC:* oil-water contact. (After Galloway et al. 1983)

Chapter 8

Faulted Surfaces

8.1
Introduction

This chapter covers how to construct maps containing faults, the integration of fault maps with horizon maps, interpreting faults from isopach maps, mapping overlapping and intersecting faults, and mapping cross-cutting faults.

8.2
Geometry of a Faulted Surface

Faults cause discontinuities in the map of a marker horizon and result either in gaps, overlaps, or vertical offsets of the marker surface. Normal faults produce gaps (Fig. 8.1a,b), reverse faults cause overlaps (Fig. 8.1c,d), and vertical faults produce line discontinuities on the map of a marker horizon (Fig. 7.17). Structure contours on the fault plane intersect the offset marker where the contours on the marker and the fault are at the same elevation. It is best practice to show the fault contours on the map, scale permitting. Structure contours on a fault surface are easily constructed by connecting the points of equal elevation where the marker horizon intersects the fault surface on opposite sides of the fault. This is called the implied fault surface (Tearpock and Bischke 2003). A fault intersects the marker surface along lines known as hangingwall and foot-wall cutoff lines. If the fault ends within the map, then the contours on the marker surface are continuous at the ends of the fault.

Heave and throw are the components of fault separation directly visible on the structure contour map of a faulted marker horizon. Throw is the vertical component of the dip separation; heave is the horizontal component of the dip separation (Fig. 8.2; also Sect. 7.4.3), both components being defined in the dip direction of the fault. The throw is the change in vertical elevation between the hangingwall and footwall cutoff lines in the direction of the fault dip. The throw and heave are related to the stratigraphic separation by Eqs. 7.5–7.8.

8.2.1
Heave and Throw on a Structure Contour Map

On a structure contour map the heave is the map distance between the hangingwall and footwall cutoff lines, measured in the direction of the fault dip (Figs. 8.3, 8.4). The fault gap (or overlap) is the horizontal distance between the hangingwall and

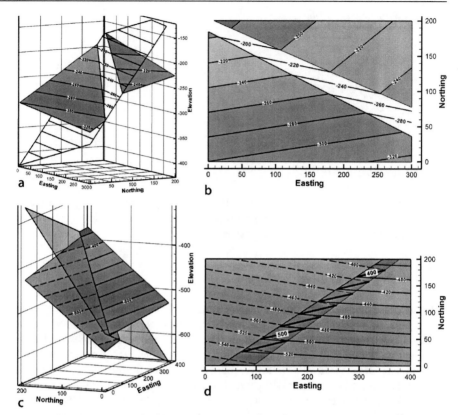

Fig. 8.1. 3-D and map views of faulted surfaces (*shaded*), fault surfaces are *unshaded*, structure contours shown on all surfaces. **a** Normal fault, oblique view. **b** Normal fault map. **c** Reverse fault, oblique view, FW contours *dashed*. **d** Reverse fault map. FW contours *dashed*, fault contours *heavy lines*

Fig. 8.2.
Vertical cross section in the dip direction of a fault showing throw and heave as components of the dip separation of an offset marker horizon

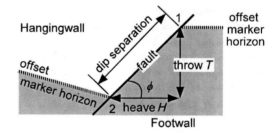

footwall cutoffs, measured perpendicular to the trace of the fault on the map. Although both are measured in the horizontal plane, the fault heave is not the same as the fault gap (or overlap) because, as seen in Figs. 8.3 and 8.4, the perpendicular distance between the hangingwall and footwall cutoff lines is not necessarily in the direction of the fault dip.

Fig. 8.3.
Marker surface offset by a normal fault. Contours on the fault are *thin lines* and contours on the marker surface are *heavier lines*. Throw on the fault is 32 m, heave is 25.6 m, and the gap is 24.9 m

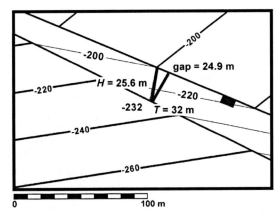

Fig. 8.4.
Marker surface offset by a reverse fault. Contours on the fault, *heavy lines*; on the hangingwall, *solid lines*; on the footwall, *dotted lines*. The throw is 3.6 m, heave 34 m, and overlap 29.1 m

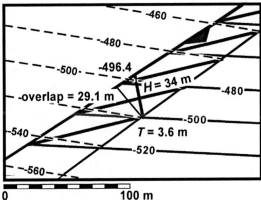

8.2.2
Stratigraphic Separation from a Structure Contour Map

Stratigraphic separation is found from the heave or throw by solving Eqs. 7.5 and 7.6 for *t*, giving

$$t = H \cos \rho / \cos \phi \ , \tag{8.1}$$

$$t = T \cos \rho / \sin \phi \ , \tag{8.2}$$

where t = stratigraphic separation, ρ = the angle between the fault dip vector and the cross-fault bedding pole, ϕ = amount of fault dip, H = heave, and T = Throw. If bedding is horizontal then the stratigraphic separation = Throw. The angle ρ is found by one of the methods in Sect. 4.1.1.1 or 4.1.1.2. The dip of the fault can be determined from the heave and throw as

$$\phi = \arctan (T / H) \ , \tag{8.3}$$

where ϕ = fault dip, T = throw, and H = heave.

Fig. 8.5.
Stratigraphic separation from
heave or throw. **a** Structure
contour map of offset marker
and fault. **b** 3-D oblique view
to NW, including base of
marker at fault cut in FW
(*dark shading*)

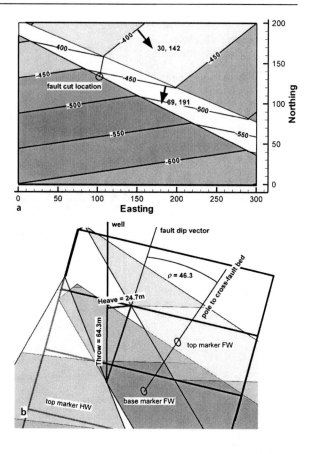

As an example, find the stratigraphic separation predicted for a fault cut at the location indicated on Fig. 8.5a. At this point the heave is 24.6 m and the throw is 64.5 m, measured as indicated in Sect. 8.2.1. The cross-fault bedding dip and the fault dip magnitudes are found from the contour spacings using Eq. 2.22. The dip directions are perpendicular to the structure contours, giving for the fault, $\phi = 69, 191$, and for the bed, $\delta = 30, 142$, and the angle $\rho = 46.3°$ from Eq. 4.3. Applying either Eqs. 8.1 or 8.2, the stratigraphic separation is $t = 47.6$ m. This result is shown in a 3-D model (Fig. 8.5b) in which all the parameters can be measured directly to illustrate and validate the relationships in Eqs. 7.5–7.6 and 8.1–8.2.

8.3
Constructing a Faulted Marker Horizon

Placing the trace of faults on the structure contour map of a marker surface generally begins with a preliminary map of the marker horizon without faults (Fig. 8.6a), then the faults are mapped and the traces of the faults are found (Sect. 2.7) on the preliminary marker surface (Fig. 8.6b). The marker surface is separated into hangingwall and

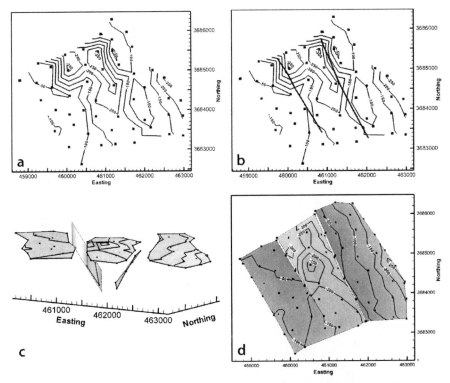

Fig. 8.6. Example of procedure for fault interpretation applied to top of Gwin coal cycle, Deerlick Creek coalbed methane field, Alabama (based on data in Groshong et al. 2003b). Contour interval is 50 ft, *squares* show well locations. **a** Top of the Gwin contoured without faults. **b** Top of the Gwin showing approximate fault traces. **c** 3-D oblique view of top Gwin broken into separate blocks at fault traces and re-contoured. **d** Top of Gwin re-mapped to fit the faults. Additional *squares* are points defining HW and FW cutoff lines

footwall blocks (Fig. 8.6c) and finally re-mapped to join the marker surface to the fault planes (Fig. 8.6d). The final re-mapping requires contouring up to and across the fault planes and must honor the known fault separations. Locating the faults, projecting beds to the fault surface, and contouring up to and across a fault will be discussed next.

8.3.1
Locating the Fault

In mapping a single marker surface with data from wells, the exact location of a fault is usually uncertain. Closely spaced, parallel contours on the marker surface provide a clue to the presence of a fault but do not necessarily give an accurate location. For example, the traces of the faults in Fig. 8.6b are in regions of closely spaced contours but do not follow the contours. Furthermore, it is rare for a well to cut a fault exactly at the marker horizon being mapped and so direct evidence of the fault location in the marker horizon is usually absent. All fault cuts at any stratigraphic level that can be correlated should be mapped, as discussed in Sect. 7.7.2. The right-hand fault in Fig. 8.6c is a typical result: the directly mappable fault usually does not extend laterally or vertically across the entire region of interest. The ambiguity of fault location in any single horizon can be greatly reduced by mapping multiple horizons in the vicinity of the fault.

Mapping multiple beds on both sides of a proposed fault will help to confirm its existence and location. As an example, the fault trace in Fig. 8.6c is not accurately located on any single horizon. Once multiple horizons have been mapped the location of the fault is much better defined (Fig. 8.7). An oblique view down the fault trace in 3-D (Fig. 8.7a) shows a narrow but clear path for the fault plane. The edge views (Fig. 8.7b,c) show that the units maintain nearly constant thickness and that they have been correctly separated into hangingwall and footwall blocks. Now it can be seen that the best-fit fault trace (Fig. 8.8) is much smoother and straighter than that suggested by the closely spaced structure contours mapped without a fault (Fig. 8.6a). If 3-D software is not available, serial cross sections across the fault provide nearly the same information about the hangingwall, footwall and fault trace.

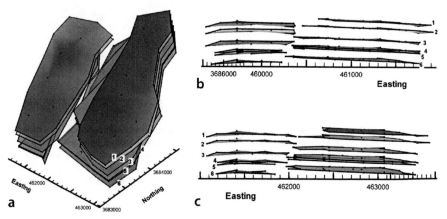

Fig. 8.7. Marker surfaces on both sides of the eastern fault in Fig. 8.6c (Deerlick Creek coalbed methane field). The markers are numbered; *1:* top Gwin, the same surface mapped in Fig. 8.6. *Small squares* are control points from wells. **a** Oblique view to NW down the fault trace. **b** View NW, parallel to bedding in footwall. **c** View NW, parallel to bedding in hangingwall block

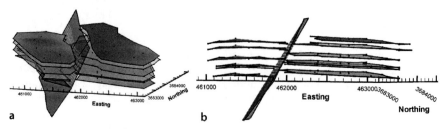

Fig. 8.8. Interpreted fault surface between hangingwall and footwall blocks of Fig. 8.7. **a** Oblique view to NW. **b** View to NW, approximately parallel to fault strike and bed dip

Faults may die out laterally or vertically into folds. During interpretation of the fault location, wells should be examined for evidence of continuity across the proposed fault trace. An unfaulted well (or outcrop) at an intermediate elevation and between the upthrown and downthrown blocks would be evidence for a fold. No such evidence is found for the fault in Figs. 8.7 or 8.8, hence it is mapped as extending across the entire stratigraphic section, e.g., across the top of the upper map horizon (Fig. 8.8) originally mapped as being continuous (Fig. 8.6a).

8.3.2
Joining Offset Marker Surfaces to a Fault

The final step in constructing a complete structure contour map of a faulted surface is to extend the marker surfaces until they join the fault. The basic principle is that the offset marker must join the fault along cutoff lines where the elevations exactly match those of the fault as exemplified by the faulted surfaces in Fig. 8.1. Rarely are there enough data close to the fault plane to eliminate the need for interpretation (Fig. 8.9). The data gaps adjacent to the faults in Fig. 8.6c are typical.

The marker-surface geometries on opposite sides of the fault may or may not be directly related, depending on the nature of the fault. If the fault is curved or if the structures on opposite sides of the fault developed independently using the fault as a displacement discontinuity, then the marker surfaces may be completely unrelated on opposite sides of the fault (Fig. 8.10a,b) and the two sides should be contoured

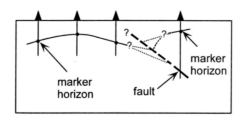

Fig. 8.9. Cross section showing typical subsurface data available for contouring a marker horizon across a fault. *Thin dotted lines* show area of uncertain marker bed location. *Thick dashed line* shows area of uncertain fault location

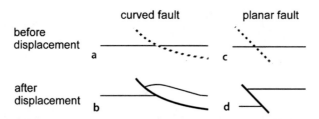

Fig. 8.10. Effect of fault curvature on the relationship between hangingwall and footwall bed geometry. **a** Listric fault before displacement. **b** Listric fault after displacement: the hangingwall shape is changed. **c** Planar fault before displacement. **d** Planar fault after displacement: the hangingwall shape is unchanged

independently. Folds produced by displacement on faults are discussed in Chap. 11. If a planar fault displaces a pre-existing structure, then the shape of the marker horizon will be unaffected by the fault (Fig. 8.10c,d) and the two sides should be contoured so that the hangingwall and footwall bed geometries are related to one another across the fault. The fault separation, if known, should be used to control the size of the fault. The next sections describe interpolation techniques designed to complete the faulted surface based on the different types of information commonly available.

8.3.2.1
Extrapolation of Marker Surface to the Fault

The following method is appropriate where there are enough data to define reasonable marker geometries in the vicinity of the fault. The marker on one side of the fault is contoured to define its shape (Fig. 8.11a). The marker surface is projected into the fault to find the line of intersection, known as the cutoff line. In 3-D software the model can be rotated until the view is along the marker trend and the cutoff line traced directly onto the fault. If a fold is present, see Sect. 5.2 for defining the fold trend and Sect. 6.6 for projection techniques. Once the cutoff line has been found it is added to the marker-horizon data set and contoured (Fig. 8.11b). If comparable information is available, the same procedure is followed on the opposite side of the fault to produce the finished map (Fig. 8.12).

Where no information is available about the dip of the fault and the fault separation is unknown, the structure contours on the fault surface can be generated from the fault geometry most reasonable for the local structural style, for example, a 60° dip for a normal fault. The marker beds are extrapolated from both the hangingwall and footwall until they intersect the fault. The structure contours on the hypothetical fault and the extrapolated marker horizon are intersected to produce the fault trace on the map. The resulting map will be internally consistent but is, of course, hypothetical.

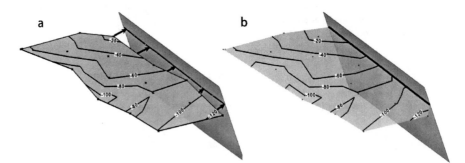

Fig. 8.11. Projection of a marker surface to a fault plane. The example is the top Gwin of the footwall on the western fault in Fig. 8.6c. **a** Faulted marker surface is projected along trend (*arrows*) to its intersection with the fault. **b** Points along the resulting footwall cutoff line are contoured with the original marker data to produce a surface that accurately meets the fault

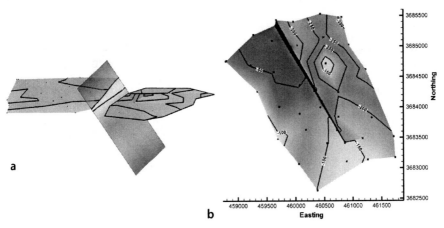

Fig. 8.12. Completed projection of a marker surface to a fault plane. The example is the top Gwin of the fault in Fig. 8.11. **a** Faulted marker surface showing projected hangingwall cutoff line on the fault surface. **b** Final structure contour map of the top Gwin including the fault cutoff lines. Fault gap is *black*

8.3.2.2
Honoring Stratigraphic Separation

If the stratigraphic separation on a fault is known, it provides an important constraint on the interpretation and provides a critical check on the size of a mapped fault. Heave and throw are calculated from the stratigraphic separation (Sect. 7.4.3) and the map is constructed to accurately include these values.

Structure contours on the best-known marker horizon are extrapolated to join the structure contours on the fault surface, defining the trace of the "known" cutoff line (Fig. 8.13). Either the heave or the throw can then be used to control the position of the other cutoff line. To use the throw (Fig. 8.13), move directly up the fault dip (or down the dip, as appropriate) to the point where the elevation is equal to the original elevation plus (or minus) the throw. This represents one point on the inferred cutoff line. The heave and throw are likely to change along the trace of the fault, thus the values closest to the points being inferred should be used. A smooth line through the inferred points is the trace of the inferred cutoff line. The elevations along the inferred cutoff line provide control points for the marker surface on that side of the fault.

The strength of this method is that it works where the marker horizon shapes are different across the fault. When fault blocks are contoured independently, use of this method will ensure that the fault separation between the blocks is correct. The weakness in the method is that the location of the fault cutoff and the cross-fault dip must be known or inferred before the cutoff can be projected across the fault. Developing the best interpretation is commonly an iterative process. The map may be used to provide the dips and then the heave and throw used to improve the map in the vicinity of the fault. Changes in the map may give different values for the fault dips, requiring another revision of the heave and throw calculations, and so on.

Fig. 8.13.
Structure contour map illustrating fault cutoff line inferred from throw. *Heavy solid
line:* known cutoff line; *heavy
dashed line:* inferred cutoff
line on opposite side of fault;
T: throw at known points,
arrows point directly up the
fault dip

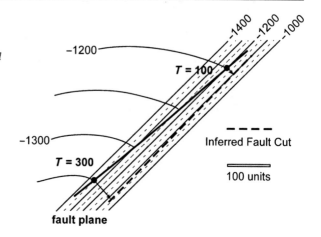

8.3.2.3
Restored Vertical Separation

Contouring based on restored vertical separation (restored tops) is appropriate where the geometry of the hangingwall and footwall beds is unchanged by faulting except for the vertical separation on the fault (Jones et al. 1986; Tearpock and Bischke 2003). This method is used to find the shape of a surface before faulting and its configuration after a displacement amount equal to the vertical separation of the fault (Fig. 8.14). Begin the interpretation by finding the vertical separation from the stratigraphic separation at a fault cut (well 4, Fig. 8.14a). To calculate the vertical separation (Eq. 7.3) it is necessary to know or assume the dip of bedding at the fault cut. For dips below about 25°, the stratigraphic separation is similar in magnitude to the vertical separation and can be used as an approximation. The stratigraphic separation is always less than the vertical separation, but below a marker dip of 25° the difference is under about 10%. Next, remove the vertical separation by adding it to all marker elevations on the downthrown side of the fault (wells 1–3, Fig. 8.14a). Then contour the marker horizon as if no fault were present to obtain the restored top (Fig. 8.14a). Then, subtract the vertical separation from all contour elevations on the downthrown side of the fault (Fig. 8.14b). Recontour the data while maintaining the shapes of the structure contours from the restored-top map. Break the map at the fault by intersecting the contoured fault surface with the marker horizon on both the upthrown and downthrown sides of the fault. This method ensures that the vertical separation on the fault is included on the final map and that the marker geometry is related across the fault.

The strength of this method is that it guarantees that the marker horizon shape is related across the fault. The weakness in the method is that in many structural styles the hangingwall and footwall geometries should not have the same shape (Fig. 8.10a,b). If there is any strike-slip component to the fault displacement, it must be removed before the points are contoured in order to correctly relate the hangingwall and footwall geometry.

Fig. 8.14.
Cross sections in the fault dip direction, showing the method of restored vertical separation to control position of the marker horizon. The dip shown at the fault cutoff is used to determine the vertical separation. **a** Fault separation is added to all marker elevations on the downthrown side of the fault and elevations are contoured to give the restored top. **b** Vertical separation is subtracted from the downthrown side and the marker horizon re-contoured to give the inferred geometry

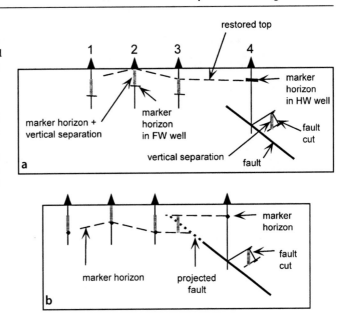

8.4
Fault Cutoff Maps and Allan Diagrams

A fault cutoff map is a map of the traces of beds where they are truncated against a fault (Fig. 8.15). Maps can be made of either or both hangingwall and footwall cutoffs. Superimposed hangingwall and footwall cutoff maps are known as Allan diagrams from their use by Allan (1989) in the prediction of fluid migration pathways and traps for hydrocarbons migrating in the vicinity of a fault zone. The cutoff lines are also termed Allan lines. Fault cutoff maps provide quality control on the bed geometries at the fault, are useful in predicting traps and spill points of hydrocarbon reservoirs formed against faults, and potentially provide a relatively unambiguous method for determining the slip on a fault without piercing points.

8.4.1
Construction

The first step in producing a fault cutoff map is to project the faulted horizons to the fault surface and find the lines of intersection between the horizons and the fault, as has been discussed in Sect. 8.3.2.1. Projections of the fault cutoffs onto either vertical or horizontal planes are convenient when the section is made directly from a completed structure contour map. A vertical map plane is suitable where the fault is steeply dipping (Fig. 8.16) and will show the throw on the fault. If the fault is curved in plan view, a vertical projection foreshortens the length of the fault. A horizontal map plane is suitable where the fault is gently dipping and will show the heave on the fault

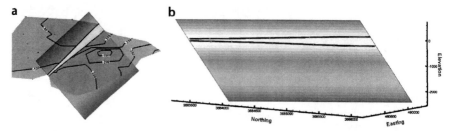

Fig. 8.15. Fault cutoff map of fault from Deerlick Creek coalbed methane field (data from Groshong et al. 2003b). **a** 3-D oblique view of marker surface truncated at fault. **b** 3-D oblique view of fault surface showing hangingwall and footwall cutoff lines (*heavy lines*)

Fig. 8.16.
Cross section in the dip direction of the fault, showing projection of fault cutoffs onto horizontal and vertical fault cutoff map planes

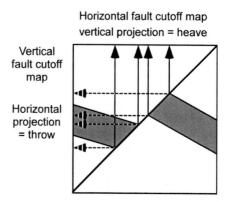

(Fig. 8.16). If the fault is curved in cross section, a horizontal projection foreshortens the width of the fault. Three-dimensional computer graphics techniques allow maps to be made directly on the fault surface (Fig. 8.15b).

The construction of a fault cutoff map is illustrated for a fault seen in an underground mine along two coal seams in the Black Warrior basin of Alabama. The fault dips about 60° to the northeast and forms one side of a full graben (Fig. 8.17). To the southeast it dies out by transferring its displacement to a parallel fault. To the northwest the fault zone continues along multiple parallel branches. The first step in constructing the cutoff map is to choose the plane of the map. In this example the fault dips steeply and so projection onto a vertical plane is suitable. The trace of the map plane is taken parallel to the trend of the fault and with a view direction to the southwest, from the hangingwall to the footwall (Fig. 8.18). The vertical scale is exaggerated by a factor of three because the maximum displacement on the fault is so small relative to its length. The elevations of the cutoffs of the top of both coal seams against the fault are transferred to the cutoff map along lines perpendicular to the trace of the map, which is the method of illustrative section construction (Sect. 6.3). The splays at the northwest end of the fault necessitate a choice of which elevations to map. The entire zone is included here so that the cutoff map (Fig. 8.18) shows the total displacement across the fault zone.

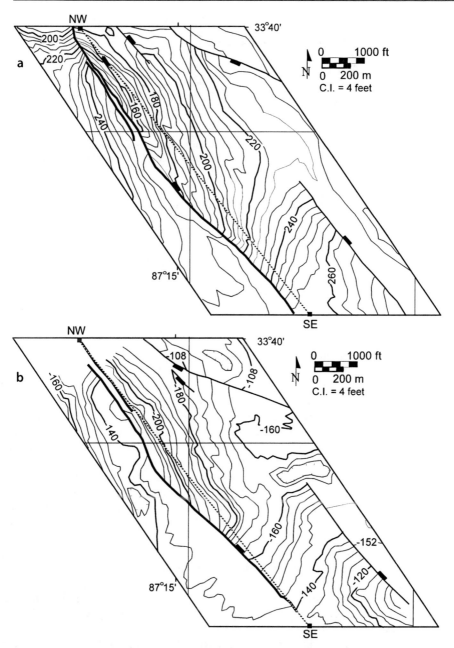

Fig. 8.17. Structure contour maps on the tops of two coal seams in the Drummond Goodsprings No. 1 Mine, central Alabama. The line of the fault cutoff map is the NW-SE *dotted line*. The fault zone represented on the cutoff map is indicated by *heavy solid lines*. **a** Top of America coal seam. Elevations are in feet above sea level. **b** Top of Mary Lee coal seam. Elevations are in feet below sea level. *C.I.* Contour interval. (After Hawkins 1996)

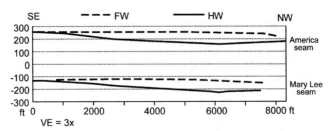

Fig. 8.18. Fault cutoff map (horizontal projection) of throw on the coal seams for the fault zone indicated by the *heavy solid lines* in Fig. 8.17. The section is along azimuth 321°. *Dashed* cutoff lines belong to the footwall (*FW*) and *solid* cutoff lines belong to the hanging-wall (*HW*)

The cutoff map (Fig. 8.18) shows that the fault separation dies out to the southeast and has a maximum at about 6 000 ft on the profile. Both seams have similar throws, with the maximum decreasing downward from 90 ft in the America seam to 80 ft in the Mary Lee seam. The dip separation on the 60° dipping fault is 104 ft in the America seam and 92 ft in the Mary Lee seam (obtained from $L = T / \sin \phi$, where L = dip separation, T = throw, and ϕ = fault dip). Assuming that the point of maximum dip separation is close to the center of the fault, the length/displacement ratio is about 8 000 / 98 or 82 to 1.

8.4.2
Determination of Fluid Migration Pathways

Fault cutoff maps are extremely useful in determining the possible routes of fluid migration in fault zones where the fault plane itself is not a barrier to the migration (Allan 1989). A fault trap is a closure in a permeable bed that is sealed by impermeable units across the fault. Multiple porous beds in the section can lead to very complex migration paths (Fig. 8.19) in which the migrating fluid crosses back and forth across the fault. Figure 8.19 shows eight fold closures against the fault, only three of which are sealed for the up-dip migration of a light fluid like oil or gas. The spill point at the base of each closure is located where permeable beds are in contact across the fault. Fluids that are heavier than water, such as man-made contaminants, may spiral downward across a fault into synclinal closures at some distance from the original contamination site.

The sequential migration of hydrocarbons through multiple traps like those in Fig. 8.19 can lead to a reversal in the expected positions of the oil and gas. The expected sequence in the filling of a trap in place is illustrated in Fig. 8.20. Thermal maturation of the hydrocarbon source leads first to the formation of oil (Fig. 8.20a) which is less dense than water and displaces the water in the trap. Gas forms later in the source bed or in the trap itself and, being less dense than the oil, displaces the oil to form a gas cap on the reservoir (Fig. 8.20b). The formation of a large volume of hydrocarbons can fill the trap to the spill point where the closure is no longer complete and allow displaced hydrocarbons to be forced out at the base of the accumulation to continue migrating up dip. The process of spilling from the base of the reservoir causes the most dense hydrocarbons to continue migrating up dip into new traps (Gussow

Fig. 8.19.
Allan diagram showing fluid
migration pathways through
permeable beds separated by
impermeable beds; footwall
shaded. Arrows give migration
routes of fluids that are lighter
than water. Oil accumulations
are *solid black*; gas accumulation is indicated by *vertical
lines*. (After Allan 1989)

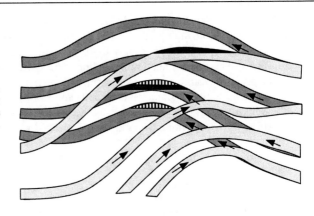

Fig. 8.20.
Sequential formation of oil
(*black*) and gas (*vertical lines*
and *open circles*) and filling
of a trap. **a** Burial to the temperature of the formation of
oil. **b** Additional burial to the
temperature of the formation
of thermal gas. The gas displaces oil in the trap

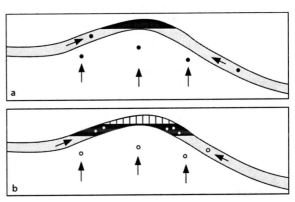

1954; Allan 1989). The lightest hydrocarbons, mainly gas, remain in the deeper traps.
Thus the presence of deep gas-filled traps and shallow oil-filled traps as shown in
Fig. 8.19 can be caused by the upward migration of hydrocarbons across multiple spill
points (assuming that oil and gas are both thermally stable at the trap depths).

8.4.3
Determination of Fault Slip

The slip on a fault is normally determined by the offset of geological lines that pierce
the plane of the fault. Geological lines may be formed by original stratigraphic features, such as linear sand bodies or paleo-shorelines; intersection lines, such as a vein-
bed intersection or a vein-vein intersection; or a structural feature like a fold hinge
line. An Allan diagram provides an excellent format for the display of the cutoff information.

An anticline-syncline pair offset by a normal fault (Fig. 8.21a) appears at first glance
to have both right-lateral and left-lateral slip, a common map pattern where faults cut
folded beds (see also Fig. 7.17). The Allan diagram (Fig. 8.21b) makes it clear that the
fault is dip slip and shows the amount of the slip from the offset of the fold hinge lines.

Structures may develop independently across some fault zones and so may not directly correlate across the fault. That the fold geometry is the same on both sides of the fault in Fig. 8.21 supports the concept that a pre-existing fold has been cut by a later fault and that the arrows connecting hinge lines (Fig. 8.21b) represent the slip.

If croscutting features are present at the fault surface, an Allan diagram will make both the slip and rotational components obvious and readily measurable. Both vein-bed and vein-vein intersections can be correlated across the fault in Fig. 8.22. As a result of the rotational displacement on the fault, each correlated point has a different net slip.

Fig. 8.21.
Anticline-syncline pair cut by a normal fault. **a** Map. The *shaded* bed is the same on both sides of the fault. *Full arrows* show bedding dips, *half arrows* show strike separations, not slip. **b** Fault cutoff map of hangingwall (*darker shading*) and footwall cutoffs. The *horizontal line* is the line of intersection of the outcrop map and cutoff map. *Arrows* show the hangingwall slip vectors of the anticlinal and synclinal hinge lines and indicate pure dip slip

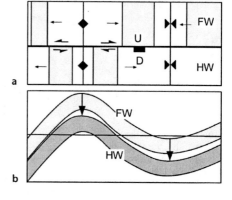

Fig. 8.22.
Rotational fault displacement on a fault cutoff map. Superimposed fault-surface sections of hangingwall (*unshaded*) and footwall (*shaded*). Horizontal footwall beds labeled *A* and *B* rotate to *A'* and *B'*; veins *C* and *D* rotate to *C'* and *D'*. *Arrows* show hangingwall slip of correlated bed-vein intersection points. Fault slip is a left-lateral reverse translation produced by a 15° clockwise rotation of the hangingwall relative to the footwall

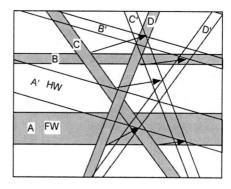

Fig. 8.23.
Allan diagrams for **a** dip-slip and **b** strike-slip dislocation-model faults

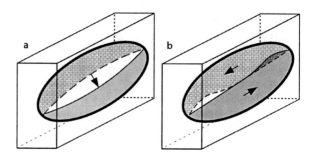

The Allan diagram of a complete fault may, by itself, provide all the information needed for a reasonable estimate of total slip. It is well established that the Allan diagram of a single bed cut by a dip-slip fault resembles that shown in Fig. 8.23a, with the arrow giving the slip. Displacement-distance graphs and the bow-and-arrow rule for fault displacement are comparable concepts (c.f., Sect. 7.5.2). Strike-slip and oblique-slip faults will doubtless show more complex geometries. Figure 8.23b is a hypothetical strike-slip bed geometry that approximately preserves bed length parallel to the fault strike between the fault tips. The maximum strike-slip displacement would be at the center and equal to the sum of the difference between the straight-line length and the curved-bed length between the fault tips for both sides of the fault.

8.5
Faults on Isopach Maps

Faults cause characteristic thickness variations on isopach maps (Hintze 1971). If unrecognized, these variations could be misinterpreted as being stratigraphic in origin. Recognized as fault related, they provide a useful tool for fault interpretation. A normal fault thins the stratigraphic unit (Fig. 8.24) and a reverse fault thickens the unit. A single fault produces two parallel bands of thickness change (Figs. 8.24, 8.25), one associated with the cutoff of the top of the unit and one with the cutoff of the base of the unit. The affected width on the isopach map is appreciably wider than the fault gap or overlap on the struc-

Fig. 8.24.
Effect of normal-fault displacement on the thickness of a unit. **a** Cross section. *T* throw on the fault. **b** Isopach map. (After Hintze 1971)

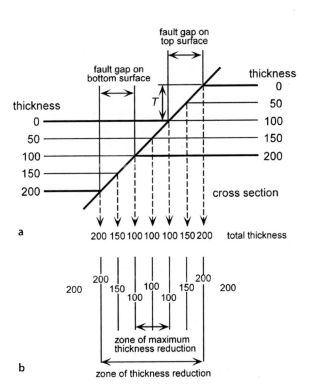

Fig. 8.25.
Effect of normal-fault displacement on the thickness of a growth unit. **a** Cross section, T_u = fault throw at upper surface, T_l = fault throw at the lower surface of the unit. **b** Isopach map

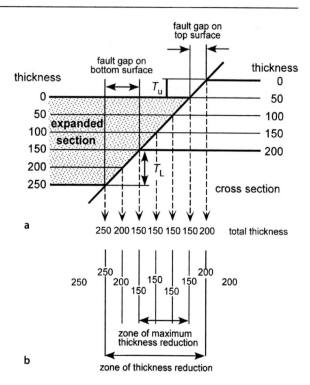

ture contour map of the top or base of the unit. The affected zone extends from the fault cutoff of the upper surface of the affected bed on one side of the fault to the fault cutoff on the lower surface of the affected bed on the other side of the fault (Fig. 8.24). The zone of thinning (normal fault) or thickening (reverse fault) is centered on the trace of the fault through the middle of the unit. Such elongate zones of thinning or thickening can indicate the presence of faults having displacements that are too small to completely separate the map unit and that might otherwise go unrecognized.

The throw on the hangingwall or the footwall of a fault that does not completely separate the unit can be determined from the maximum and minimum thicknesses of the unit as

$$T = t_{max} - t_{min} \quad , \tag{8.4}$$

where T = fault throw, t_{max} = maximum thickness inside (reverse fault) or outside (normal fault) the zone of thickness change, and t_{min} = minimum thickness inside (normal fault) or outside (reverse fault) the zone of thickness change.

For a growth unit (Fig. 8.25) the thickness change is different on opposite sides of the fault. Less thickness change occurs on the upthrown side of the fault because the displacement is partly erased by sedimentation. The asymmetry can be used to infer the dip direction of the fault. A normal fault dips toward the thicker side and a reverse fault dips away from the thicker side.

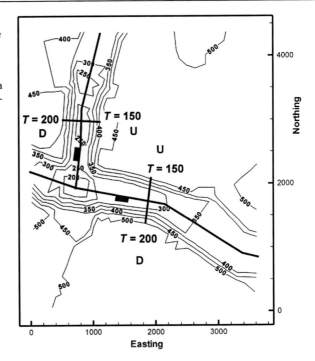

Fig. 8.26.
Interpreted isopach map of the top of the lower Taylor Formation on part of the Hawkins salt dome, Texas. The contour interval is 50 ft. *D* Downthrown block, *U* upthrown block. (Isopach map after Hintze 1971)

A map on a portion of a salt dome provides an example of the effect of normal faults on an isopach map. Elongate zones of isopach thinning are developed along faults (Fig. 8.26). On a traverse across the north-south fault, the throw is 150 ft on the east (Eq. 8.4: 400–250) and 200 ft (Eq. 8.4: 450–250) on the west. The thicker section on the west of the normal fault indicates that it dips to the west. A similar analysis indicates that the east-west fault is a growth fault, down to the south with 150 ft of throw at the top of the unit and 200 ft at its base.

8.6
Displacement Transfer

This section describes the geometry of linked faults. As a fault dies out it usually transfers its displacement to another fault (Fig. 8.27). Displacement transfer between nearby faults is an important process in both map and cross section (Crowell 1974; Childs et al. 1995; Walsh et al. 1999). Faults that transfer displacement but do not intersect are termed soft linked and are separated by a continuous bed segment called a relay ramp (Kelly 1979). Faults that intersect are said to be hard linked (Walsh and Watterson 1991; Childs et al. 1993) and join along a branch line (Boyer and Elliott 1982). Conjugate faults (Sect. 1.6.4) that overlap will form convergent or divergent links (Fig. 8.27). Bends or overlaps that trend oblique to the slip direction are either restraining or releasing (Crowell 1974; Mansfield and Cartwright 1996) depending on their overlap direction relative to the slip direction (Fig. 8.28).

Fig. 8.27.
Displacement-transfer geometries on normal faults in map view. (After Morley et al. 1990)

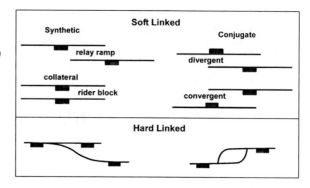

Fig. 8.28.
Displacement-transfer geometries on normal faults in cross section

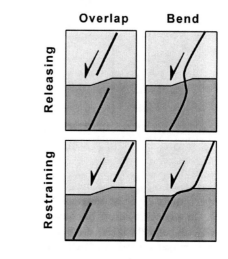

Fig. 8.29.
Possible evolution of a relay zone into a hard link. *Dashed arrows* indicate fault propagation directions

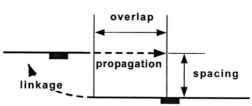

Soft links may evolve into hard links as displacement increases and the faults grow longer. Faults may propagate in their own plane or curve to intersect (Fig. 8.29). In general, a wide spacing between the propagating faults favors overlap and a small spacing favors linkage.

In the following sections the basic patterns of fault-to-fault displacement transfer are examined. All types of faults show the same forms of displacement transfer. The examples given will mainly be normal faults because they are easy to visualize, but the same principles and terminology can be applied to reverse and strike-slip faults.

8.6.1
Relay Overlap

Faults that transfer displacement from one to the other without intersecting constitute a relay pattern (Fig. 8.27). The displacement is said to be relayed from one fault to the other. The displaced horizon in the zone of fault overlap forms a ramp that joins the hangingwall to the footwall (Figs. 8.30, 8.31a). The ramp may be unfaulted or may itself be broken by faults. An unbroken ramp provides an opportunity for pore fluids to migrate across the main fault zone. Second-order faults within a ramp typically trend at a low angle to the strike of the ramp and are therefore at a high angle to the relay faults. Faults exhibiting a relay pattern may appear to be unrelated on a map because a single fault will have the displacement pattern of an isolated fault that dies out along strike (Figs. 7.22, 7.25). The covariation of displacement on the two faults reveals their relationship to be that of displacement transfer (for example, Fig. 7.25c). Detailed examination of large fault zones may reveal that the main fault consists of multiple segments with relay overlaps between the segments.

On the structure contour map of a horizon displaced in a relay zone, the ramp is the region between the two faults where the contours on the marker horizon are at a high

Fig. 8.30.
Displacement transfer at a relay overlap. *Arrows* indicate amount of dip separation on the relay faults

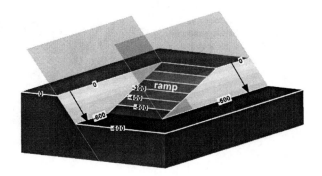

Fig. 8.31.
Structure contour maps based on the relay overlap geometry in Fig. 8.30. Both maps are at the same scale. **a** Displaced stratigraphic horizon. *Solid contours* are on the offset marker, *dashed contours* are on faults *A* and *B*. **b** Portions of the two fault planes; contours on fault *A* are *dashed* where they lie below fault *B*

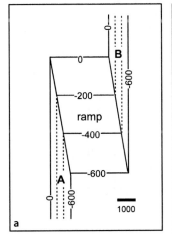

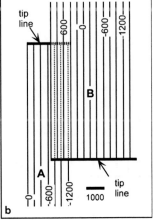

angle to the trend of the faults (Fig. 8.31a). Structure contour maps on the faults show them to overlap in the relay zone (Fig. 8.31b). Both faults end at tip lines that define the ends of the relay zone. The extent of the faults may be large, both updip and downdip, and the tip lines may be straight or curved.

8.6.2
Branching Fault

The intersection of two faults (Fig. 8.32) occurs along a branch line (Boyer and Elliott 1982). Some of the displacement is transferred from the main fault to the branch fault at the intersection. Displacement is conserved at the branch line (Ocamb 1961). As can be seen from Fig. 8.32, at the branch line, the throw of the largest fault will be equal to the sum of the throws on the two smaller faults. All the faults may change throw independently away from the branch line.

Fig. 8.32.
Displacement transfer at a branch line on a normal fault. Fault surfaces are *shaded*. The total throw is constant at the branch line. In this example faults *A* and *C* are a single plane; fault *B* is the branch

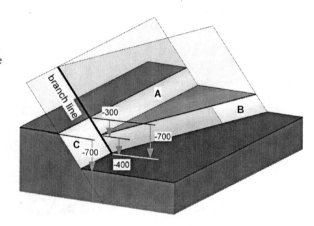

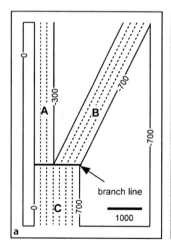

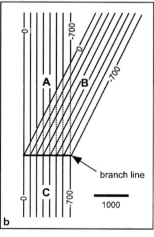

Fig. 8.33.
Structure contour maps of a branching normal fault, based on the geometry of the structure in Fig. 8.32. Both maps are at the same scale. **a** Structure contours on the displaced marker horizon (*solid lines*) and on the faults (*dashed lines*). **b** Structure contours on portions of the three fault planes; contours on fault *A* are *dashed* where the plane lies below fault *B*. Fault planes *A* and *C* are the same

The structure contour map of a marker horizon displaced by a branching fault (Fig. 8.33a) shows two intersecting faults and an abrupt change in throw at the intersection. The throw of fault A plus that of fault B is equal to that of C at the branch line. In this example, the fault plane of A is continuous with the plane of fault C. Contours on the faults (Fig. 8.33b) show a straight branch line and a region of fault overlap. All three faults at the branch line could have different strikes.

A good test of the fault interpretation on a map is to measure the throws in the vicinity of all branch lines and show that the throws of the two smaller faults are equal to that of the larger fault. If this is not true, the map is wrong. An abrupt change in the throw of a fault is probably caused by an undetected fault branch (Ocamb 1961) or relay fault that carries the missing throw.

8.6.3
Splay Fault

Splay faults are minor faults at the extremities of a major fault (Bates and Jackson 1987). The master fault splits at one or more branch lines to form the splay faults (Figs. 8.34, 8.35). The splays then die out along strike at tip lines. Splay faults distribute the total

Fig. 8.34.
Splay faults (*B* and *C*) at the end of a master normal fault (*A*)

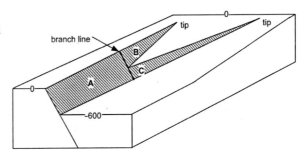

Fig. 8.35.
Structure contour maps of a splaying normal fault based on geometry of the structure in Fig. 8.34. The master fault is *A*, its continuation into the region of splay faulting is *B*, and the other splay fault is *C*. Both maps are at the same scale. **a** Offset marker horizon. *Solid contours* are on the marker horizon and *dashed contours* are on the faults. **b** Portions of the faults. Contours are *dotted* where they lie below another fault

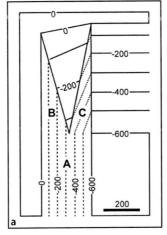

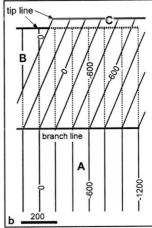

offset over a wider area, which makes it possible for more of the displacement to be accommodated by folding rather than faulting. This helps a large fault die out into a fold. The master fault may continue straight in the region of fault splays (Fig. 8.34), or all the splays may have different orientations from that of the master fault.

Splay faults are ultimately bounded in all directions by branch lines and tip lines. The leading edge of a structure is its termination in the transport direction, and the trailing edge is in the opposite direction (after Elliott and Johnson 1980). For a thrust fault, the leading edge of a splay is a tip line and the trailing edge is a branch line (Fig. 8.36a). The leading hangingwall and footwall cutoff lines of a unit carried on a splay join and end at the tip line of the leading fault (Fig. 8.36b). The trailing bed cutoff line continues along the trailing thrust beyond the tip of the splay.

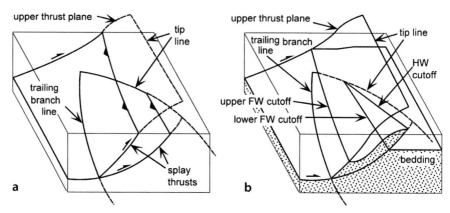

Fig. 8.36. Splay faults at the leading edge of a thrust fault. **a** Fault surfaces in three dimensions. **b** Bedding cutoff lines of a unit transported on the lower splay. (After Diegel 1986)

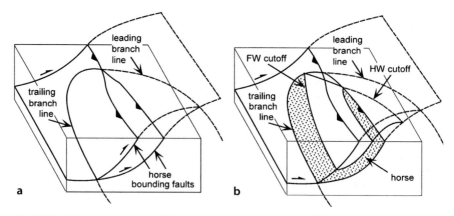

Fig. 8.37. Fault horse bounded on all sides by fault surfaces that end in all directions at a closed branch-line loop. **a** Main fault and branch lines. **b** Bedding cutoff lines for a stratigraphic unit within the fault horse. (After Diegel 1986)

8.6.4
Fault Horse

A fault horse (Fig. 8.37) is a body of rock completely surrounded by fault surfaces (Dennis 1967; Boyer and Elliott 1982). The geometry can be thought of as a splay fault that rejoins the main fault. A horse is completely bounded by a branch-line loop along which the two fault surfaces join (Fig. 8.37a). A stratigraphic unit within a horse will have leading and trailing cutoff lines at the two bounding fault surfaces (Fig. 8.37b). Multiple fault horses form a duplex (Boyer and Elliott 1982).

8.7
Crossing Faults

Two intersecting faults may both continue past their line of intersection, a relationship here termed crossing faults. Where one fault offsets the other, the line of intersection between the two faults is a cutoff line of one fault by the other, not a branch line. Crossing faults may be sequential or contemporaneous. Sequential faulting means that an older fault is cut and displaced along a younger fault. Contemporaneous faulting means that the faults form at approximately the same time and as part of the same movement picture.

8.7.1
Sequential Faults

Crossing dip-slip normal faults are illustrated in Fig. 8.38. Both faults have constant (but different) displacement, but the shaded marker horizon is displaced to four different elevations, reflecting the four different combinations of displacement from the original elevation (no displacement, fault 1 alone, fault 2 alone, fault 1 + 2). At first glance the structure contour map (Fig. 8.39a) appears to imply that fault 1 is younger and is a strike-slip fault because its trend is straight whereas the trend of fault 2 appears to be offset. Comparison with Fig. 8.38 shows that fault 2 is the through-going plane. It is the different elevations of the marker horizon along fault 2 that make this fault appear to bend at the intersection. The fault-plane map (Fig. 8.39b) shows fault 2 to be unbroken and fault 1 to be displaced. In the area near the intersection of the two faults, both faults would be penetrated by a vertical well (Fig. 8.39b).

The footwall cutoff line formed by the intersection of fault 1 with the marker horizon forms a geological line that can be used to determine the net slip on fault 2. In the example in Fig. 8.38, the slip on fault 2 is 300 units of pure dip slip, as indicated by the arrow on the fault surface. No geological line is present that predates the displacement of fault 1; hence the displacement can only be constrained as having a dip separation of 400 units.

Where fault 2 displaces fault 1, fault 2 carries the combined stratigraphic separation of both faults (Figs. 8.38, 8.39), making this area have a much larger separation fault than either fault 1 or 2 alone. The zone of combined separation for a particular horizon is restricted to a small region (Fig. 8.39a) where the second fault crosses the first fault. In three dimensions, the zone of combined separation (Dickinson 1954)

Fig. 8.38.
A young normal fault (*2*) cuts
and displaces an old normal
fault (*1*). An offset marker surface is *shaded*. Displacement
on each fault is dip slip of a
constant amount. The displacement of the offset geological
line is the slip on fault 2

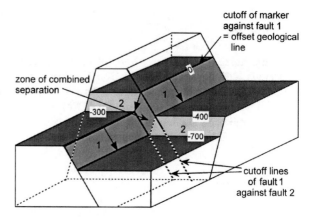

Fig. 8.39.
Structure contour maps based
on the geometry of the structure in Fig. 8.38. Both maps
are at the same scale. Fault 2 is
through-going and displaces
fault 1 down dip 300 units to
the south. **a** The marker horizon. **b** Portions of both faults
in the vicinity of the offset
marker horizon

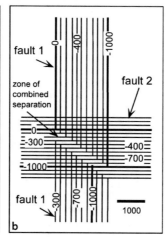

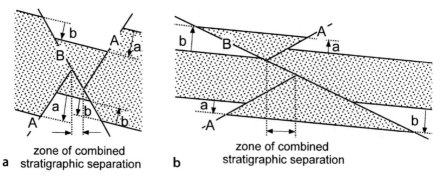

Fig. 8.40. Zone of combined stratigraphic separation on younger fault *B*. Fault *A* is displaced by fault *B*.
a Two normal faults. **b** Two reverse faults

continues along the plane of the younger fault, between the traces of the cutoff lines of the older fault against the younger fault (Figs. 8.39b, 8.40). Other horizons will have their zones of combined separation along the same trend, but offset from one another. Given low bedding dips, a vertical well drilled into the zone of combined separation for two normal faults will cut one fault that carries the combined separation (Fig. 8.40a). A vertical well drilled into the zone of combined separation for two reverse faults will have three fault cuts (Fig. 8.40b), the middle one of which will carry the combined separation.

The amount of the combined separation is given by a relationship from Dickinson (1954). If the younger fault (B, Fig. 8.40a) is normal,

$$t = -b + (\pm a) \quad , \tag{8.5}$$

and if the younger fault (B, Fig. 8.40b) is reverse,

$$t = +b - (\pm a) \quad , \tag{8.6}$$

where t = combined stratigraphic separation, a = stratigraphic separation on the older fault, b = stratigraphic separation on the younger fault, the "+" sign indicates reverse separation = thickening, and the "−" sign indicates normal separation = thinning. The simplicity of this result is due to the fact that all the beds have the same dip. If dip changes occur across the faults but are small, then Eqs. 8.5 and 8.6 will still provide good estimates.

The map pattern produced by crossing normal faults depends on the angle of intersection of the faults, the dip of the faults and the marker horizon, and the magnitude and sense of slip of the faults. The trace of the fault on the marker horizon depends on the dip of the marker as well as on the attitude of the fault surface. The following examples illustrate some of the possibilities. In the three parts of Fig. 8.41, the orientation of the older fault (A) is changed while the orientation of the younger fault (B) and the attitude of the marker horizon remain constant. The structure changes from a geometry that could be described as tilted steps (Figs. 8.41a,c) to a horst that changes into a graben along a northwest-southeast trend (Fig. 8.41b). Note that the direction of the fault-bedding intersection is not parallel to the fault strike in any of the examples because the strike of bedding is not parallel to the strike of the faults.

The width of the zone of combined stratigraphic separation is reduced to a line for certain combinations of the displacement directions (Fig. 8.42a,b). An apparent strike separation on the older fault will be caused by slip on the younger fault that is in the strike direction of the younger fault. In Fig. 8.42c the later, through-going fault (B) appears to be offset by left-lateral strike-slip on fault A, although it is, in fact, fault B that displaces fault A.

The development of the map pattern of cross-cutting reverse faults is illustrated with a forward model (Fig. 8.43). The first fault (A) trends obliquely across the north-westerly regional dip of the marker (Fig. 8.43a). The trace of the second fault (B) is found by intersecting the contours of the fault with the displaced marker surface

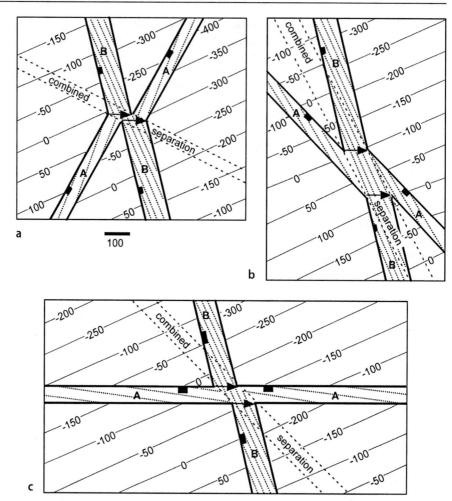

Fig. 8.41. Structure contour maps showing the effect of the angle of fault intersection on the geometry of sequential dip-slip normal faults. Fault *A* is older and has a pure dip-slip of 100 units; fault *B* is younger, strikes 335°, and has a slip of 200 units down to the east. *Arrows* give the slip direction of fault *B*. Structure contours on the displaced marker horizon are *thin solid lines*; contour interval is 50 units. *Dotted contours* are on the fault planes. Note that the strikes of the faults, as shown by the structure contours, are not the same as the trend of the fault cutoff lines of the marker horizon. **a** Fault *A* strikes 020°. **b** Fault *A* strikes 326°. **c** Fault *A* strikes 276°

(Fig. 8.43b). In the final step, the hangingwall of fault B is displaced up to the west and the location of the hangingwall cutoff and the contours are drawn as an overlay on the footwall (Fig. 8.43c). Three fault cuts are present in the vertical zone of combined stratigraphic separation where the older fault is repeated by the younger fault (Dickinson 1954).

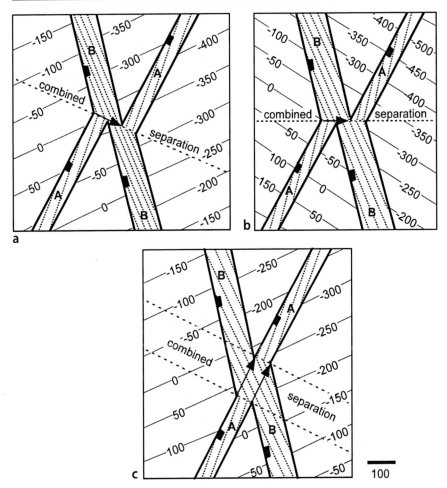

Fig. 8.42. Structure contour maps showing effect of the slip direction of the later fault (*B*) on the intersection geometry between normal-separation faults. *Thin solid contours* are on the marker horizon, *dotted contours* are on the faults. Fault *A* strikes 020°, is older, and has a throw of 100 units. Fault *B* strikes 335° and has a throw of 200 units. The slip is downthrown in the direction of the *arrows*. Note that the strikes of the faults, as shown by the structure contours, are not the same as the trend of the fault cutoff lines. **a** The slip direction of fault *B* is obliquely down to the southeast, in the dip direction of *A*. The attitude of the marker horizon is the same as in Fig. 8.41. **b** The slip direction of fault *B* is obliquely down to the east. **c** The slip direction of fault *B* is obliquely down to the northeast, in the strike direction of *A*. The attitude of the marker horizon is the same as in Fig. 8.41

A cross section (Fig. 8.44) shows the evolution of two reverse faults having the same dip direction, like those in the previous map (Fig. 8.43c). In the zone of combined stratigraphic separation between the dashed lines, a vertical well would cut three faults, but the top of the shaded unit would be penetrated only twice. Inside this zone the throw on the top of the shaded unit is the combined value for both thrusts.

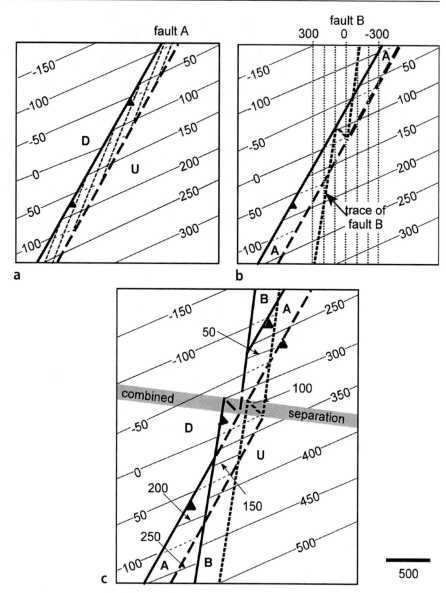

Fig. 8.43. Structure contour maps showing evolution of intersecting reverse faults having similar dip directions and sense of displacement. Contours on the fault planes are *dotted*. Contours on hidden surfaces are *dashed*. *Thin solid lines* are contours on the marker horizon. D: Downthrown block, U: upthrown block. **a** First displacement. The northeast striking thrust (A) has a displacement of 100 units up on the southeast. The hangingwall cutoff is a *thick solid line* and the footwall cutoff is a *thick dashed line*. **b** Structure contours on incipient fault B are added. The trace of thrust B on the marker surface is *dotted*. **c** Second displacement, enlarged to show detail. Thrust B displacement is 200 units up on the east. *Dotted bed contours* are in the footwall of fault A; contours indicated by *arrows* are in the footwall of fault B

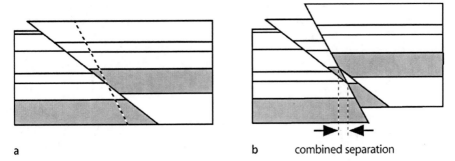

a   b   combined separation

Fig. 8.44. Evolution of crossing reverse faults having the same dip direction. The cross sections approximate the evolution along an east-west cross section through the center of Fig. 8.43c. **a** First fault displacement showing the *dashed* trace of the second fault. **b** Displacement on the second fault. *Dotted lines* bound the zone of combined stratigraphic separation for the *shaded* horizon

Fig. 8.45.
Structure contour map of crossing reverse faults. Fault *A* is older, strikes 273°, and has a heave of 100 units, up on the south; fault *B* has a heave of 200 units, strikes 347° and is up on the east with displacement parallel to the trace of fault *A* (*arrow*). *Thin solid lines* are contours on the marker horizon. The intersection lines of faults with the contoured horizon are *wide lines. Dashed contours* are hidden below faults. (After Dickinson 1954)

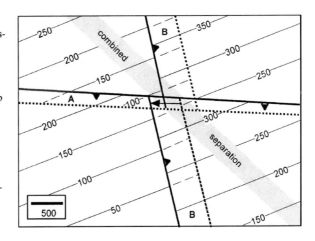

Just as for normal faults, a variety of different geometries are produced by the intersection of reverse faults of varying attitudes, amounts and directions of slip, and that cut marker horizons of differing attitudes. The map of Fig. 8.45 is the result of displacement of a southeast dipping bed by two orthogonal reverse faults, fault A being the older. Figure 8.46 shows the evolution of two parallel thrusts having opposite dips. The evolution of an east-west cross section of a structure approximately like that of Fig. 8.46c is shown in Fig. 8.47. A vertical well in the zone of combined separation would have three fault cuts and penetrate the top of the shaded unit twice. Even though the two faults thicken the section, a unit within the zone of combined separation can be reduced in thickness by the second fault. In the region labeled "T" (Fig. 8.47b) the shaded unit is thinned by the crosscutting reverse faults and the middle fault cut could be mistaken for a normal fault downthrown to the west.

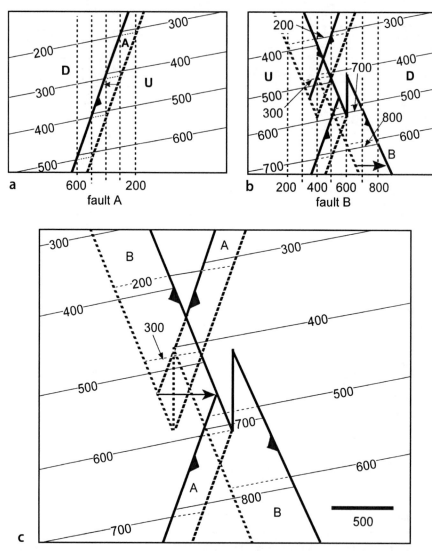

Fig. 8.46. Evolution of the structure contour map of reverse faults having equal and opposite dips. North-south *dashed contours* are on faults A and B. *Thin solid lines* are contours on the marker horizon. *Wide lines* are fault traces. Contours are *dotted* where hidden. **a** Fault A strikes north–south and is displaced 100 units up due west. **b** Fault B strikes north–south and is displaced 200 units up due east. **c** Final map, enlarged to show detail. *Arrow* shows the displacement direction of fault B

It should be noted that all the previous examples of crosscutting faults are based on the simplest possible fault geometries. The faults are planar and so the attitude of the marker horizon is not changed by displacement on the faults. The displace-

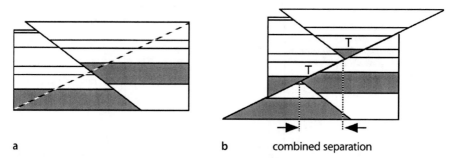

a   b   combined separation

Fig. 8.47. Evolution of crossing reverse faults having opposed dips, similar to those in Fig. 8.46c. **a** Displacement on the first fault, with location of the second fault *dashed*. **b** Displacement on the second fault. *Dotted lines* bound the zone of combined stratigraphic separation for the *shaded* horizon. *T* indicates regions of local structural thinning of the shaded unit

ment on each fault has been assumed to be constant, a reasonable assumption over a small portion of a fault, but not likely to hold over a long distance along the fault surface.

8.7.2
Contemporaneous Faults

Intersecting faults may be of the same age, that is, contemporaneous. (The term contemporaneous faulting has also been used for faults in which the displacement is contemporaneous with deposition (Hardin and Hardin 1961) but the term growth fault is now widely used for that concept.) According to the Andersonian theory of faulting (Sect. 1.6.4) a biaxial state of stress is expected to produce two conjugate fault trends of the same age that intersect with a dihedral angle of 40–65°. A triaxial stress state may cause three or four fault trends (Oertel faults) to form simultaneously.

The fault pattern in the zone of intersection of contemporaneous faults tends to be more complex than the patterns for sequential faults discussed in the previous section. Multiple small faults may occur in the zone of intersection (Fig. 8.48). In the experimental studies of normal faults by Horsfield (1980), two crossing conjugate faults formed initially and with continuing extension the initial horst and graben were segmented by smaller conjugate faults. Many of the crosscutting relationships described above are probably the result of contemporaneous conjugate faults that cut one another as their displacement increases and they grow along strike and down dip.

As yet there is little published on the geometry of crossing contemporaneous faults. This style of structure may have been overlooked because of the incorrect assumption that crossing faults cannot be contemporaneous or because the geometric relationships are complex. Where timing information, such as growth strata, is available, the time relationships of the fault sets can be determined and used to substantiate the interpretation of fault contemporaneity.

Fig. 8.48.
Crossing conjugate normal
faults from a naturally deformed outcrop of Pleistocene
sand. (After Walsh et al. 1996,
from a photo by Dietmar
Meier)

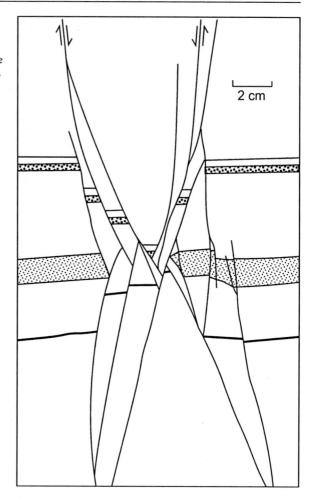

8.8
Exercises

8.8.1
Heave and Throw from a Map

Find the heave and throw on the fault at the point shown on the map in Fig. 8.49.

8.8.2
Construct the Fault Trace

Construct the hangingwall and footwall cutoff lines on both maps in Fig. 8.50. What is
the heave, throw, and gap for a point near the middle of each map?

Fig. 8.49.
Structure contour map of
faulted surface

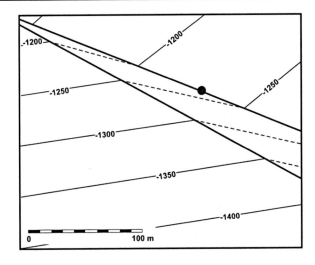

Fig. 8.50.
Unfinished structure contour
maps of a fault (*dashed lines*)
and a marker surface on both
sides (*solid lines*). **a** Fault dips
north. **b** Fault dips south

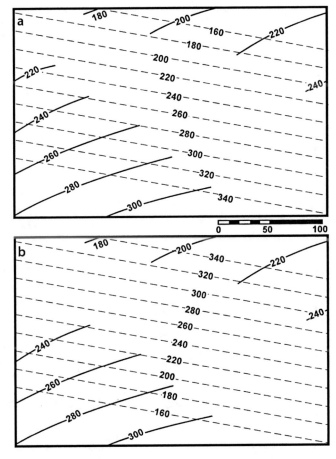

8.8.3
Construct the Fault Trace

Construct the missing fault trace and missing marker horizon on both maps in Fig. 8.51 for a fault throw of 40. Does the marker surface on the map belong to the hangingwall or footwall? Construct one map to show one to be a normal fault and the other to be reverse. What is the heave and gap for a point near the middle of each map?

8.8.4
Reservoir Structure

Use the subsurface information in Fig. 8.52 to make a structure contour map of the top of the Hamner Sandstone reservoir. What type of fault or faults are present? Are there any faults in the reservoir? What is the dip of the fault or faults? What is the calculated heave and throw at each fault cut? Do the fault map and fault separation data agree with the map?

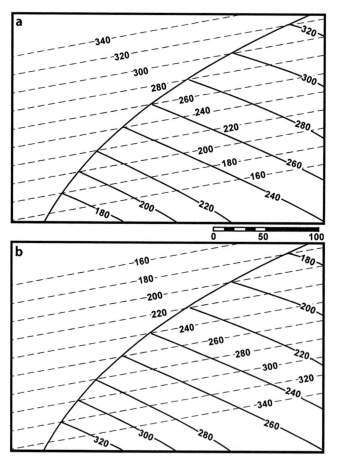

Fig. 8.51.
Structure contour maps of a fault (*dashed lines*) and a marker surface on one side (*solid lines*). **a** Fault dips south. **b** Fault dips north

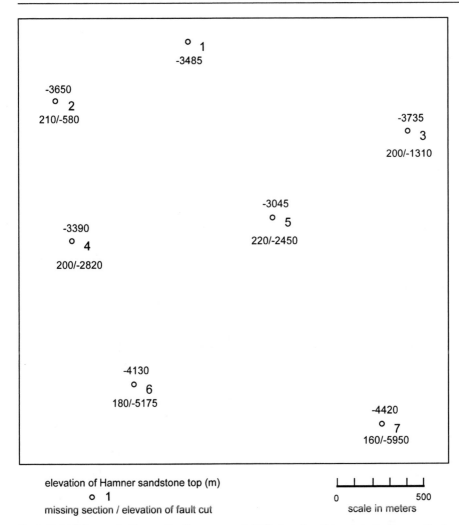

o 1
-3485

-3650
o 2
210/-580

-3735
o 3
200/-1310

-3045
o 5
220/-2450

-3390
o 4
200/-2820

-4130
o 6
180/-5175

-4420
o 7
160/-5950

elevation of Hamner sandstone top (m)
o 1
missing section / elevation of fault cut

0   500
scale in meters

Fig. 8.52. Well data on the Hamner Sandstone reservoir. Wells (*numbered*) penetrate the Hamner Sandstone (*single number* is the elevation of the top contact). Fault cuts are indicated by a *pair of numbers* (amount/elevation). Elevations are in meters, negative below sea level

8.8.5
Normal Fault

Determine the structure of the faulted Oil City Sandstone in Fig. 8.53. Is a single fault present? What kind? What is the evidence? What is the attitude of the fault? of bedding in the hangingwall? of bedding in the footwall? What is the heave and throw on the fault? Does the map agree with the attitude information? Explain the reason for the hydrocarbon trap.

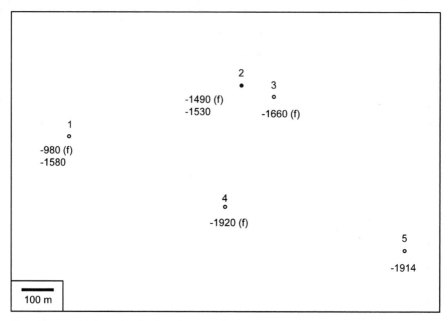

Fig. 8.53. Map of information for the Oil City Sandstone. Posted on the map are the elevations of fault cuts (*f*) and the top of the sandstone. Well *2* is an oil well. Dipmeters in well *1* indicate a bedding attitude of 10, 315 and a fault strike of 045; in well *5* the bedding attitude is 8, 315. Elevations are in meters

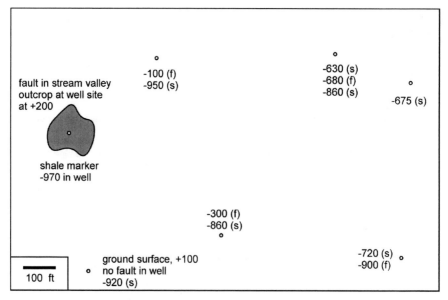

Fig. 8.54. A shale marker and fault cuts in a groundwater basin. Posted on the map are the elevations of fault cuts (*f*) and of the shale tops (*s*) in each well where present. Elevations are in feet, negative below sea level

8.8.6
Reverse Fault

Map the fault (f) and the shale marker horizon (s) using the data in Fig. 8.54. Is a single fault present? What kind? What is the evidence? What is the attitude of the fault? of bedding in the hangingwall? of bedding in the footwall? What is the heave and throw on the fault? If

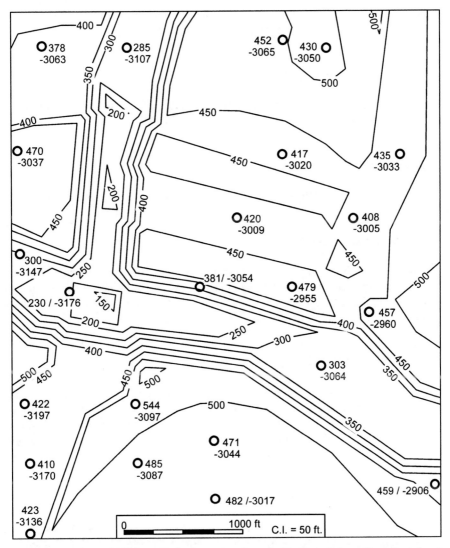

Fig. 8.55. Isopach map of the lower Taylor Formation above the Hawkins salt dome, Texas. Posted next to the wells (*circles*) are the thickness (upper or left-hand *number*) and the elevation of the formation top, in feet below sea level. (After Hintze 1971)

a heavy liquid is spilled in the stream valley in the shaded area, could the fault provide a barrier to the fluid movement in porous units above the shale marker? Explain the reason behind your answer. Where would a fault trap for fluids lighter than water be located?

8.8.7
Faults on an Isopach Map

Figure 8.55 is an isopach map. Locate the faults that explain the thickness changes. Indicate the upthrown and downthrown sides of each fault. Determine the throw on each fault. Make a structure contour map of the faults.

8.8.8
Cutoff Map of Normal Fault

Construct a fault cutoff map for the northern fault on the structure contour map in Fig. 8.56. Project the cutoff lines to the blank profile below the map.

8.8.9
Cutoff Map of Reverse Fault

Construct a fault cutoff map for fault A across the structure contour map in Fig. 8.57.

8.8.10
Fluid Migration across a Fault

Suppose a toxic liquid that is heavier than water is spilled onto the surface in the center of the structure illustrated by the Allan diagram in Fig. 8.58. Where will the liquid go? Will the liquid be trapped at a location on the cross section? If liquid is trapped, will it all be in the same location?

8.8.11
Thrust-Faulted Fold

Based on the map in Fig. 3.29, answer the following questions. What is the 3-point dip of the fault at the surface? Construct a structure contour map of the fault from its surface dip. Intersect the previously-constructed structure contour map of the top Fairholme (Exercise 7.7.4) with the map of the fault. Does the projected structure contour map agree with the drilled depths to the top of the Fairholme?

8.8.12
Relay Zone

Map the faults and the top of the Northriver Sandstone on the map of Fig. 8.59. Where is the relay zone? What is the attitude of the sandstone away from the faults? What is the attitude of the sandstone between the faults? What are the attitudes of the faults? What is the maximum throw and heave on the faults?

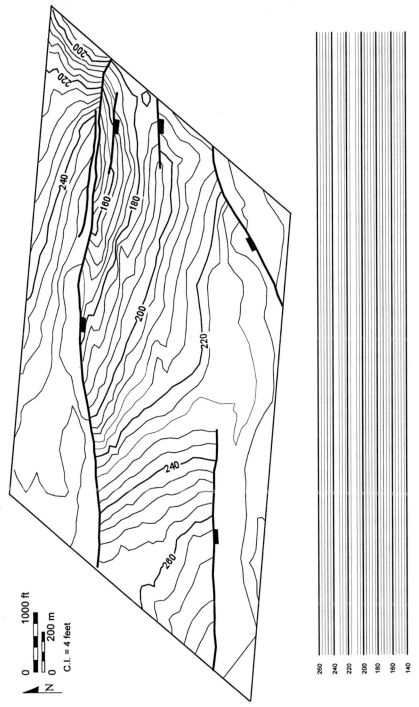

Fig. 8.56. Structure contour map with faults and blank cross-section template

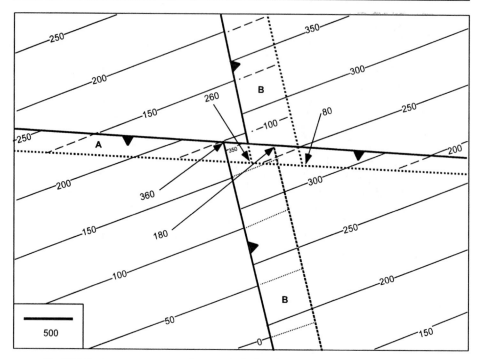

Fig. 8.57. Structure contour map of reverse-faulted marker horizon (*thin lines*). Faults are *thick lines*, hidden contours are *dashed*

spill

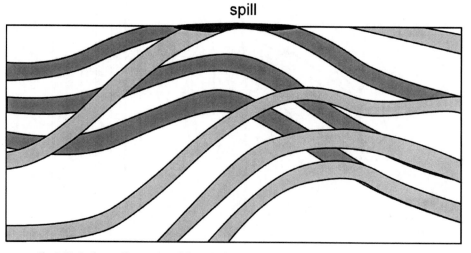

Fig. 8.58. Fault cutoff map, viewed from the hangingwall toward the footwall. The footwall beds have *darker shading* than the hangingwall beds. *Shaded* units are porous and permeable; *unshaded* units are impermeable

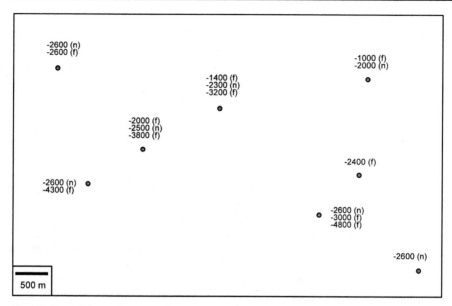

Fig. 8.59. Top of the Northriver Sandstone (*n*) and fault-cut elevations (*f*) in wells. Elevations are in meters, negative below sea level

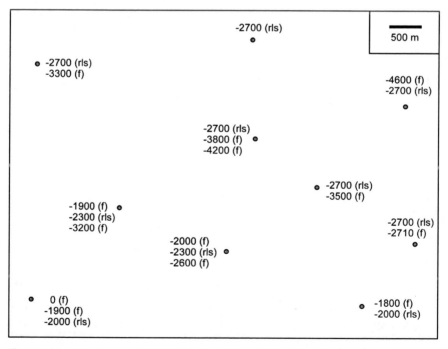

Fig. 8.60. Map giving the top of the Reef Limestone (*rls*) and faults (*f*) in wells. Elevations are in meters, negative below sea level

8.8.13
Branching Fault

Map the faults and the top of the Reef Limestone on the map of Fig. 8.60. Where is the branch line? What is the attitude of the limestone away from the faults? What is the attitude of the limestone between the faults? What are the attitudes of the faults?

8.8.14
Splay Faults

The water-well map of Fig. 8.61 shows a distinctive clay seam to be absent in some wells due to faulting. Map the faults and the top of the clay seam. Where is the branch line? What are the attitudes of the faults? What is the maximum throw and heave on the clay seam? If the clay seam is a barrier to ground water flow from the surface, where is this barrier absent? Is a spill of toxic heavy liquid in the southwest corner of the map area likely to sink below the clay seam? Why or why not? In which direction will a spill of heavy liquid in the southeast corner of the map migrate?

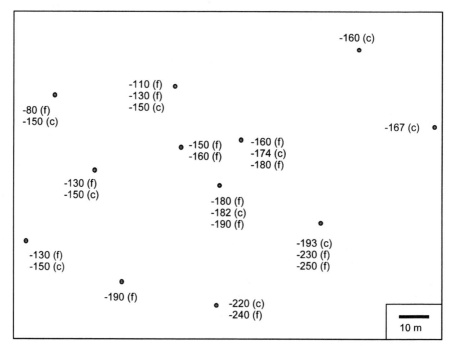

Fig. 8.61. Map of the top of a clay seam in water wells drilled into an alluvial aquifer. Elevations below sea level are negative

8.8.15
Sequential Faults 1

Two different fault trends occur in the area of Fig. 8.62. Map the faults and the A sand. What is the reason for the hydrocarbon trap in the A sand? What are the attitudes of the faults? What is the throw and heave on each fault? Which fault is older? If the hydrocarbons migrated before the formation of the younger fault, would the trapping potential of the structures be the same?

8.8.16
Sequential Faults 2

Contour the Northport Dolomite in the map of Fig. 8.63, being careful to explain the fault cuts and the oil trap(s). Is there one oil field or two? What are the attitudes of the faults? What is the throw and heave on each fault? Which fault is older? If the hydrocarbons migrated before the formation of the younger fault, would the trapping potential of the structures be the same? Are there any additional hydrocarbon prospects?

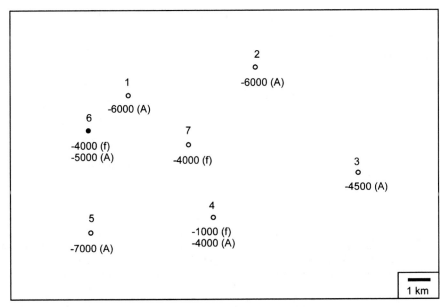

Fig. 8.62. Map of the top of the A sand (A) and faults (f) in wells drilled for oil. The *solid circle* is an oil well, *open circles* are dry holes. Everywhere away from the faults clear bedding dips are recorded on the dipmeter; they are about 27, 334. Close to the fault in well 4 the bedding dip is at azimuth 062. In well 6 the bedding dip close to the fault is at azimuth 189. In well 7 the dips of bedding are in all directions near the fault. Elevations are in meters, negative below sea level

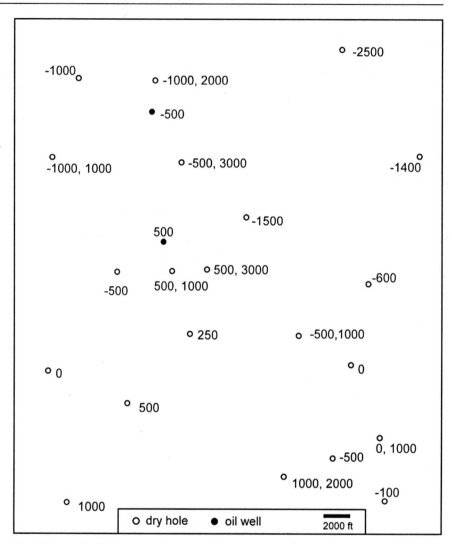

Fig. 8.63. Map of well information from the Northport Dolomite. *Two numbers* together next to a well give fault cut information: 500, 100: depth of fault cut, amount of fault cut. A *single number* by the well is the top of the dolomite. Where only fault-cut information is given, the dolomite is faulted out. The fault trends are generally northwest–southeast. Elevations are in feet, negative below sea level

Chapter 9

Dip-Sequence Analysis

9.1
Introduction

The three-dimensional geometry of a structure can be determined from the bedding attitudes measured in a single well bore or on a traverse through a structure. The method of dip sequence analysis presented here was developed for the structural analysis of dipmeter logs by Bengtson (1981a) but is equally informative whether the traverse is down a well or along a stream. Major problems with the structural interpretation of dip data are the high stratigraphic noise content and the complexity of the structures to be interpreted. Dip sequence analysis techniques, called Statistical Curvature Analysis Techniques (SCAT or SCAT analysis) by Bengtson (1981a), are particularly good for extracting the structural signal from the noise. Using SCAT it is possible to determine the plunge of folds, the locations of fold axial surfaces, crests, and troughs, to infer the strike and dip directions of faults, and to separate regional fold trends from local fault trends. The power of the technique derives from (1) the noise-reduction strategy of examining the data as dip components in both the strike and dip directions of folding and (2) providing models for the SCAT responses of the geometry to be interpreted.

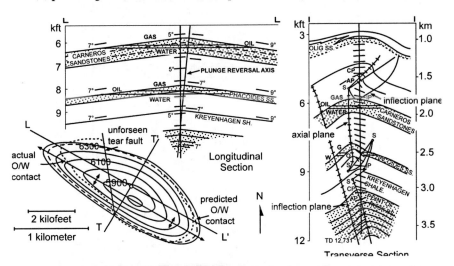

Fig. 9.1. Railroad Gap Field, California, predicted longitudinal and transverse cross sections and structure contour map on the top Carneros sandstone, based on the SCAT analysis of a single well at the crest of the anticline. *O/W*: oil-water contact. (After Bengtson 1981a)

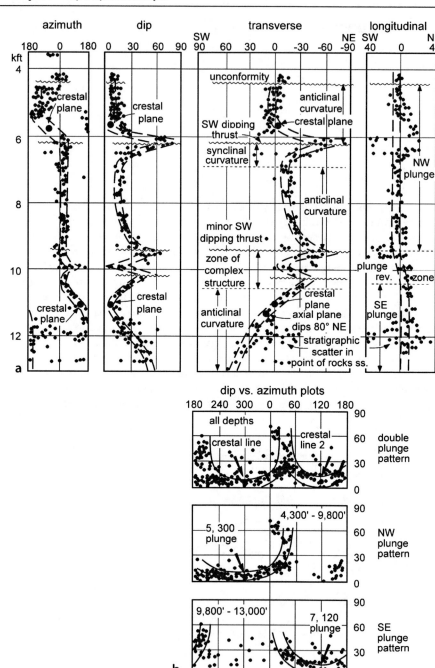

Fig. 9.2. SCAT plots for the discovery well of Railroad Gap Field, California. For the map and cross sections see Fig. 9.1. **a** Dip component vs. depth plots. **b** Dip vs. azimuth plots. (After Bengtson 1981a)

The potential of the method is indicated by the interpretation of the Railroad Gap oil field on the basis of the SCAT analysis of a single, favorably located well (Figs. 9.1, 9.2). SCAT analysis (Bengtson 1981a) was used to predict the structure on perpendicular cross sections from which the map was generated. The map view shows the close correspondence between the predicted and observed oil-water contact.

9.2
Curvature Models

Figure 9.3 illustrates the basic curvature geometries. The first step in the analysis is to differentiate a monoclinal dip sequence from a fold. This is accomplished with an azimuth histogram and/or with a tangent diagram. An azimuth histogram is a plot of the azimuth of the dip versus the amount of the dip (Fig. 9.4). The natural variation of

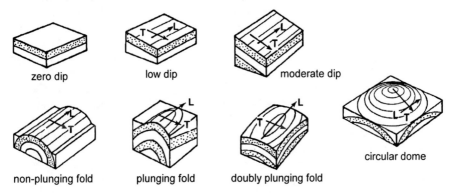

Fig. 9.3. Models of structural curvature geometries. *L:* Longitudinal; *T:* transverse. (Bengtson 1981a)

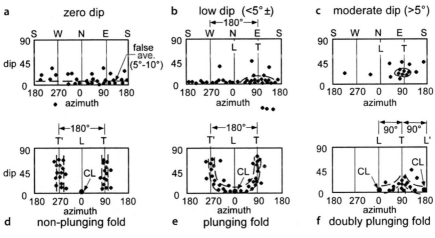

Fig. 9.4. Dip vs. azimuth patterns corresponding to the models of Fig. 9.3. *CL:* crestal line (after Bengtson 1981a). **a** Zero dip. **b** Low dip. **c** Moderate dip. **d** Non-plunging fold. **e** Plunging fold. **f** Doubly plunging fold

dips around a monoclinal dip gives a horizontal distribution of noise. Thus a zero true dip gives a small false positive average on the azimuth histogram because all dips are recorded as positive (Fig. 9.4a). As the dip increases the dips form point concentrations that become better defined as the true dip increases (Fig. 9.4b,c). Non-plunging and uniformly plunging folds give vertical concentrations of points corresponding to the limbs (Fig. 9.4d,e) and a doubly plunging fold produces an arrow-head-shaped distribution of points (Fig. 9.4f). On a tangent diagram a monocline plots as a point concentration of dips, and a fold (Figs. 5.3, 5.5) will produce a linear or curvilinear concentration of points.

9.3
Dip Components

A key step in a SCAT analysis is to determine the dip components in the transverse direction (T = transverse = regional dip) and the longitudinal direction (L = longitudinal = regional strike) which is at right angles to it. These dip components represent the dips on vertical cross sections in the T and L directions and are used to produce the SCAT histograms (Fig. 9.2) and cross sections (Fig. 9.1) in the T and L directions. The T and L directions are found from the dip vs. azimuth histogram (Fig. 9.4) or from the plot of bedding dips on a tangent diagram (Fig. 9.5). For monoclinal dip, the center of the point concentration on the dip-azimuth histogram is the T direction and the L direction is 90° away from it (Fig. 9.4b,c). For a fold, the center of the limb concentrations on an azimuth histogram is the T direction and the midpoint between the concentrations is the L direction (Fig. 9.4d–f). On a tangent diagram (Fig. 9.5), the orientation of the crest (or trough) line is the L direction and the T direction is at right angles to the crest (or trough) line. Both the T and L lines go through the center of the tangent diagram regardless of the fold plunge.

The dip components can be found either graphically or analytically. In the graphical method, the T and L lines are drawn on a tangent diagram. The T and L components are the projections of the dip vectors onto the T and L axes (Fig. 9.6). The dip com-

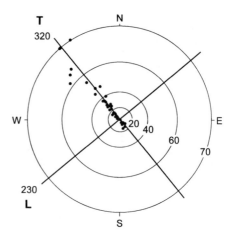

Fig. 9.5.
Determination of T and L directions on the tangent diagram of a fold. *Solid dots* represent dip vectors of bedding

Fig. 9.6.
Dip components in T and L directions. Here the T direction is NE–SW and the L direction is NW–SE. Bed attitude is 55, 082. The dip components are the lengths found by orthogonal projection of the dip vector onto the T and L lines. The T component is 50°NE and the L component is 40°SE

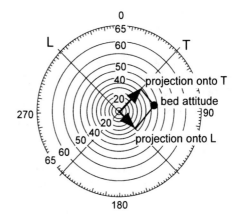

Fig. 9.7.
Geometry of the T and L components. α: angle between dip vector and T direction; θ_T: azimuth of T direction; θ: azimuth of dip vector; T_c: T component; L_c: L component

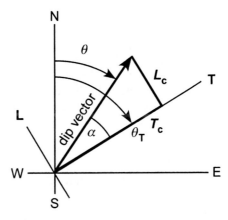

ponents are themselves vectors and have both magnitude and direction. The quadrant of the component, as well as its magnitude, must be recorded.

The dip components can easily be found analytically. Based on the geometry of Fig. 9.7,

$$\alpha = \theta_T - \theta \ , \tag{9.1}$$

$$T_c = \delta \cos \alpha \ , \tag{9.2}$$

$$L_c = \delta \sin \alpha \ , \tag{9.3}$$

where $T_c = T$ component, $L_c = L$ component, $\alpha =$ angle between dip vector and T direction, $\theta_T =$ azimuth of T direction, $\theta =$ azimuth of dip vector, $\delta =$ dip. Computer programs for the preparation of SCAT diagrams have been published by Elphick (1988). SCAT analysis can be performed entirely on a spreadsheet. Plot the tangent diagram as described in Sect. 2.8, use Eqs. 9.1–9.3 to find the T and L components, and plot the dip-component diagrams as xy graphs.

9.4
Analysis of Uniform Dip

The dip component diagrams are the primary noise reduction strategy in SCAT analysis. For zero dip the azimuth of the dip is random (actually stratigraphic scatter) and the dip amount shows a false positive average (Fig. 9.8). As the amount of homoclinal dip increases (Figs. 9.9, 9.10) the concentration of points becomes sharper. The component plots for zero dip show the correct zero average (Fig. 9.8). Low and moderate planar dips (Figs. 9.9, 9.10) show their true dip values on the transverse component plots because these are in the dip direction. The longitudinal dip components average zero because they are in the strike direction. The zero L component average (Figs. 9.9, 9.10) confirms the choice of the L and T directions.

9.5
Analysis of Folds

Folds produce distinctive curves on the dip vs. depth plots. The azimuth vs. depth plot shows the reversal of azimuth at the crest of the fold, CP (Figs. 9.11–9.13). The dip component plots are the most informative. The transverse component plots all cross the zero dip line at the crest of the anticline (CP), show an inflection point at the axial plane (AP) and show a dip maximum at the inflection plane (IP) that separates anticlinal curvature from synclinal curvature. Any variations in plunge are apparent on the longitudinal component plot. The non-plunging fold (Fig. 9.11) is defined by a straight line on the plot of

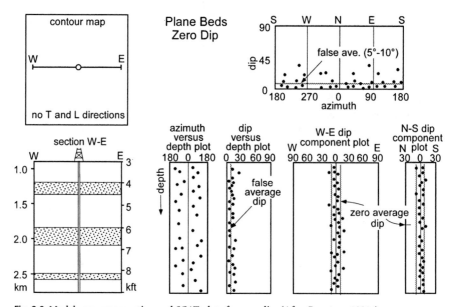

Fig. 9.8. Model map, cross section and SCAT plots for zero dip. (After Bengtson 1981a)

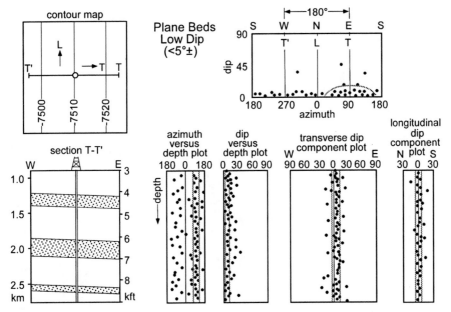

Fig. 9.9. Model map, cross section and SCAT plots for low monoclinal dip. (After Bengtson 1981a)

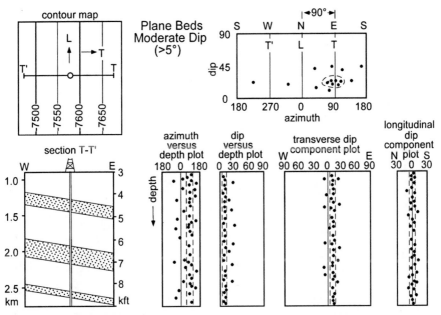

Fig. 9.10. Model map, cross section and SCAT plots for moderate to steep monoclinal dip. (After Bengtson 1981a)

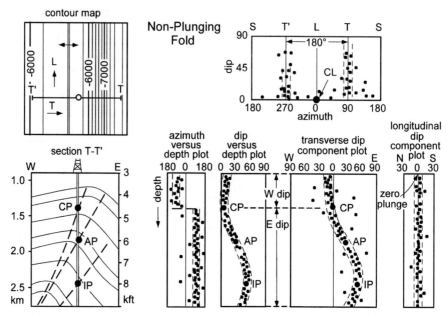

Fig. 9.11. Model map, cross section and SCAT plots for a non-plunging fold. *AP:* axial plane; *CL:* crestal line; *CP:* crestal plane; *IP:* inflection plane. (After Bengtson 1981a)

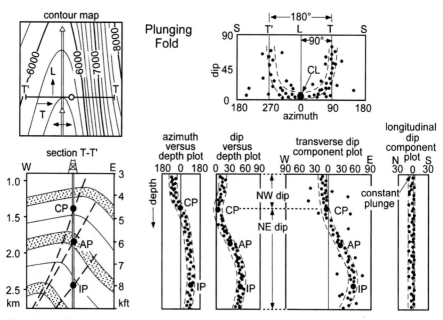

Fig. 9.12. Model map, cross section and SCAT plots for a plunging fold. *AP:* axial plane; *CL:* crestal line; *CP:* crestal plane; *IP:* inflection plane. (After Bengtson 1981a)

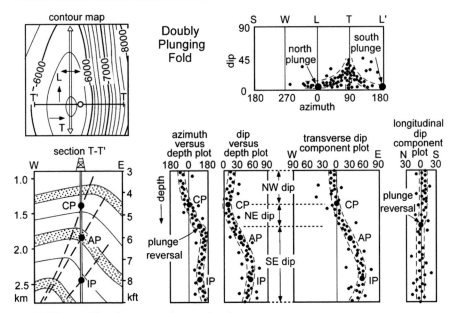

Fig. 9.13. Model map, cross section and SCAT plots for a doubly plunging fold. *AP:* axial plane; *CP:* crestal plane; *IP:* inflection plane. (After Bengtson 1981a)

L dip with depth that gives the average plunge of zero. A uniformly plunging fold (Fig. 9.12) plots as a line of constant plunge with depth. A doubly plunging fold (Fig. 9.13) shows the plunge reversal with depth on the *L* component plot.

Dip-sequence analysis can be performed on a traverse in any direction through a structure. As an example, the method is applied to a horizontal traverse across the map of the Sequatchie anticline originally presented in Fig. 2.4. The traverse (Fig. 9.14) runs from northwest to southeast at right angles to the fold axis along a stream valley that provides the best exposure and therefore the most data. The traverse is broken into three straight-line segments at the dashed lines in order to follow the valley. The attitudes of bedding are located on the SCAT diagrams according to their distance from the northwest end of the traverse. The numerical values are given in Table 9.1.

The *T* and *L* directions are determined from the tangent diagram and the dip-azimuth diagram. The linear trend of dips on the tangent diagram (Fig. 9.15a) is the trend of *T*, and *L* is at right angles to it. On the dip-azimuth diagram (Fig. 9.15b) the two vertical lines of points indicate, by comparison to Fig. 9.4, a non-plunging fold with a crest that trends 230. The dip-azimuth diagram should always be checked against the tangent diagram before finally deciding on the plunge direction and amount. Here the trend of the crest and the lack of significant plunge agrees with the tangent diagram. The direction of the crest line, here equal to the fold axis direction, is the *L* direction to be used in the next stage of the analysis. The *T* direction is at 90° to *L*, parallel to the azimuth of the limb dip.

The SCAT diagrams reveal the details of the structure. The bedding azimuths and dip components are plotted in Fig. 9.16. The azimuth and dip diagrams (Fig. 9.16a,b) show the locations of the crestal plane, axial plane, and inflection plane (compare

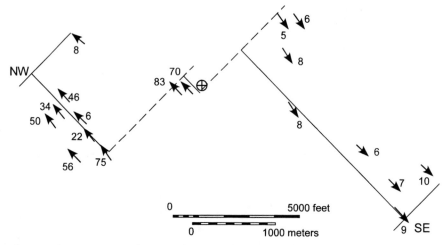

Fig. 9.14. Dip traverse across the Sequatchie anticline in the Blount Springs area, showing locations of bedding attitude measurements. *Dashed lines* are offsets in the line of traverse

Table 9.1.
Dip traverse across Sequatchie anticline

Distance from NW end of traverse (ft)	Dip, azimuth	T component (from 320/140)	L component (from 230/50)
256	8, 308	8 NW	1 SW
1384	46, 315	45 NW	7 SW
1640	34, 316	34 NW	2 SW
1660	50, 320	50 NW	0
2328	6, 320	6 NW	0
3143	22, 316	22 NW	1 SW
3261	56, 318	56 NW	1 SW
4096	75, 330	75 NW	25 SW
4253	Break		
4253	83, 315	83 NW	~45 SW
4528	70, 315	70 NW	13 SW
4891	0, 000	0	0
5147	Break		
5323	5, 145	5 SE	1 SW
5815	6, 144	6 SE	1 SW
6404	8, 145	8 SE	1 SW
8005	8, 144	8 SE	1 SW
10942	6, 127	6 SE	1 NE
12789	7, 136	7 SE	1 NE
13466	10, 136	10 SE	1 NE
13692	9, 136	9 SE	1 NE

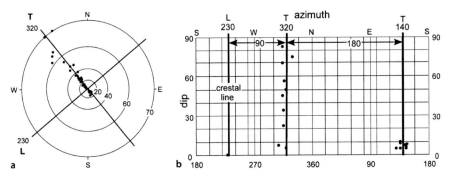

Fig. 9.15. Finding the *T* and *L* directions for the traverse across the Sequatchie anticline. **a** Tangent diagram. **b** Azimuth-dip diagram. *T* transverse dip direction; *L* longitudinal dip direction; crest line is at 0, 230

Fig. 9.16.
SCAT analysis of the Sequatchie anticline. **a** Azimuth-distance diagram. **b** Dip-distance diagram. **c** *T* component dip-distance diagram. **d** *L* component dip-distance diagram. *AP*: axial plane; *CP*: crestal plane; *IP*: inflection plane

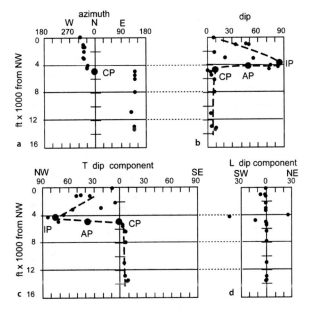

Fig. 9.16 with 9.11). The locations of the crestal plane and inflection plane are well defined in Fig. 9.16b,c and the axial plane falls between the two. Note that in the dip-depth (distance) plot (Fig. 9.16b) all dips plot to the right, whereas in the *T*-component plot the dips are plotted by their quadrant direction. The dip data for the northwest limb is noisy, even on the *T*-component plot, although the signal remains clear. Most of the dips on the *L*-component diagram (Fig. 9.16d) are zero or close to zero, confirming the choice of the plunge direction and the interpretation that the plunge is zero. Significant plunge aberrations occur between the inflection plane and the crest plane which is the location of the steep limb of the structure. This suggests that the structure of the steep limb is complex, perhaps containing obliquely plunging minor folds, not just a simple monoclinal dip or curvature around a single axis.

9.6
Analysis of Faults

The drag geometry (Sect. 7.2.7) provides the basis for fault recognition by SCAT analysis. The presence of a fault is recognized from the distinctive cusp pattern on the transverse dip component plot (Figs. 9.17–9.20). A cusp is also present on the dip vs. depth plot but may not be as clearly formed. The cusp is caused by the dips in the drag fold adjacent to the fault and is expected to occur within a distance of meters to tens of meters from the fault cut. A traverse perpendicular to the fault plane will show the minimum affected width, whereas a traverse at a low angle to the fault plane, such as a vertical well drilled through a normal fault, will show the maximum width. The fault cut is at the depth indicated by the point of the cusp. The azimuth vs. depth plots distinguish between steepening drag that occurs where the faults dip in the direction of the regional dip of bedding (Fig. 9.17) and flattening drag that occurs where the fault dip is opposite to the regional dip of bedding (Fig. 9.18). Steepening drag maintains a constant dip direction whereas flattening drag may produce a reversal in the dip direction. A drag-fold axis that is oblique to the regional fold axis produces multiple fold axes on the dip-azimuth diagram (Figs. 9.19, 9.20). Both the regional dip and the drag-fold axis appear on the dip-azimuth diagram and the tangent diagram, allowing both directions to be determined.

If either the dip direction of the fault or its sense of slip is known, the other property of the fault can be determined from the direction the cusp points on the T-component

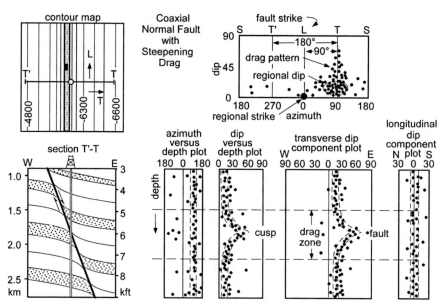

Fig. 9.17. Structure contour map, cross section, and SCAT plots for a normal fault with drag that steepens the regional dip. Fault strike is parallel to the regional strike. *L*: regional strike; *T*: regional downdip direction; *T'*: regional up-dip direction. (After Bengtson 1981a)

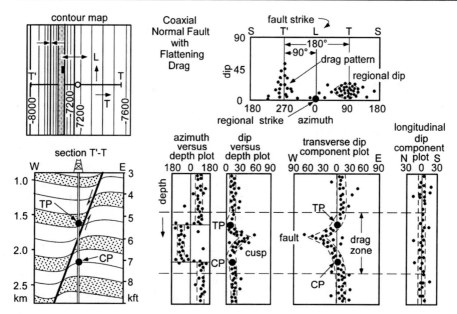

Fig. 9.18. Structure contour map, cross section, and SCAT plots for a normal fault whose drag flattens the regional dip. Fault strike is parallel to regional strike. *L:* regional strike; *T:* regional down-dip direction; *T':* regional up-dip direction; *CP:* crestal plane; *TP:* trough plane. (After Bengtson 1981a)

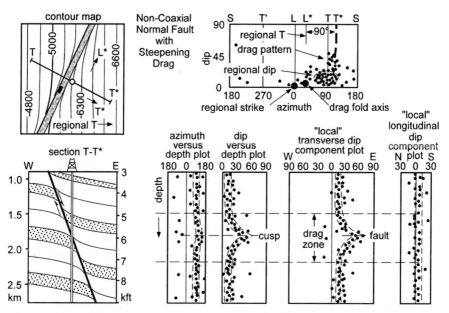

Fig. 9.19. Structure contour map, cross section, and SCAT plots for a normal fault striking oblique to regional dip with drag that steepens the regional dip. *L:* regional strike; *T:* regional down-dip direction; *T':* regional up-dip direction; *L*:* fault strike; *T*:* normal to fault strike. (After Bengtson 1981a)

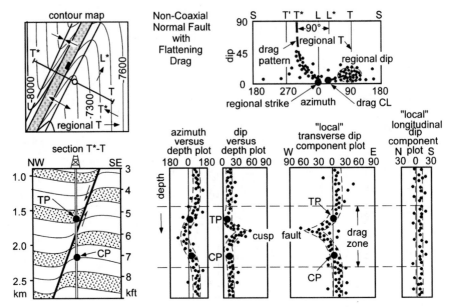

Fig. 9.20. Structure contour map, cross section, and SCAT plots for a normal fault striking oblique to regional dip with drag that flattens the regional dip. *L*: regional strike; *T*: regional down-dip direction; *T'*: regional up-dip direction; *L**: fault strike; *T**: normal to fault strike; *CP*: crestal plane; *TP*: trough plane. (After Bengtson 1981a)

diagram. For a normal fault, the cusp points in the direction of the fault dip. For a reverse fault, the cusp points opposite to the direction of the fault dip. Note that the cusp on the dip vs. depth plot always points in the same direction because the dips are not plotted according to direction.

A drag fold may be present on one side of the fault but absent on the other, resulting in a half-cusp pattern. As indicated by Fig. 9.21a, this geometry may be present at the map scale as well as at the drag-fold scale. Folds of this type produce a half-cusp pattern on the transverse dip component plot (Fig. 9.21b). In association with a reverse-fault, the dips may increase to vertical and then become overturned. On the *T*-component plot (Fig. 9.21b), the half cusp curves smoothly to the left to a 90° dip, then reappears where dips are 90° to the right. The isolated group of dips near 90° on the right represents overturned beds, providing a method for recognizing overturning from the dip sequence alone.

A synthetic example of a dipmeter run across a normal fault will serve to illustrate the method. The example also illustrates SCAT analysis using a spreadsheet. The traditional paper-copy dipmeter (Fig. 9.22) is a graph of dip versus depth in a well. The "tadpole" heads indicate the amount of dip and the tails the direction of dip. Solid heads represent the best data and open heads the worst. In the case of a four-armed dipmeter a solid head represents a dip based on correlation of all four arms and an open head means three of the four arms can be correlated. If only two arms can be correlated, the dip cannot be calculated and no point is plotted. The numerical data set is given in Table 9.2.

Fig. 9.21.
Drag geometry on a reverse fault showing a half-cusp transverse dip component plot and overturned beds. (After Bengtson 1981a)

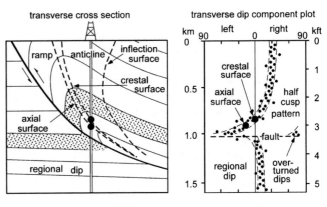

Fig. 9.22.
Synthetic dipmeter log representing a well containing a fault cut. Reference level for well is ground surface at 350 ft elevation. Quality ranking of data: *solid head:* best, *open head:* lower

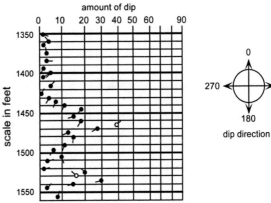

Table 9.2.
Attitudes from dipmeter log in Fig. 9.22

Depth	Dip	Azimuth	Depth	Dip	Azimuth
−1350	2	115	−1465	39	46
−1360	4	320	−1470	27	242
−1365	2	41	−1475	13	250
−1375	3	350	−1480	15	190
−1385	3	91	−1490	11	204
−1395	2	206	−1495	6	219
−1400	5	227	−1505	10	155
−1405	2	66	−1510	4	72
−1415	6	45	−1520	2	79
−1425	1	29	−1525	20	322
−1430	5	213	−1530	16	314
−1435	7	200	−1535	30	259
−1440	11	212	−1540	15	260
−1445	18	223	−1545	4	42
−1455	15	258	−1555	8	24
−1460	18	209			

Interpretation begins with the azimuth-depth and dip-depth plots (Fig. 9.23). At shallower elevations in the well, the dip magnitude is consistently low and the azimuth highly variable. These are the characteristics expected for a low regional dip (c.f., Fig. 9.9). The next deeper interval appears to be a cusp on the dip-depth diagram. In the context of a cusp, the azimuth-depth diagram suggests flattening drag (c.f., Fig. 9.18). The lower portion of the well contains no clear structural pattern and may represent stratigraphic noise, perhaps a unit containing disparate dips like a conglomerate (where the pebble boundaries would produce dip readings) or a reef (where individual corals might be producing the dips).

Having isolated the cusp as an interval of interest, further analysis will be performed on that part of the well log alone. The T and L directions are most clearly found on the tangent diagram (Fig. 9.24a). There is a substantial amount of scatter but the least squares best-fit line does a good job of locating the T direction. Where it can be checked against other geological data, the best fit line has proved to be remarkably reliable, even where the scatter is large. If a quadratic best-fit line approximates a hyperbola and fits the data reasonably well, then the fold is probably conical. Where a quadratic best-fit is not hyperbolic, the best fit is linear and the fold is cylindrical. The dip-azimuth diagram (Fig. 9.24b) shows substantial scatter but can be interpreted with reference to Fig. 9.18 as showing a regional dip component and a drag-fold component. The results of this stage of the analysis give $T = 052$ and $L = 322$. Additional valuable information is the strike of the fault, which must be approximately parallel to the L direction, 322°.

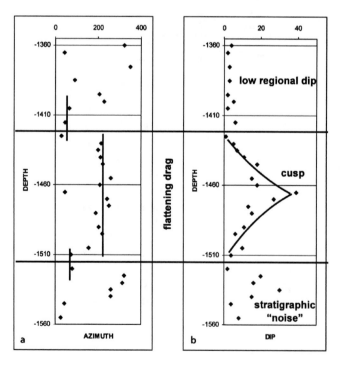

Fig. 9.23.
Azimuth-depth and dip-depth diagrams for data in Table 9.2. **a** Azimuth versus depth. **b** Dip versus depth

Fig. 9.24.
Determination of *T* and *L* directions for data in Table 9.2.
a Tangent diagram. **b** Dip-azimuth diagram

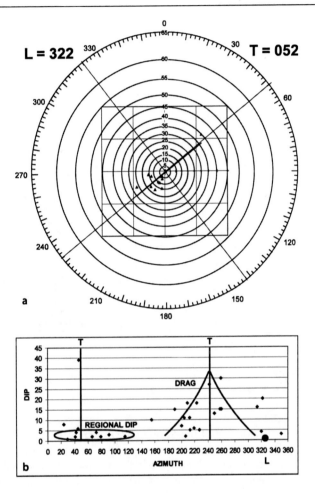

The final part of the interpretation is based on the *T* and *L* component plots. A normal fault dips in the direction the cusp points on a *T* component plot, which is to the southwest for this example (Fig. 9.25a). At this stage, the effect of the choice of the *T* and *L* directions on the component plots should be examined. Vary their directions and watch for the effect on the *L*-component plot. The best result is one which shows the points falling the closest to the zero line, indicating that the *L* direction has been correctly chosen. The result in Fig. 9.25b is the best that can be obtained from this data set. Note that the data point that was at the tip of the cusp on the dip-depth plot (Fig. 9.23b) lies on the wrong (NE) side of the *T*-component plot (Fig. 9.25a). Re-examination of the original data (Fig. 9.22) shows this to be a point with poor data quality. It might represent a dip on a fracture or be a spurious result on fractured rock in the fault zone. No data at all might be expected from a fault zone in which the bedding has been highly disrupted by the deformation.

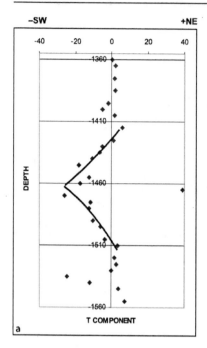

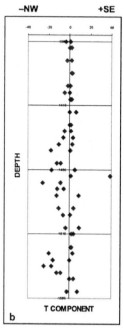

Fig. 9.25.
T and L component plots for data in Table 9.2, given $T = 052$ and $L = 322$. Plus and minus values assigned to the compass directions are for spreadsheet plotting purposes. **a** T component versus depth. **b** L component versus depth

9.7
Exercises

9.7.1
SCAT Analysis of the Sequatchie Anticline

Use the data in Table 9.1 to perform a complete SCAT analysis on the dip traverse across the Sequatchie anticline. Plot the azimuth-distance and the dip-distance diagrams. What are the T and L directions? What are the dip components in the T and L directions? Plot them on the dip-component diagrams.

9.7.2
SCAT Analysis of Bald Hill Structure

Use the data in Table 9.3 to perform a complete SCAT analysis on the dip traverse across the Bald Hill structure to see if a fault is present and its location and orientation, given that the faults in the area are reverse.

9.7.3
SCAT Analysis of Greasy Cove Anticline

Perform a complete SCAT analysis on the Greasy Cove anticline (Table 9.4). Consider both fold and fault geometry. The anticline is part of the southern Appalachian fold-thrust belt.

Table 9.3.
Bald Hill bedding attitudes

Distance from the northwest (km)	Attitude Dip, azimuth	T component	L component
0.54	60, 310		
0.70	90, 289		
0.90	55, 311		
1.10	40, 295		
1.30	14, 124		
2.38	12, 319		
2.68	26, 281		

Fig. 9.26.
Geologic map of the Bald Hill area. Topographic contours (in feet) are *thin lines*; geologic contacts are *wide gray lines*. Data have been projected parallel to strike onto NW-SE traverse line. (Modified from Burchard and Andrews 1947)

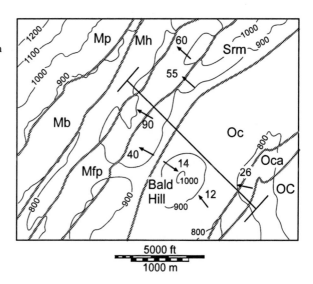

Table 9.4.
Southeastern Greasy Cove anticline bedding attitudes

Distance (ft)	Bedding attitude	T component	L component
0	10, 300		
200	35, 302		
500	43, 293		
600	85, 300		
1 200	90, 329		
1 500	70, 130		
1 600	45, 133		
2 000	70, 133		
2 100	30, 130		
2 200	50, 130		
3 500	60, 133		
4 000	40, 150		

Quality Control

10.1
Introduction

This chapter focuses on the quality control of completed interpretations as expressed in maps and cross sections. Quality control means locating and correcting errors in the data and in the interpretation. Problems can arise from data transcription errors, incomplete exposure in the field, interpolation uncertainties between wells and seismic profiles, and missing or misleading information in seismic interpretations. The quality-control issues discussed in this chapter can be broadly categorized as data errors and contouring artifacts, inconsistency of trends, bed thickness anomalies, and impossible fault shapes. Additional techniques for detecting and correcting errors involving the restoration and balancing of cross sections are covered in Chap. 11. Those topics are covered separately because they also include related methods for extracting additional geological information from the data and for making model-based predictions of the geometry. The single best quality control for an interpretation is to build an internally consistent, 3-D model of the entire structure. Many of the individual problems noted in the following sections would be obvious in 3-D. The following sections discuss methods that are commonly applied in 2-D, that is, to geological outcrop maps, structure contour maps, and to cross sections.

10.2
Data Errors and Contouring Artifacts

Before a map is finalized, it should always be examined for data errors, edge effects, and contouring artifacts. Some of these problems are nearly inevitable because of human error in data input and the intrinsic behavior of computer contouring algorithms.

10.2.1
Data Errors

Data errors are a likely possibility where single points fall far from the average surface. Problem points may be recognized by the presence of small closed highs or lows, usually defined by multiple contours that surround an individual point. Such errors commonly arise from mistakes in the interpretation of the unit boundaries or as transcription errors in transferring data to the map. This type of error tends to produce very local highs and lows on a preliminary structure contour map (Fig. 10.1a). Point 1 (Fig. 10.1a) is almost certainly a bad data point as it forms a small, deep depression in the map surface. Point 2 is a closure at an elevation consistent with other elevations on the map

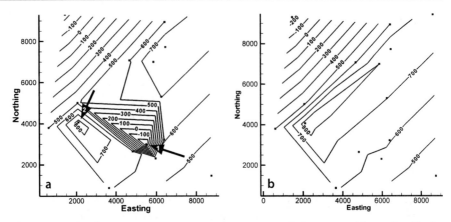

Fig. 10.1. Locating data errors on a structure contour map. **a** Preliminary map with questionable closures. **b** Map recontoured after removal of point 1

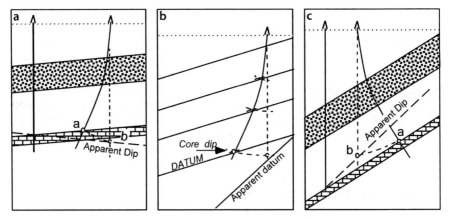

Fig. 10.2. False apparent dips caused by unrecognized well deviation. **a** Apparent dip determined between two wells, one unknowingly deviated down dip. **b** Dip determined from core dip in the unknowingly deviated well. **c** Apparent dip determined between two wells, one unknowingly deviated up dip. (After Low 1951)

and could be real or a bad data point. Removal of point 1 leads to a more structurally reasonable map (Fig. 10.1b) and shows point 2 to be consistent with the revised map. Maps produced by triangulation are the best for recognizing data errors because gridded maps smooth the values and thereby reduce the effect of anomalous values. It is always worth checking a point that produces a one-point closure. Local closures, if actually present, are important because, for example, they might form hydrocarbon traps.

Bad points can arise from a deviated well that is mistakenly interpreted as being vertical, causing both location and thickness anomalies. The log depth to a formation boundary is larger in a well that deviates down dip than in a vertical well at the same surface location (Fig. 10.2a). If the formation boundary is plotted vertically beneath the well location, its depth will be too great. If an apparent dip is then determined between this well and a correctly located formation top in another well, the apparent dip will be wrong, perhaps even in the wrong direction (Fig. 10.2a). If the dip is determined from a core in a deviated well (Fig. 10.2b) that is mistakenly thought to be vertical, then the inferred dip will be too large. The apparent thicknesses are too large in both situations. A well that deviates up dip will result in an apparent steepening of the dip between two wells and thicknesses (if mistakenly corrected for dip) that will be too small (Fig. 10.2c).

10.2.2
Edge Effects

Both humans and computer-contouring algorithms appear to prefer contours that close within the map area. In Fig. 10.3a, the triangulation algorithm has closed the contours on southwest end of the anticline. Based on the available control, open contours to the southwest (Fig. 10.3b) are equally valid and perhaps more geologically reasonable.

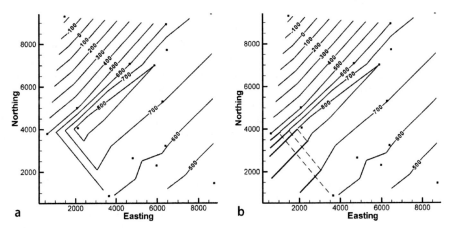

Fig. 10.3. Is the closure real? **a** Anticline closed to the southwest in triangulated map from data in Fig. 3.5. **b** The same structure opened to the southwest (*heavy lines*), deleted contours *dashed*

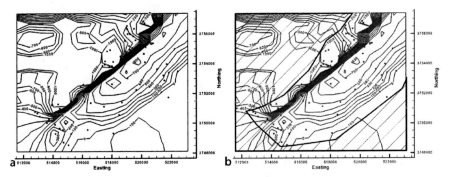

Fig. 10.4. Edge effects on a map of the Sequatchie anticline. Control points are *solid squares*. **a** Map from Fig. 3.23a. **b** Same map with region of no data *shaded* (diagonal lines)

Gridding algorithms typically extend the contours to the edges of the grid, regardless of whether or not data are present to support the extrapolation. Figure 10.4a is a map produced by a computer kriging algorithm over a rectangular region. The region without any data to control the contours is shaded in Fig. 10.4b. The closed highs and lows within the shaded area are completely spurious.

A TIN network may contain extremely elongated triangles along the edges of the data, implying a relationship between widely separated points. Such widely separated points are not necessarily geologically related and probably should not influence the shape of the surface between them. The TIN network should be examined for such problems and such long-distance connections might be removed. Some computer programs eliminate triangles for which one or two of the angles are smaller than a threshold value. This tends to reduce the problem but does not eliminate it.

10.2.3
Excessive Detail

Excessively wiggly contours or areas containing multiple, small, structural closures may be contouring artifacts (Krajeweski and Gibbs 1994). Closed contours that do not contain control points should be viewed with suspicion as artifacts of the gridding algorithm (Fig. 10.5). This type of artifact only occurs with gridding algorithms and is more likely when using high-order surfaces (e.g., kriging with quadratic drift and grid-node densities much greater than the control-point density).

Another type of artifact may arise if data from multiple sources, such as different seismic surveys, are contoured together, because their datums may be different. Each

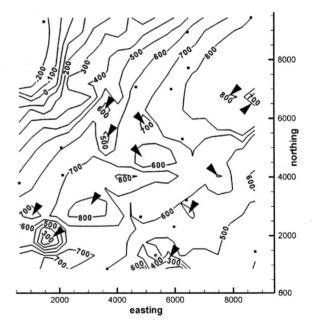

Fig. 10.5.
Structure contour map from
Fig. 3.11d. Control points are
small squares. *Triangles* point
to local closures unjustified by
control points

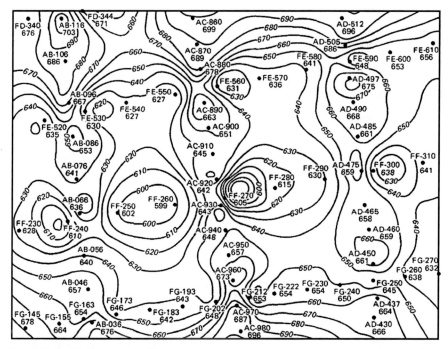

Fig. 10.6. Cloverleaf structure contour pattern produced by data along three north-south seismic lines mis-tied to three east-west lines. Control points are indicated by *dots*. Contours are concentrated along lines of data collection. (Jones et al. 1986)

data set can be internally consistent, but one may have elevations systematically shifted with respect to the other. Cloverleaf patterns of highs and lows (Fig. 10.6) characterize this type of problem (Jones et al. 1986; Jones and Krum 1992). Two-dimensional seismic lines in areas of dipping units have the additional problem that the reflections on lines parallel to strike may be shifted laterally from the surface location of the line, leading to a similar datum shift between lines that are at right angles to one another (Oliveros 1989). Mis-ties between reflectors and incorrect stacking velocities can lead to similar cloverleaf patterns.

10.3
Trend Incompatibilities

The basic principle of trend compatibility is that the trends obtained from different parts of the total data set, which may include structure contour maps, geological contact locations and bedding attitude measurements, must agree, or a reason for any discrepancies must be found.

Structure contours are a valuable aid in evaluating the validity of the contact locations on a geological map. To test the preliminary geologic map of Fig. 10.7a, find the orientations of the structure contours implied by both the top and bottom

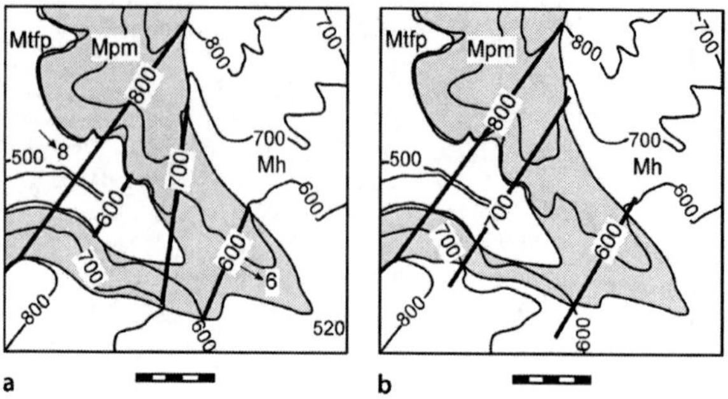

Fig. 10.7.
Geologic map. Units from youngest to oldest are Mh, Mpm, Mtfp. Topographic elevations are in feet and the scale bar is 1 000 ft. **a** Structure contours (*heavy lines*) determined from the intersection of mapped formation boundaries with the topographic contours. **b** Revised geologic and structure-contour map

Fig. 10.8.
Structure contours on the top and base of an outcropping unit on a topographic map. Structure contours drawn between closest corresponding elevations across the V formed by the outcrop trace. *Solid dots* are control points; *long dash contours* are on top of the bed; *dotted contours* are on the base of the bed. *Question marks* show where structure contours indicate that the bed surface should intersect the topographic surface but no intersection is observed

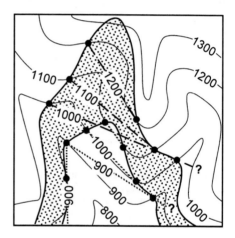

boundaries of the Mpm (Fig. 10.7a). The 800-ft and 600-ft contours are parallel, but the 700-ft contour is quite different, suggesting that the contact locations producing the 700-ft contour could be wrong. The map is improved by changing the least certain outcrop trace to make the 700-ft contour to be parallel to the others (Fig. 10.7b). The revised map becomes a new working hypothesis which should be field checked.

The consistency between a structure contour map and bedding attitudes from outcrop measurements or dipmeters provides a powerful test of the interpretation. The fold axis trend must be consistent with the structure contour trend. The potential problem is illustrated with the map in Fig. 10.8. If the formation boundary elevations are sampled at individual points around the structure, they will be contoured as shown (previously discussed in Sect. 5.5). A problem will be recognized only if compared to the trend of the fold axis, which trends north–south with zero plunge. The correct structure contours are shown in Fig. 10.9a. This interpretation would be clear with good exposure in the field (Fig. 10.9b), but in the subsurface could not be interpreted without knowledge of the fold axis.

A few locations with dip information can be used to test and control the structures. The information provided by a small number of formation tops alone can be contoured in numerous ways as evidenced by the two maps (Fig. 10.10a and 10.10b)

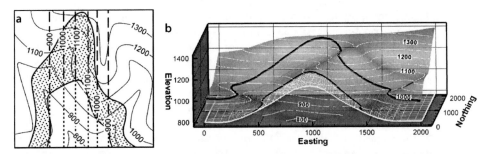

Fig. 10.9. Correct interpretation of the map in Fig. 10.8. **a** Structure contours on the base of the bed, based on knowledge that the fold axis trends north–south with zero plunge (after Weijermars 1997). *Long dashes* indicate contours below ground, *short dashes* indicate contours projected above ground. **b** 3-D view of fold. Trace of the upper boundary is shown on the topography. The lower boundary is rendered as an outcrop trace and as a 3-D mesh

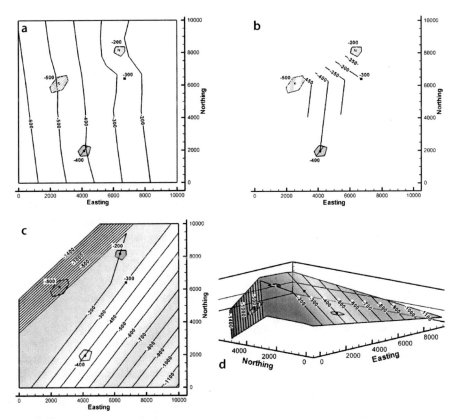

Fig. 10.10. Structure contour maps based on the four labeled points. Dips are known at the 3 points surrounded by *polygons*. **a** Kriged contouring based on elevations. **b** Triangulation contouring based on elevations. **c** Dip-domain map based on contour direction and spacing from 3 known bedding attitudes. **d** 3-D oblique view NE of **c**

derived by applying different computer contouring techniques to identical data. If dip information is available from even a few locations, the derived contour directions and spacings (Sect. 3.6.1) greatly constrain the interpretation. The map in Fig. 10.10c has been produced by extending the known dips in 3-D until they intersect to define boundaries with the adjacent domains, then contouring the resulting structure by triangulation. The map so constructed (Fig. 10.10c,d) agrees perfectly with the known dips.

Any conflict between the structure contours and the local bedding attitudes should be investigated. Local attitude measurements with a compass in outcrop or from a dipmeter in the subsurface can be accurate yet not reflect the map-scale structure. Small-scale bedding variations, cross bedding, or minor folding, can all lead to bedding attitudes that diverge from the trend that should appear on the structure contour map. The regional trend will be clearer, in spite of any small-scale variability of the bedding attitudes, if the fold axis is found using multiple attitude measurements (Sect. 5.2).

10.4
Bed Thickness Anomalies

Many quality-control measures are ultimately based on the concept that bed thicknesses should remain constant or be smoothly varying. This concept is applied as a quality control tool to the compatibility between multiple horizons on maps and cross sections, the internal consistency of composite-surface maps, and the geological likelihood of the growth history given by the expansion index.

10.4.1
Compatibility between Structure Contour Maps

The structure of nearby stratigraphic horizons is usually fairly similar. If unit thicknesses are approximately constant, the shapes of nearby horizons will be nearly the same. The compatibility of maps on closely spaced horizons is indicated by structure contours that are nearly parallel (Fontaine 1985) and separated by approximately constant distances (in gently dipping beds). A significant difference in trends on adjacent surfaces suggests a possible misinterpretation of one or both of the surfaces. The structure contours on different horizons cannot intersect without implying a structural or stratigraphic discontinuity. Only faults and unconformity surfaces can cut across stratigraphic boundaries.

Mapping different horizons independently can easily lead to incompatible surfaces. Figure 10.11 shows structure contour maps on two closely spaced horizons on the same part of the Sequatchie anticline. In this example more control data were generated for the lower surface (Fig. 10.11b) by means of vertical projection (Sect. 3.6.2). To determine if the maps are compatible, they are superimposed (Fig. 10.12). The superposed contours are approximately parallel and so are compatible by this criterion, except in the shaded region of second-order folding on the southeast limb of the major anticline (Fig. 10.12a). Disharmonic folding is implied by the folds confined to the Mtfp. This may be geologically reasonable if the mechanical stratigraphy is appropriate (Sect. 1.5.1) but should be checked.

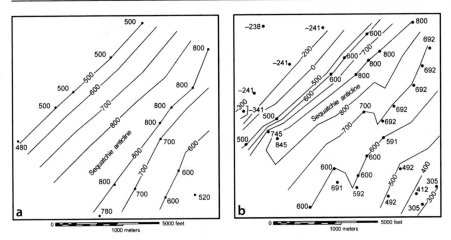

Fig. 10.11. Structure contours on adjacent horizons. Control points are labeled. Contour interval is 100 ft. **a** Top of the Mpm. **b** Structure contour map on the top of the Mtfp, 108 ft stratigraphically below the Mpm

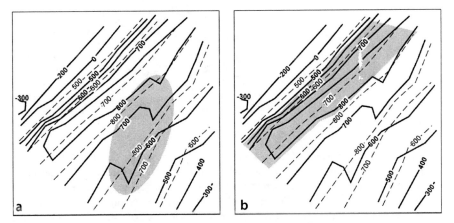

Fig. 10.12. Testing for compatible surfaces by superimposing structure contour maps. Structure contour map of the top Mpm (*dashed contours*) overlain on the composite map of the top Mtfp (*solid contours*). **a** *Shaded area* is a zone of contour shape incompatibility. **b** *Shaded area* shows approximate area where the stratigraphically higher surface (Mpm, *dashed lines*) lies below the stratigraphically lower surface (Mtfp, *solid lines*)

A comparison between elevations of the two maps in Fig. 10.11 reveals a fatal flaw in the interpretation: the upper stratigraphic surface lies below the lower stratigraphic surface in the shaded area of Fig. 10.12b. This indicates a glaring lack of compatibility between the two surfaces. Usually it will be the map with the lesser amount of control that should be corrected. A cross section would also reveal this problem and make it easy to visualize, one of the reasons that cross sections should always be part of the quality control process.

Fig. 10.13.
Compatibility of geological
contacts in outcrop. **a** Geo-
logical map of the southern
Sequatchie anticline at Blount
Springs, Alabama on a 30 m
DEM base from Fig. 2.4. Ob-
lique view to NE. Distance
between grid lines, 1 km.
Thick black lines are geologic
contacts. **b** Composite struc-
ture contour map of the top
Mtfp, oblique view to NE (from
Fig. 3.23b)

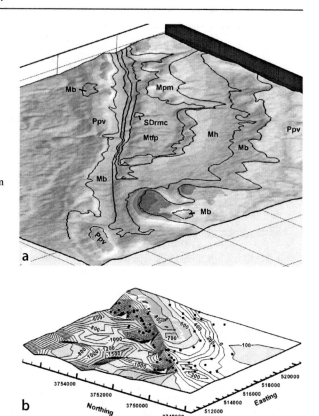

A composite surface map (Sect. 3.6.2) provides a method for the simultaneous testing
of a complete geological map for internal consistency. It is very difficult to determine by
inspection alone whether the outcrop traces on a geological map (e.g., Fig. 10.13a) are
consistent with one another. The composite surface map of a representative horizon provides such a test. If the composite-surface map shows a realistic structure, without unex-
plained peaks and valleys, as in Fig. 10.13b, then the combined data are internally consistent. Because outcrop trace locations, dips and thicknesses all work together to produce
the final map, all of this information is tested by the composite surface map.

10.4.2
Compatibility of Thicknesses on Cross Sections

A cross section is a powerful tool for checking and improving the structural interpretation presented on an outcrop map or on multiple structure contour maps. First be sure
the cross section matches the map from which it was constructed. In section view, poorly
controlled map horizons may be improved with information from other horizons
(c.f. Sect. 6.4). Errors in mapping on one horizon can be recognized as incompatibilities
with the geometry of horizons above or below. Many of the mapping pitfalls noted pre-

Fig. 10.14.
Preliminary map of top Mary
Lee structure contours in the
southeast Deerlick Creek coalbed methane field, Black
Warrior Basin, Alabama. Data
from Groshong et al. (2003b).
Control points are *squares*.
Easting and northing in
meters, elevations in ft. Trace
of cross section in Fig. 10.15
is shown

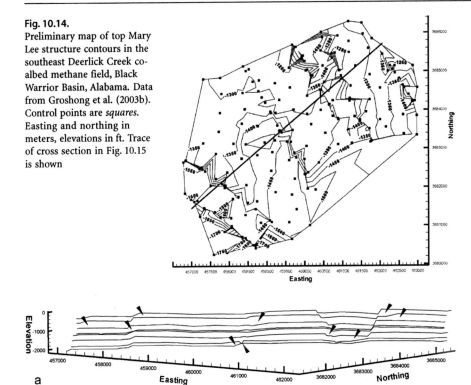

Fig. 10.15. Cross section of SE Deerlick Creek coalbed methane field. Horizontal view to NW. No vertical exaggeration. Easting and northing in meters, elevations in ft. **a** Preliminary interpretation of all
horizons. *Arrows* point to thickness anomalies. **b** After locating faults and re-mapping

viously and their correct interpretations are immediately obvious on a cross section. A
valid cross section must be compatible with the structure laterally adjacent to the section
as well. Accurate interpretation in three dimensions is demonstrated by means of maps
and cross sections that are each internally consistent and consistent with each other.

The preliminary structure contour map in Fig. 10.14 shows excessively wiggly contours and some trends that might represent faults. A cross section of the preliminary
contour maps on multiple horizons is constructed by slicing the maps. The section
(Fig. 10.15a) shows numerous thickness anomalies and discontinuities that can be resolved
by inserting faults (Fig. 10.15b). Once the faults are mapped and the marker surfaces re-
mapped (Sect. 8.3.2), the thickness anomalies disappear, and the new marker-surface map
is much simpler (Fig. 10.16).

Fig. 10.16.
Top Mary Lee structure contour
map (same data as Fig. 10.14)
after reinterpretation of all
map horizons to include faults.
Easting and northing in meters,
elevations in ft

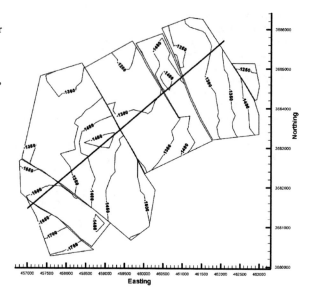

10.4.3
Realistic Growth History

The growth sediments, deposited during deformation, contain a record of the structural evolution. The implied evolution must be compatible with the regional structural style and evolution. Apparent inconsistencies in the indicated geological evolution may indicate errors in the interpretation. The expansion index plot across a fault (Sect. 7.7.4) is very helpful in recognizing and fixing this type of problem. The appropriate form of the plot is determined from the local structural style. For example, in an extensional region, a reasonable expectation is that the sense of displacement on normal faults does not reverse and, consequently, that the expansion index must always be greater than one. A value of less than one means that the downthrown side received less sediment than the upthrown side which, in turn, implies that the downthrown side was high (upthrown) during the interval having an $E < 1.0$. In this situation, an $E < 1.0$ could indicate a miscorrelation, either of the units or of the fault cuts. In Fig. 10.17, if the top three units are correlated, E is less than one in unit 2. A better interpretation is that the thin unit 2 on the hangingwall is part of a thicker interval that correlates with unit 2 on the footwall (interpretation 2). Another possibility is that the abnormally thin unit in well B on the downthrown side of a fault has been truncated by another fault (Fig. 10.18a). Inserting fault f_2 at the proper location (Fig. 10.18b) results in the correct form of the expansion-index diagram. The expansion index increases steadily downward for the reinterpreted fault blocks.

The growth history should agree with the stratigraphic separation on the fault. Within the growth interval on a normal fault, the separation is expected to increase down the dip. The throw across the fault can be plotted against depth (Fig. 10.18). A misinter-

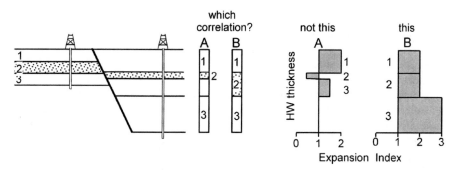

Fig. 10.17. Use of the expansion index to test unit correlations. An expansion index of less than one implies reverse fault movement or a miscorrelation across a normal fault

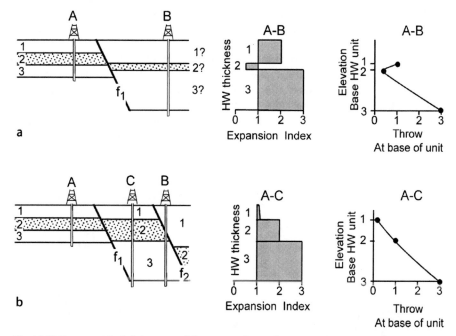

Fig. 10.18. Presence of a fault interpreted from anomalies in the expansion index and vertical separation. **a** A misinterpretation of the hangingwall of fault f_1 that gives an expansion index of less than one for unit 2 and a fault throw that decreases and then increases with depth. **b** The anomalous thinness of unit 2 in well B is explained by normal fault f_2. C is a well that would penetrate an unfaulted section in the hangingwall of the fault f_1

preted growth stratigraphy (Fig. 10.18a) shows the separation to decrease downward and then to increase again. The correct interpretation (Fig. 10.18b) shows a continuing increase of vertical separation downward.

10.5
Unlikely or Impossible Fault Geometries

It is surprisingly easy to create interpretations that include impossible or at least very unlikely 3-D fault geometries. Faults placed on single-horizon structure-contour maps or on geological outcrop maps without regard to the implied 3-D fault shape are the most likely to cause problems. The quality of a fault interpretation can be assessed using the implied fault shape and the cutoff-line geometry of hangingwall and footwall. A final quality-control issue is the distinction between a fault and an axial surface.

10.5.1
Fault Shape

Fault surfaces should always be mapped. The shape of an undeformed fault should be planar, smoothly curved, or have a stair-step shape. A very irregular fault (Fig. 10.19) probably cannot move and so is unlikely to be a correct interpretation unless the fault has been folded by a later event.

Structure contour maps of faulted horizons are commonly shown without contours on the faults. Contouring the implied fault surface, by matching elevations across the fault, is a very simple and effective test of the interpretation. Incorrect interpretations of the fault-bedding relationships will show unreasonable or impossible contour patterns for the fault surface. The implied fault surface in Fig. 10.20 has a spiral shape, possible perhaps, but not very likely. In more complex situations, crossing contours may occur on the implied fault, an even less likely result.

Implied structure contours may reveal problems in addition to unlikely fault shapes. In Fig. 10.21a the implied structures on the fault are parallel and so are reasonable. But once the fault contours are extended across the map (Fig. 10.21b) a problem becomes

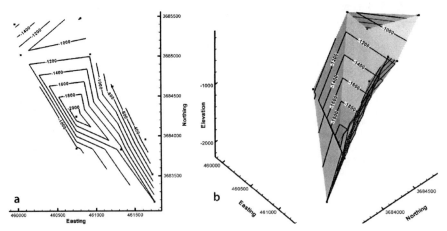

Fig. 10.19. Map of two faults from the Deerlick Creek area (Fig. 7.32) interpreted as being a single fault. *Squares* are the positions of fault cuts, *contours* are on the fault surface. **a** Structure contour map, synclinal shape is unreasonable. **b** 3-D view to the NW of the inferred synclinal fault

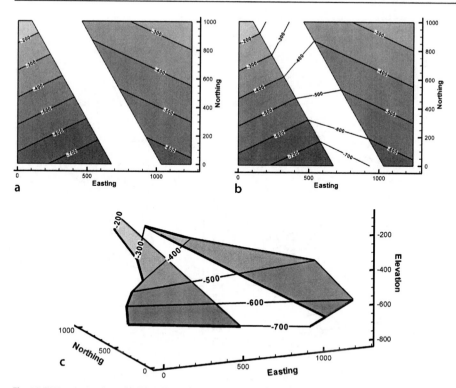

Fig. 10.20. Interpretation of unlikely fault geometry from implied structure contours. **a** Structure contour map of offset marker without contours on fault. **b** The implied fault contours found by connecting equal elevations of the hangingwall and footwall bed cutoffs. **c** 3-D view to NE showing the implied spiral fault shape

apparent. The map shows the marker bed to be above the fault on both sides of the fault gap. Normally an offset bed will occur on both sides of the fault. This is not required for a fault where erosion has removed beds from one side, but otherwise is anticipated. The 3-D view (Fig. 10.21c) makes it clear that both segments of the marker bed are in the hangingwall of the fault. Either the map in Fig. 10.21a is wrong or the fault has truncated a steeply plunging synform, the trough of which has yet to be located. Finding the synclinal hinge of the marker-bed in the footwall would validate the map.

10.5.2
Fault Separation

The stratigraphic separation from a fault cut in a well must agree with the heave and throw shown by the structure contour map of the faulted surface. Suppose the well that cuts the normal fault at the location shown in Fig. 10.22 has a stratigraphic separation of 75 m, is this consistent with the map? The map indicates a heave of 24.6 m and a throw of 64.5 m (Sect. 8.2). From Eqs. 8.1 or 8.2, the stratigraphic separation on the fault should be 47.6 m, a significant difference. The alternative strategy is to find

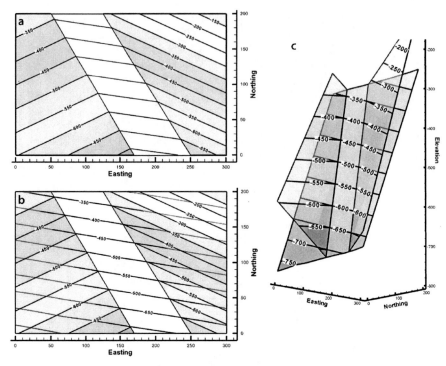

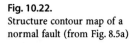

Fig. 10.21. Questionable relationship between marker bed and fault. **a** Structure contour map of marker bed with contours across fault gap. Fault resembles a simple normal fault with missing section. **b** Fault contours extended across the map. **c** 3-D oblique view shows both segments of the marker belong to the hangingwall

Fig. 10.22.
Structure contour map of a
normal fault (from Fig. 8.5a)

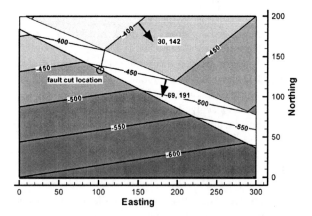

the heave and throw from Eqs. 7.5–7.6 or 7.7–7.8 and compare to the values on the map. The discrepancy is revealed by either method and requires that the fault cut interpretation be changed or the map reinterpreted.

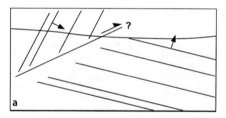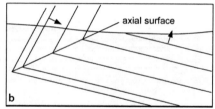

Fig. 10.23. Axial surface vs. fault, cross-section view. *Full arrows* indicate stratigraphic facing (up) direction. **a** Dip change used to infer location of a fault, *half arrow* indicates possible displacement direction. **b** Extrapolated markers meet, showing an axial surface is present, not a fault

The defining feature of a fault is the separation of a marker surface. It is tempting to map a fault anywhere the dip changes rapidly (Fig. 10.23a) in either outcrops or on seismic profiles. At such locations, the continuity of beds or reflectors may be difficult to establish, leading to the inference of a fault. If extrapolation of the markers towards each other shows that they can meet without offset (Fig. 10.23b), then the probability that the feature is an axial surface must be considered. Dip-domain fold hinges can be very tight. Faults with no stratigraphic separation are probably axial surfaces with large dip change. Of course strike-slip and oblique-slip faults cutting folds may have locations where the stratigraphic separation is zero even though the slip may be significant. In such situations it is expected that fault separation will appear elsewhere along the fault surface, demonstrating that displacement has, in fact, occurred.

10.5.3
Fault Cutoff Geometry

The geometry of the cutoff lines of marker surfaces against the fault provides a test of the quality of the interpretation. The relationships are nicely shown on an Allan diagram (Sect. 8.4). The Allan diagram of the normal fault shown in Fig. 10.24 represents a reasonable throw distribution along a fault. The throw increases from SE to NW and is approximately constant down the dip of the fault at any one location.

The Allan diagram from Fig. 10.24 has been modified in Fig. 10.25 to show typical fault-separation problems. At point 1 the throw on marker 5 is significantly more than on the markers above and below it. At point 3 the throw is significantly less than on markers above and below. The lack of consistency along the fault and up and down the fault, makes the interpretation of marker 5 suspect. Lesser throw on one horizon, as at point 3, might be explained by a change in lithology to a rock type that favors folding. Nevertheless the interpretation should be checked. The reversal of separation at point 2 implies a significant problem. Only a strike-slip fault would be expected to show reversals of separation along strike. But the strike slip would not be confined to a single marker horizon, suggesting an error in the interpretation.

A stratigraphic separation diagram shows the fault separation in terms of the units juxtaposed across the fault along a single line, for example along the map trace of the fault (Sect. 7.3.3). The curves for the hangingwall and footwall are not expected to cross (Fig. 10.26), because this implies that either (1) the fault changes from a thrust to a

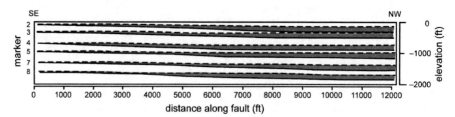

Fig. 10.24. Allan diagram of throw (projection to a vertical plane) for a normal fault in Deerlick Creek coalbed methane field (data from Groshong et al. 2003b). *Dashed line* is FW cutoff, *gray shading* shows fault throw. Base of the gray is the HW cutoff

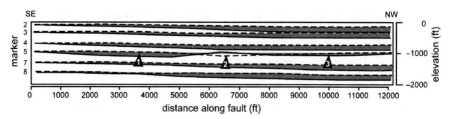

Fig. 10.25. Allan diagram showing inconsistent throw at numbered locations. *Dashed line* is FW cutoff, *gray shading* shows fault throw

Fig. 10.26.
Stratigraphic separation diagram showing a separation anomaly. *FW curve:* Stratigraphic position of the thrust in the footwall; *HW curve:* stratigraphic position of the fault in the hangingwall

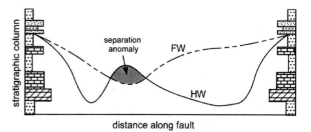

normal fault or from a normal fault to a thrust, or (2) the fault is out of the normal evolutionary sequence and cuts an older fold or fault. A separation anomaly requires careful attention to the correlation of the fault cuts and to the interpretation of the fault if the correlation is accepted. A strike-slip fault might have an apparent separation anomaly like that in Fig. 10.26.

10.6
Exercises

10.6.1
Cross-Section Quality

Critique the cross section in Fig. 10.27. What might be the origin of any problems?

10.6.2
Map Validation

The structure contour map of the top of a faulted limestone (Fig. 10.28) fits the well information and explains the hydrocarbon trap. Is the interpretation valid? Based on the map as presented, what kind of faults are present? In which direction do they dip? Where are the structural closures on the map? Could any of them be new hydrocarbon traps? Why or why not? Draw implied structure contour maps on the faults. Is the original map correct? Construct an improved map that honors all the well data.

Fig. 10.27.
Cross section of a compressional anticline (after Brown 1984). Elevations are in feet, *dashed wells* are projected to the line of section, *solid circles* represent oil producers, *open circles* are dry holes

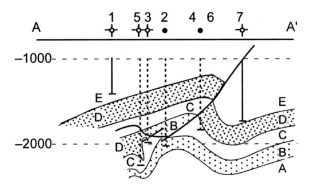

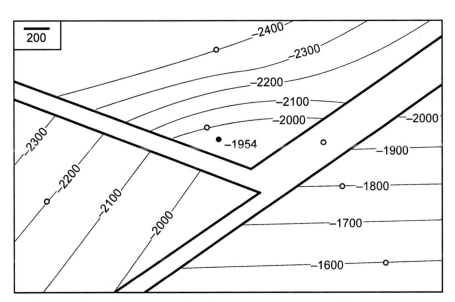

Fig. 10.28. Hypothetical structure contour map of the top of the Appling Bend Limestone gas reservoir. Contours are depths below sea level. The *solid circle* is a gas well, *open circles* are dry holes. The limestone is missing in the well in the fault gap. Choose the measurement units to be in either feet or meters. Elevations below sea level are negative

10.6.3
Map and Fault Cut Validation

Is the fault in Fig. 10.29 valid? If not, can it be corrected? The well near the center of the map is thought to have a stratigraphic separation of 15.0 m. Is that correct? If not, what is stratigraphic separation is consistent with the map?

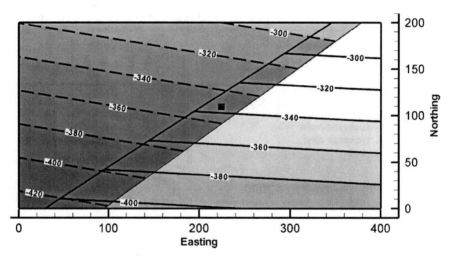

Fig. 10.29. Structure contour map of an offset horizon. Scale in m, *small square* is a well location

Chapter 11

Structural Validation, Restoration, and Prediction

11.1
Introduction

This chapter presents techniques for validating structural interpretations and for extracting additional information, such as the shape of the structure beyond the data, predicting the presence of structures too small to be seen at the resolution of the data, and determining the structural evolution. Most of the techniques of structural restoration, balance and prediction are related to one another by use of a common set of kinematic models, which is why they will be discussed together here.

A restorable structure can be returned to its original, pre-deformation geometry with a perfect or near-perfect fit of all the segments in their correct pre-deformation order (Fig. 11.1). Restoration is a fundamental test of the validity of the interpretation. A restorable structure is internally consistent and therefore has a topologically possible geometry. An unrestorable structure is topologically impossible and therefore is geologically not possible (Dahlstrom 1969). An interpretation based on a large amount of hard data, such as a complete exposure, many wells, or good seismic depth sections controlled by wells, is nearly always restorable, whereas interpretations based on sparse data are rarely restorable. This is the empirical evidence that validates restoration as a validation technique.

Fig. 11.1.
A three-dimensional structural interpretation (**a**) and its restoration (**b**). The *shaded surface* is a fault

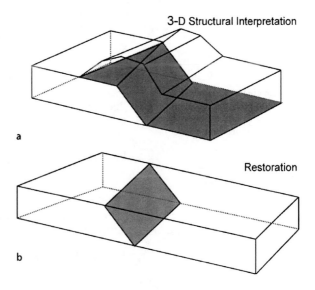

3-D Structural Interpretation

a

Restoration

b

The original concept of a balanced cross section (Chamberlin 1910) is that the deformed-state and restored cross sections maintain constant area and so would balance on a beam balance (Fig. 11.2). This concept was generalized by Dahlstrom (1969) to a constant volume criterion. In many structures there is little or no deformation along the axis of the structure, and so in practice the third dimension can often be temporarily ignored and constancy of volume can be applied to a cross section as a constant-area rule. Units which maintain constant bed length are said to be length balanced and units that maintain constant area but not constant bed length or bed thickness are said to be area balanced. A balanced cross section is generally understood to be one which is restorable to a geologically reasonable pre-deformation geometry, as well as maintaining constant area.

The techniques for the restoration of a structure are necessarily based on models for the evolution of the geometry. A kinematic model defines the evolution through time of the geometry of a structure. Four basic kinematic models are commonly used for restoration (Fig. 11.3). The most appropriate model for a given structure will be determined by the mechanical stratigraphy and the boundary conditions that produced the structure. The simplest model is rigid-body displacement (Fig. 11.3a) which may include both translation and rotation. Layer-parallel slip (Fig. 11.3b) implies slip between closely spaced layers that maintain constant thickness unless otherwise specified. If the slip is between layers that are visible at the scale of observation, the folding mechanism is known as flexural slip (Donath and Parker 1964) and so this is called the

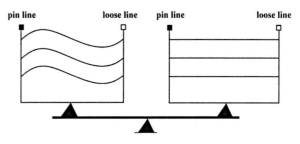

Fig. 11.2.
The concept of a balanced cross section (after Woodward in Woodward et al. 1989). Deformed-state section on the left, restored section on the right

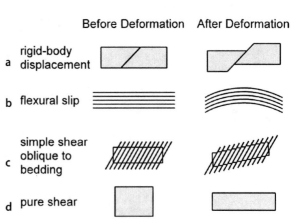

Fig. 11.3.
Basic kinematic models.
a Rigid-body displacement.
b Flexural slip. **c** Simple shear oblique to bedding. **d** Pure shear

flexural-slip model. Simple shear is the geometry produced by slip on closely spaced, parallel planes with no length or thickness changes parallel or perpendicular to the slip planes. Simple shear parallel to bedding is the mechanism of flexural slip. Simple shear oblique to bedding (Fig. 11.3c) is a kinematic model that causes bed length and bed thickness changes. Pure shear is an area-constant shape change (Fig. 11.3d) for which the shortening in one direction is exactly balanced by extension at right angles to it. Restorations are produced by applying one or more of these kinematic models to the deformed-state structure to return it to its pre-deformation configuration.

A valid map or cross section may be restorable by more than one kinematic model, and different models will produce somewhat different restored geometries. It follows that any given restoration does not necessarily represent the exact pre-deformation geometry or the specific path followed by the structural evolution. Nevertheless, the internal consistency of the restoration by any technique constitutes a validation of the interpretation. If a restoration is possible, it shows that the structure is internally consistent even if the restoration technique is not a perfect model for the deformation process.

Kinematic models contain the relationships needed to predict the geometry and evolution of a structure. The predictive capabilities of the models are the basis of techniques for utilizing very limited amounts of information to predict the geometry in areas of very sparse data or no data (Fig. 11.4). This chapter introduces the basic kinematic models and their predictive capabilities.

Kinematic models represent simplified descriptions of the mechanical processes that form structures. The deformation in some structures is more complex than can be fit by one of the simple kinematic models. For these structures the more general area-balancing methods can be appropriate. Using the relationship between displaced area and depth, a structure can be tested for area balance and its lower detachment predicted without performing a restoration or a model-based prediction. Layer-parallel strain is treated here because it is an intrinsic part of both the kinematic models and the area-depth relationship and because it provides a tool for predicting sub-resolution structure (i.e., folds and faults too small to be seen at the resolution of the data) and is another tool for validating the structural interpretation.

The chapter begins with the most general concepts, an overview of balance and restoration followed by a discussion of strain and strain partitioning. Then the model-independent area-balance methods and area-depth technique are given, followed by the individual kinematic-model-based restoration and prediction techniques.

Fig. 11.4.
Prediction of fault shape from geometry of a key bed using the flexural-slip kinematic model (after Geiser et al. 1988). **a** Key bed required for prediction. **b** Predicted complete hangingwall geometry and fault shape

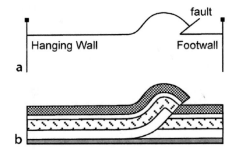

11.2
Restoration and Balance

This section covers general concepts and terminology used in all the balancing and restoration techniques to follow.

11.2.1
Boundaries

Ideally, the boundaries of a section to be restored are chosen so that the section will restore to a rectangle (Fig. 11.5a). The side boundaries are pin lines, and the upper boundary is a reference bed that will be returned to its original depositional geometry (Dahlstrom 1969; Elliott in Geiser 1988; Marshak and Woodward 1988). The original position of a horizon, including both its shape and elevation, is known as the *regional datum*, commonly shortened to just the *regional* (McClay 1992). The restored positions of all other horizons are determined with respect to the reference horizon. The base of the section is normally either a stratigraphic marker or a detachment horizon (Fig. 11.5), but may be simply the lowest visible unit. The bounding pin lines are leading and trailing pins, according to their position in the structure with respect to the transport direction. A locally balanced structure is one in which bed-normal pins on either side of the structure of interest define a region in which the area has remained

Fig. 11.5.
Pin lines bounding a region of interest. *LPL:* leading pin line; *TPL:* trailing pin line; *WPL:* working pin line. **a** Undeformed. **b** Locally balanced. **c** Regionally balanced with transport of material out of the structure. **d** Regionally balanced with simple-shear transport of material into the structure

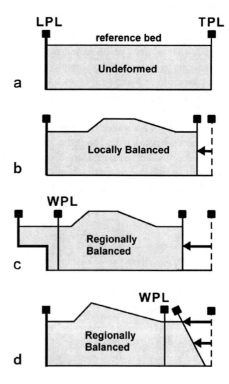

constant (Fig. 11.5b). If material has been transported across the chosen pin lines by displacement on an upper detachment (Fig. 11.5c) or into or out of the region by layer-parallel simple shear (Fig. 11.5d), the structure should be regionally balanced, even if the region of interest is not locally balanced. Thus for a regionally balanced structure a vertical pin lines may represent a *working* pin line (Fig. 11.5b,c) that will restore with an offset or a tilt. A valid cross section might fail to restore to a rectangle because of unrecognized transport across a working pin line. The restoration reveals these otherwise hidden displacements.

11.2.2
Palinspastic vs. Geometric Restoration

Restoration is a purely geometric manipulation of the cross section according to a specific set of rules. A geometric restoration (Fig. 11.6) is a restoration that is not specifically related to the direction of transport that formed the structure. A palinspastic restoration is the restoration of the units to their correct pre-deformation configuration by exactly reversing the displacements that formed the structure. The independence of the restoration from the transport direction is illustrated in Fig. 11.6. Two different cross sections through a fold have been restored. Normally it is assumed that

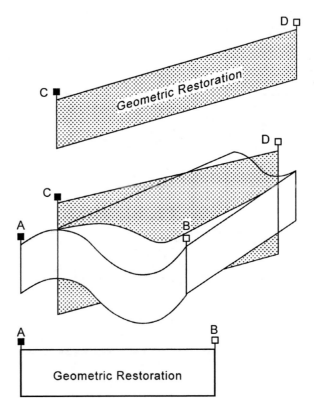

Fig. 11.6.
Geometric restoration of a cross section in any direction

the transport direction is perpendicular to the fold axis, and so the palinspastically restored section would be AB. If the transport is in the CD direction, however, the palinspastic restoration would be in the CD cross section. Either restoration serves to validate the structure but only the palinspastic restoration returns the units to their actual pre-deformation locations.

The only restriction on the choice of the orientation of the line of section is that the transport direction should be approximately constant in the vicinity of the cross section (Elliott 1983; Woodcock and Fischer 1986). The transport direction need not be parallel to the line of section as long as faults with transport oblique to the section are not crossed. A cross section oblique to the transport direction is the same as a section having vertical or horizontal exaggeration (Cooper 1983, 1984; Washington and Washington 1984). This exaggeration can be removed by projecting the section into the transport direction (Sects. 6.5, 6.6). Restoration of a cross section that crosses oblique-transport faults will probably result in a structural or stratigraphic discontinuity at the fault. The discontinuity might be removed by a lateral shift of the cross section at the fault.

As an example of the difference between geometric and palinspastic restoration, consider restoring the Säntis anticline, a major fold above the Säntis thrust in the Helvetic fold and thrust belt of Switzerland (Fig. 11.7). The anticline is oblique to the

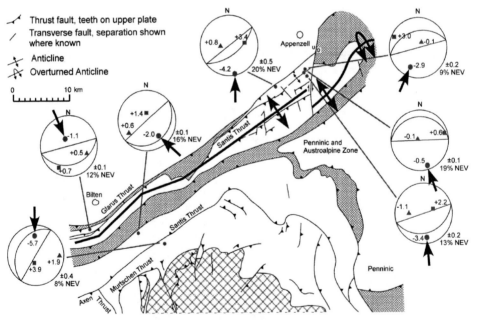

Fig. 11.7. Geologic map of the Säntis anticline in the Helvetic fold-thrust belt, eastern Switzerland constructed at a 1500 m datum (modified from Groshong et al. 1984). Calcite twin-strain axes are shown on lower-hemisphere, equal-area stereograms: *dot:* maximum shortening axis; *square:* maximum extension axis; *triangle:* intermediate strain axis. *Large arrows* show transport direction interpreted as being parallel to the direction of internal shortening strain

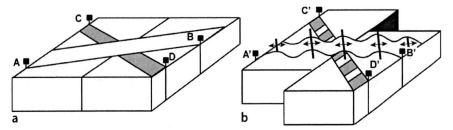

Fig. 11.8. Deformation in beds above a strike-slip fault. **a** Before deformation. **b** After right-lateral strike-slip displacement. Both cross sections A'B' and C'D' can be geometrically restored without removing the rotation caused by the strike-slip displacement

regional northward transport direction documented by the north-south pebble elongation direction observed in exposures of the basal thrust and the mainly north-south twinned-calcite shortening strains in the fold itself (Fig. 11.7). The fold axis orientation may be controlled by an underlying oblique thrust ramp. A palinspastic restoration of this anticline requires a north-south cross section. The section perpendicular to the fold axis can be restored, but it would be a geometric, not a palinspastic, restoration.

The distinction between a palinspastic and geometric restoration is important to the use of balancing techniques along wrench faults. Figure 11.8 shows representative portions of beds deformed by movement along a wrench fault. Strike-slip displacement on the fault produces shortening and rotation of section AB to the final position A'B'. Deformation dies out away from the master fault and so the pin lines at A' and B' are both in undeformed beds. The cross section A'B' could be restored, as could any other line of section. A correct palinspastic restoration requires the vertical-axis rotation to be restored as well. The line CD (Fig. 11.8) will be extended and rotated during wrench deformation into the position C'D'. Extensional structures such as normal faults will develop perpendicular to this line and could also be geometrically restored. Restorations across small-displacement wrench faults can be expected to be relatively free from major discontinuities. Restorations across large-displacement wrench faults are likely to show large discontinuities at the master fault but otherwise be restorable.

11.2.3
Sequential Restoration

A sequential restoration shows the intermediate stages between the fully deformed and fully restored stages. Where growth stratigraphy is present, the geometry of the intermediate stages is controlled by the thickness changes (Fig. 11.9). A sequential restoration is valuable because of the insight it provides into the structural evolution and because it represents an even more rigorous test of the validity of the interpretation. A sequential restoration of a section consisting entirely of pre-growth units is possible, but must be based on the kinematic model, rather than on hard data.

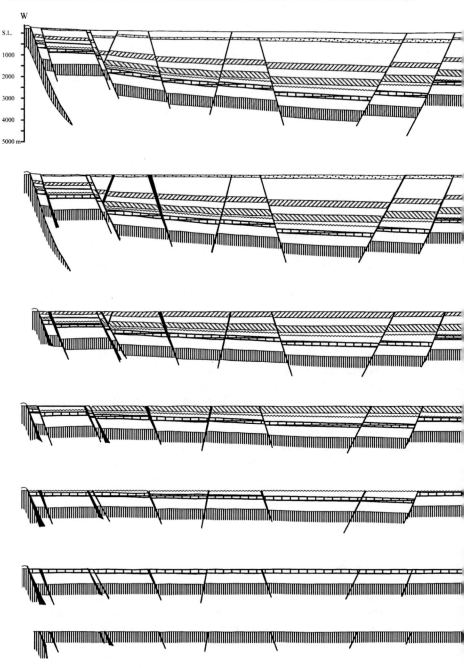

Fig. 11.9. Sequential restoration of a profile across the Rhine Graben, Germany. The upper cross section is redrawn from Doebl and Teichmüller (1979). Gaps between blocks are *black*

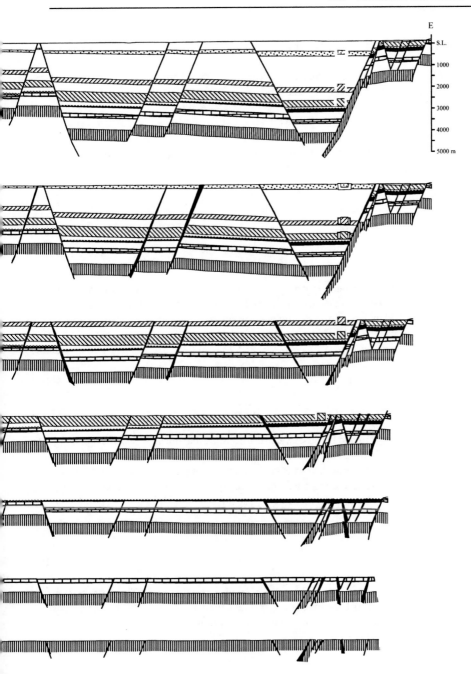

11.3
Strain and Strain Partitioning

Strain can play a significant role in structural balancing. Presented here are the practical strain measurements that can be applied to field examples at the map and cross-section scale. The most readily quantifiable macroscopic measures of strain are bed length and bed thickness (Fig. 11.10). Separate equations are given for the strain parallel and perpendicular to bedding, even though the forms are identical, in order to better emphasize the origins of the interpretations that will appear in subsequent sections.

Bed length change can be quantified as

$$\Delta L = L_1 - L_0 \ , \tag{11.1}$$

$$e_L = (L_1 - L_0)/L_0 \ , \text{or} \tag{11.2}$$

$$e_L = (L_1/L_0) - 1 \ , \tag{11.3}$$

where ΔL = change in bed length, e_L = the infinitesimal normal strain parallel to bedding, L_0 = the bed length before deformation, L_1 = the bed length after deformation. Exactly equivalent equations can be written for bed thickness change:

$$\Delta t = t_1 - t_0 \ , \tag{11.4}$$

$$e_t = (t_1 - t_0)/t_0 \ , \text{or} \tag{11.5}$$

$$e_t = (t_1/t_0) - 1 \ , \tag{11.6}$$

where Δt = change in bed thickness, e_t = the infinitesimal normal strain perpendicular to bedding, t_0 = the bed thickness before deformation, t_1 = the bed thickness after deformation. With the equations in the forms given, extension is positive and contraction is negative. The value of e is a fraction but is commonly given as a percent by multiplying by 100.

The strain can be also be measured with the stretch, β, (McKenzie 1978):

$$\beta_L = (L_1/L_0) \ , \tag{11.7}$$

$$\beta_t = (t_1/t_0) \ , \tag{11.8}$$

where β_L = layer-parallel stretch and β_t = layer-normal stretch. The stretch is always positive, greater than 1 for extension and less than 1 for contraction. For a constant area rectangle the stretch has the convenient property that

$$\beta_L = 1/\beta_t \ . \tag{11.9}$$

From Eqs. 11.7, 11.8 and 11.9, the stretches can also be given as

$$\beta_L = t_0/t_1 \ , \tag{11.10}$$

Fig. 11.10.
Bed length and bed thickness change in an area-constant deformation

Fig. 11.11.
Deformation of a bed of original thickness t_0 by oblique simple shear (after Groshong 1990). α: angle between shear direction and bedding, ψ: angle of shear, e_L: layer-parallel extension

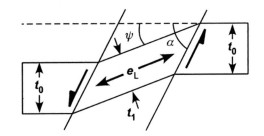

$$\beta_t = L_0 / L_1 \quad, \tag{11.11}$$

and from Eqs. 11.3, 11.6 and 11.7, 11.8,

$$e_L = \beta_L - 1 \quad, \tag{11.12}$$

$$e_t = \beta_t - 1 \quad. \tag{11.13}$$

Simple shear oblique to bedding (Fig. 11.11) changes the bed length and bed thickness. The layer-parallel strain can be determined from the shear amount and the shear angle (Groshong 1990) as

$$e_L = [\sin \alpha / \sin (\alpha - \psi)] - 1 \quad, \text{ or} \tag{11.14}$$

$$e_{90} = (1 / \cos \psi) - 1 \quad, \tag{11.15}$$

where α = angle between shear direction and bedding and ψ = angle of shear. Equation 11.15 is for vertical simple shear ($\alpha = 90°$) and Eq. 11.14 is for all other angles. To find the thickness change from the length change for constant area deformation, substitute Eq. 11.3 into 11.6 to give

$$e_t = -e_L / (e_L + 1) \quad. \tag{11.16}$$

Simple shear parallel to bedding in flexural-slip deformation has no effect on bed length or bed thickness. If a bed-normal marker is present, or assumed, then its length strain is given by Eq. 10.15 where the angle of shear is the amount of rotation of the original bed-normal marker.

Strain is commonly partitioned between deformation features that are visible at the scale of the map and cross section and deformation at smaller scales, termed sub-resolution strain. In the low-temperature deformation of lithified rocks (i.e., where hydrocarbons can be preserved), the crystal-plastic strains are usually less than

Fig. 11.12.
Representative segment of a
bed deformed by sub-resolution strain. **a** Undeformed.
b Extension on small normal
faults. **c** Contraction on small
thrust faults. **d** Low-resolution
observation of **b**. **e** Low-resolution observation of **c**

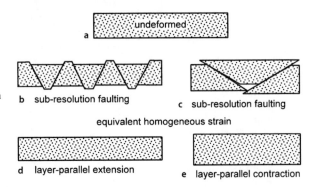

4–6%, and perhaps closer to 1–2% in very brittle lithologies such as dolomite or
in extensional settings (Groshong 1988). If the deformation is brittle, additional
strain within the layers will commonly be accommodated by small-scale faulting
(Fig. 11.12b,c). Strain measured at the scale of the oil field or a seismic line may appear
homogeneous (Fig. 11.12d,e) but represent the combined effects of sub-resolution
faults and/or folds.

11.4
Area-Balance Methods

Area balance methods include the restoration of irregularly shaped deformed regions,
area-based depth-to-detachment calculations and the area-depth relationship. These
methods do not depend on specific kinematic models and in that sense represent the
most general approaches to validation, restoration, and prediction. All the structures
discussed in Sects. 11.4.1–11.4.3 are locally balanced (Fig. 11.5b). Regionally balanced
structures (Fig. 11.5c,d) will be treated in Sect. 11.4.4.

11.4.1
Area Restoration

Area restoration is used for structures in which deformation has produced significant
changes in the original bed lengths and thicknesses. The technique is based on the
area of the deformed-state cross section (Fig. 11.13a). It is assumed that the area has
remained constant:

$$A_0 = t_0 L_0 \ , \tag{11.17}$$

where A_0 = original area, t_0 = original bed thickness, and L_0 = original bed length. The
area between the pin lines is measured and then divided by either the original bed
thickness or the original bed length, whichever is better known, and Eq. 11.17 solved
for the unknown dimension. The original bed length might be known from that of an
adjacent key bed that has not changed thickness (Mitra and Namson 1989) or the original thickness might be known from a location outside the deformed region. The shape

Fig. 11.13.
Area restoration. A_0: Original area; t_0: original bed thickness; L_0: original bed length. Shape of the restored area depends on assumed original orientations of the pin lines. **a** Deformed-state cross section. **b** Section restored to vertical pin lines. **c** Section restored to tilted pin lines

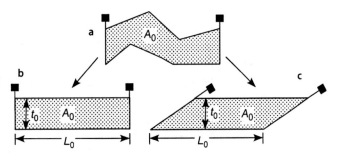

of the restored area depends on the assumed original shape (Mitra and Namson 1989). Ordinarily the unit is restored to horizontal, leaving only the orientations of the pin lines to be determined. A folded area might be appropriately restored to a rectangular prism (Fig. 11.13b), whereas a block bounded by faults would be restored to fit between the presumed original fault shapes (Fig. 11.13c).

11.4.2
Depth to Detachment and Layer-Parallel Strain

The first predictive uses of the concept of area balance were to determine the depth to detachment of the structure (Chamberlin 1910; Hansen 1965). Area uplifted above the regional as a result of compressional deformation is termed the excess area (Fig. 11.14). Area that drops below regional as a result of extensional deformation is termed lost area. More generally, these can be called the displaced areas. The classical displaced-area method is designed to find the detachment depth from the excess or lost area of one horizon in a structure. The displaced area is produced by displacement along the lower detachment such that

$$S = D H \; , \tag{11.18}$$

where S = area above or below the regional, D = displacement, and H = depth to detachment from regional (Fig. 11.14). A unique depth to detachment can be calculated from a measurement of the excess area *if* the displacement that formed the structure is known. Chamberlin (1910) and many subsequent authors have assumed that bed length remains constant and so the displacement is the difference between the curved-bed length of the marker horizon and its length at regional:

$$D = L_0 - W \; , \tag{11.19}$$

where D = displacement, L_0 = curved-bed length (assumed equal to original bed length) and W = width of structure at regional. Substituting Eq. 11.19 into 11.18:

$$H_c = S / (L_0 - W) \; , \tag{11.20}$$

where H_c indicates the detachment depth if bed length is constant.

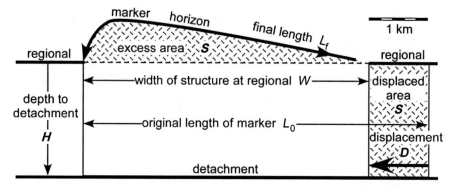

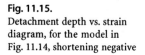

Fig. 11.14. Area balanced model. $H = 2.08$ km, $L_f = 5.422$ km, $S = 2.141$ km^2, $W = 4.879$ km, $D = 1.028$ km. Theoretical detachment for constant bed length is $H = 3.943$ km (Eq. 11.20)

Fig. 11.15.
Detachment depth vs. strain diagram, for the model in Fig. 11.14, shortening negative

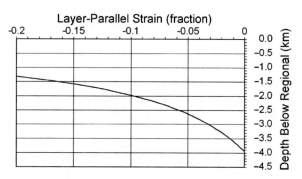

Calculations based on constant bed length typically show detachments that are extremely deep for low-amplitude structures and indicate shallower detachments for higher-amplitude structures, even in the same area (eg., the Appalachian profile of Chamberlin 1910). One significant problem is the assumption of constant bed length. Previous work (Groshong 1988, 1994) has shown that bed length is not likely to remain constant and that length changes can have a significant effect on the calculated detachment depth. For example, the theoretical depth to detachment of the area-balanced model in Fig. 11.14 is 3.943 km (Eq. 11.20), about 90% greater than the correct value of 2.08 km. The sensitivity of the calculated detachment depth to unmeasured changes in bed length (sub-resolution strain) can be demonstrated quantitatively. If the bed length changes during deformation,

$$L_f = L_0 + \Delta L \quad , \tag{11.21}$$

where L_f = observed length = final bed length (Fig. 11.14) and ΔL = change in bed length during deformation (shortening negative). Equation 11.21 is solved for L_0 and substituted into Eq. 11.19, giving the relationship between D and the changing bed length. This relationship is substituted into Eq. 11.18 and solved for H to give

$$H = S / (L_f - \Delta L - W) \quad . \tag{11.22}$$

If $\Delta L = 0$, then $L_f = L_0$, and this equation reduces to the constant bed-length relationship (Eq. 11.20).

The effect of bed-length change on the depth-to-detachment relationship for the example in Fig. 11.14 is seen by substituting the measured values of S, L_f, W, and a range of ΔL values (including zero) into Eq. 11.22 and graphing the result (Fig. 11.15). The most general result is obtained by plotting layer-parallel strain (Eq. 11.2) versus detachment depth. The maximum possible detachment depth (3.943 km) occurs for zero strain. The detachment at 2.0 km shown in Fig. 11.14 requires a layer-parallel shortening of –0.1 (–10%) in the marker horizon. A shallower detachment is possible if the layer-parallel strain is sufficient.

11.4.3
Area-Depth Relationship of Locally Balanced Structures

The relationship between the areas of multiple horizons on a cross section and their depths allows a cross section to be tested for area balance and internal consistency without restoration. Detachment depth can be determined without knowledge of the layer-parallel strain, and the layer-parallel strain can be determined once the detachment location is known.

Displacement causes an amount of the original area to be pushed above the regional (excess area) or dropped below the regional (lost area) as shown in Fig. 11.16. Measurements are made of the excess or lost areas above or below the regional and the elevation of the regional with respect to the reference level for each horizon. This information is plotted on a graph of area versus depth (elevation) (Fig. 11.17a). For a locally balanced structure, the data from multiple horizons give points that define a straight line, the equation of which is

$$h = (1/D)\,S + H_e \ , \tag{11.23}$$

where h = elevation of a surface above or below the reference level, D = displacement on the lower detachment, S = net displaced area, either excess or lost, and H_e = the elevation at which the area S goes to zero, which represents the position of the detach-

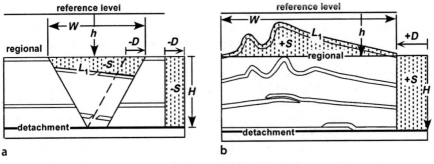

Fig. 11.16. Area-balance terminology. *S:* Excess or lost area; *D:* displacement on the lower detachment; *H:* distance from the lower detachment to the regional; *h:* elevation of the regional above or below the reference level; *L$_1$:* bed length after deformation. **a** Extension (after Groshong 1996). **b** Contraction (after Groshong and Epard 1994)

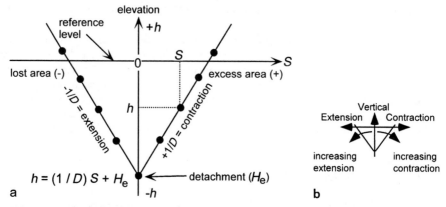

Fig. 11.17. Area-depth relationships for compressional and extensional detachment structures. **a** Area-depth graph. **b** Relationship between the slope of the area-depth line and the amount and type of deformation

ment. The elevation of the reference level is arbitrary. If the reference level is at the detachment, $H_e = 0$. If the reference level is chosen to be sea level, then H_e will be the elevation of the detachment relative to sea level. The sign convention for area is that excess area and contractional displacement are positive and lost area and extensional displacement are negative. If deformation causes both uplift and subsidence from the regional, the correct value of S is the algebraic sum or net area (Groshong 1994). The slope of the area-depth line, $1/D$, is the inverse of the displacement on the lower detachment. To be a valid cross section, the points on the area-depth graph (Fig. 11.17a) must all fall on or close to the area-depth line.

Each horizon has its own separate regional. The reference level should be at least approximately parallel to the directions of the regionals and the elevations measured perpendicular to it. If the regionals are tilted with respect to the reference level, the elevation measurements should be taken at the center of the structure (equivalent to the average value).

Equation 11.23 and the graph in Fig. 11.17a are modified from the forms originally given by Epard and Groshong (1993) in order to show elevation as the vertical axis on the graph. This makes the graphical relationship between area and depth more intuitively obvious, with the area on the area-elevation line decreasing downward on the graph to zero at the lower detachment (Fig. 11.17a), as it does in the structure itself (Fig. 11.16). In this form, a vertical area-depth line represents differential vertical displacement (Fig. 11.17b) and the lower the slope of the area-depth line, the greater the amount of extension or contraction. A positive slope of the line indicates contraction and a negative slope indicates extension. The slope of the area-depth line is the inverse of the displacement, instead of the displacement itself as in the previous formulation (Epard and Groshong 1993). For any given area-elevation data point, the range of h values caused by variations in measurement and interpretation will usually be significantly less than the range of the S values, and so a least-squares fit to the data should be a regression of S onto h, rather than h onto S as the form of Eq. 11.23 might suggest.

A potential pitfall with this technique is in the selection of the correct regional. The problem is that this level must be chosen on the deformed-state cross section. If undeformed units are preserved outside the structure of interest, they should record the correct elevation of the regional. The bottoms of synclines are good candidates for the elevations of the regionals in vertical and compressional structures and the elevations of beds in flat horsts are good candidates for the regionals in extensional structures. The effect of an incorrect choice of the regional is to produce an incorrect elevation for the lower detachment. The goodness of fit of the area-depth points to a line on the graph is unaffected by the choice of the regional, and so the method remains valuable in judging the internal consistency of the cross section, even if the regionals are unknown. Because the area-depth relationship is not dependent on the kinematic model, it applies as an additional constraint on model-based cross-section restorations and can provide additional information about the depth to detachment, strain and total displacement.

11.4.3.1
Effect of Growth Stratigraphy

The area-depth relationship described above applies to beds that were deposited before deformation. Displacement on the lower detachment of a locally balanced structure produces the same displacement in every pre-growth unit and results in a straight-line area-depth relationship. Growth beds are deposited during deformation and each bed may have a different displacement, the amount of which decreases upward in younger beds. Some of the lost area in the case of a graben (Fig. 11.18a) is filled with sediment. These effects result in an area-depth curve that is straight in the pre-growth units and curves upward toward zero area in the growth beds (Fig. 11.18b). The boundary between the growth and pre-growth intervals is clearly identified by the inflection point in the area-depth graph (horizon 6, Fig. 11.18b).

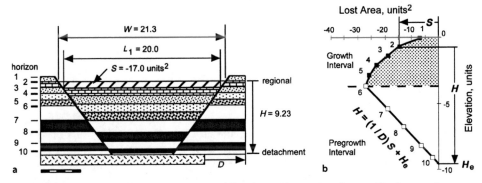

Fig. 11.18. Area-balanced model of a pure-shear full graben with growth sediments (Groshong et al. 2003a). **a** Cross section. *Solid black* and *solid white* layers are pre-growth units, *patterned* layers are growth beds. *Diagonal pattern* is the lost area of horizon 2. **b** Area-depth graph pointing out the values for horizon 2. *S:* lost area; *H:* depth to detachment for a particular horizon; *D:* displacement of pre-growth units; H_e: elevation of the lower detachment

11.4.3.2
Strain

Layer-parallel strain can be determined from the area-depth relationship if either the displacement is known (from the area-depth graph), or the detachment location is known (by direct observation or from the area-depth relationship). The method applies to growth beds as well as to pre-growth beds.

The length of a particular horizon as seen on a cross section is its final length L_1 which may include bed-length changes. From the geometry of Fig. 11.16, the original length of the horizon (L_0) is

$$L_0 = W + d = W + S/H \ , \tag{11.24}$$

where W = the width of the structure at the regional of the horizon and d = the displacement of any horizon, including a growth horizon (Groshong et al. 2003a). A lowercase d is used here for a growth-bed displacement and an uppercase D for the total displacement of a pre-growth unit. Either value can be substituted into Eq. 11.24. W is always measured parallel to the regional and is tilted if the regional is tilted. The value of L_0 calculated from the displaced area is independent of the stratigraphic growth of the unit because it depends only on the length along the upper surface of the unit, not on the thickness. The bed-length change is

$$\Delta L = L_1 - L_0 = L_1 - (W + S/H) \ . \tag{11.25}$$

The length difference can be converted to layer-parallel strain, e, by dividing Eq. 11.25 by Eq. 11.24:

$$e_L = (H \ L_1 / (H \ W + S)) - 1 \ . \tag{11.26}$$

If the total displacement, D, is known from the area-depth relationship, an alternative form of the equation is convenient:

$$e_L = (L_1 / (W + D)) - 1 \ . \tag{11.27}$$

11.4.4
Area-Depth Relationships of Regionally Balanced Structures

For regionally balanced structures, material is transported across vertical pin lines bounding the structure of interest (Fig. 11.5c,d). A representative example of a regionally balanced structure is a fault-bend fold (Figs. 11.5c, 11.19a). In a fault-bend fold, one of the original pin lines is offset along an upper detachment horizon (Fig. 11.19a). Part of the area displaced above the upper detachment is within the anticline and part is outside the anticline. Material may be displaced toward the foreland or toward the hinterland; the area-balance is the same. Two area-depth lines are produced, one for the stratigraphic levels below the upper detachment and one for the stratigraphic levels above the

Fig. 11.19.
Area-depth relationships in a
fold having displacement on
an upper detachment. **a** Fault-
bend fold model (after Suppe
1983). **b** Area-depth diagram
of **a** (modified from Epard
and Groshong 1993). D: displacement on lower detachment; d^*: slope of line for horizons stratigraphically above
the upper detachment; d: displacement on upper detachment; h_u: elevation of upper
detachment

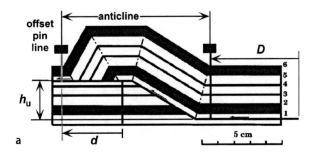

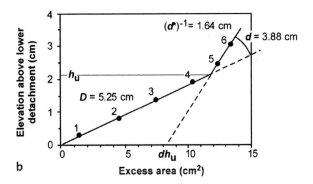

upper detachment. Below the upper detachment, no material is translated out of the
section and the area-depth line is the same as for a locally balanced structure (Eq. 1.23).
For units stratigraphically above the upper detachment, the area-depth line is

$$h_2 = [1/(D-d)]\,S_2 + k \quad , \tag{11.28}$$

where h_2 = the distance from the reference level for units above the upper detachment,
S_2 = the excess area of units stratigraphically above the upper detachment, d = displacement on the upper detachment, $k = d\,h_u/(D-d)$, and h_u = the elevation of the upper
detachment. This is the equation of a line of slope $d^* = 1/(D-d)$ that is valid for $h > h_u$.
The excess areas of a fault-bend fold are thus distributed along two line segments
(Fig. 11.19b). The lower-slope segment is valid for $h < h_u$ and corresponds to Eq. 11.23.
The upper segment is valid for $h > h_u$ and corresponds to Eq. 11.28. The displacement
on the upper detachment is the difference in the slopes of the two area-depth lines:

$$d = D - (d^*)^{-1} \quad . \tag{11.29}$$

Failure to recognize the presence of an upper detachment and taking into consideration only measurements from above the upper detachment will lead to a significant
overestimation of the depth to detachment (lower dashed projection line on Fig. 11.19b).
If the lower detachment is known to be shallower than the prediction, then an upper
detachment is indicated.

Fig. 11.20.
Length and displacement measurements for strain calculation

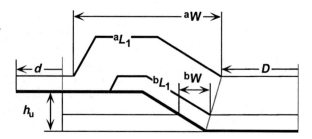

Sub-resolution strains are determined from the displacement and the width of the structure at regional. The appropriate relationship depends on whether the bed is stratigraphically below or above the level of the upper detachment horizon (Fig. 11.20). The final bed lengths are the observed lengths between the points where the structure departs from regional. The original bed lengths are given by the following equations:

$$^{b}L_0 = {}^{b}W + D \quad , \tag{11.30}$$

$$^{a}L_0 = {}^{a}W + (D - d) \quad , \tag{11.31}$$

where $^{i}L_0$ = original bed length, ^{i}W = width of structure parallel to regional at regional, i = a or b, D = displacement on the lower detachment, and d = displacement on the upper detachment. The layer-parallel strain is

$$e = ({}^{i}L_1 / {}^{i}L_0) - 1 \quad , \tag{11.32}$$

where e = layer-parallel strain as a fraction, $^{i}L_1$ = measured bed length, and $^{i}L_0$ = corresponding original bed length from Eqs. 11.30 or 11.31.

Other regionally balanced styles of importance are those produced with simple-shear transport of material into the closed structure (Fig. 11.5d). A linear shear profile has been treated by Epard and Groshong (1993). The total displacement is a combination of the constant component, D, and the distributed simple-shear component, D_s. The excess area is the sum of the constant-displacement component, Dh and the distributed shear component, S_s:

$$S = Dh + S_s \quad . \tag{11.33}$$

Assuming linear simple shear:

$$S_s = \tfrac{1}{2}D_s h \quad , \tag{11.34}$$

where

$$D_s = h \tan \psi \quad , \tag{11.35}$$

and ψ = angle of shear. Replacing D_s in Eq. 11.34 by 11.35 and substituting into Eq. 11.33 gives

$$h^2 \ \tfrac{1}{2}\tan\psi + h\ D - S = 0 \ . \tag{11.36}$$

This is a quadratic equation in h of a curve through the origin. The curvature of the area-depth line is a constant that depends on the shear angle. If the shear angle is low, the value of $\tan\psi$ is small, the curvature is not very great, and therefore could easily go undetected.

11.4.5
Applications

The first application is to the full graben model in Fig. 11.18. The reference level is at the regional elevation of horizon 1. The points representing the pre-growth interval fall on a straight line because they all have the same displacement. The inverse slope of the best-fitting straight line (–4.5 units) is the total displacement on the lower detachment used to generate the model. The point at which the lost area goes to zero is the location of the lower detachment (–9.62 units below the reference level) which is the position of the lower detachment in the model. The growth beds (horizons 1–5) have lost areas that decrease upward. The lost area of the youngest growth bed is not zero because an increment of extension has occurred after the deposition of this unit. Layer-parallel strain is found with Eq. 11.27 and thickness changes by substituting the layer-parallel strains into Eq. 11.16. The pre-growth sequence is significantly stretched horizontally and thinned vertically by the deformation (Table 11.1), as is obvious on the cross section. The total displacements of the growth beds have been determined from Eq. 11.18 (Table 11.1), given the position of the lower detachment, and match the

	Unit top	Total displacement	Layer-parallel extension (%)	Layer-normal shortening (%)
Table 11.1. Structural data calculated from the full graben model in Fig. 11.18a (Groshong et al. 2003)	1	1.0	+0.9	–0.9
	2	1.5	+2.6	–2.5
	3	2.5	+4.6	–4.4
	4	3.0	+7.0	–6.5
	5	4.0	+10.3	–9.3
	6	4.5	+15.2	–13.2
	7	4.5	+24.0	–19.4
	8	4.5	+37.5	–27.3
	9	4.5	+66.2	–39.8
	10	4.5	+104.2	–51.0

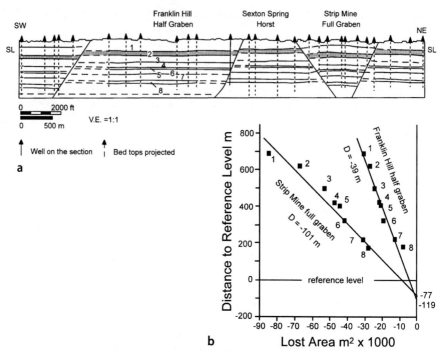

Fig. 11.21. Area-depth interpretation of the Deerlick Creek coalbed methane field. **a** Cross section perpendicular to fault strike showing the calculated lower detachment at the base of the section. The *dashed line* near the base of the section is the reference level. Units *1–8* are coal cycles. **b** Lost-area diagram for the Franklin Hill and Strip Mine grabens. *Solid lines* are the least-squares best fits; *V.E.*: vertical exaggeration. (Modified from Groshong 1994, after Wang 1994)

input values. The growth sequence includes depositional thickening which is greater than the structural thinning, giving a net thickness increase in each growth unit. The structural thinning (Table 11.1) can be recognized from the area balance (Eq. 11.27) in spite of the net thickness increase due to deposition.

A typical field application to an extensional structure is illustrated by a pair of grabens from the Pennsylvanian coal measures of the Black Warrior basin of Alabama. The cross section (Fig. 11.21a) is based on multiple horizons from closely spaced wells. Is this a valid cross section and how deep do the faults extend? The lost area is measured for each coal cycle boundary. The positions of the cycle boundaries at their footwall cutoffs coincide with regional dip across the area, from which it is inferred that footwall uplift is negligible. Lines joining the footwall cutoffs are therefore chosen as the regionals. The lost area for a given cycle boundary is the area bordered by the position of the boundary in the graben, the faults, and the regional for that cycle. The reference level was chosen to be at the base of the well control and parallel to regional dip. The distance from each regional to the reference level is measured in the center of each graben.

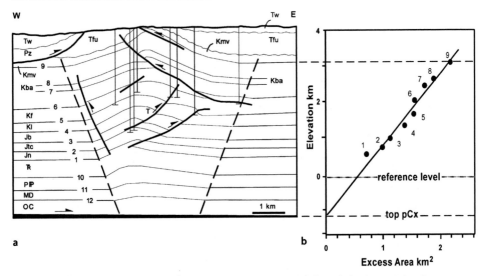

Fig. 11.22. Area-depth interpretation of the Tip Top anticline, Wyoming thrust belt, western U.S. **a** Cross section, areas below horizon 1 match the best-fit area-depth line (Groshong and Epard 1994, after Webel 1987). **b** Best-fit area-depth line (after Epard and Groshong 1993)

The area-depth curves for both grabens are straight lines with a modest amount of scatter (Fig. 11.21b). This amount of scatter is normal and so the section is validated. The best-fit lines go to zero area at depths below the reference level of –199 and –77 m, representing the predicted depths of the detachment. The agreement between the depths to detachment, calculated independently for two different grabens of different structural styles, is additional evidence of a valid result. The displacement on the lower detachment is 39 m for the Franklin Hill half graben and 101 m for the Strip Mine full graben.

The Tip Top anticline provides a field example of a compressional fold formed by shortening at the front of the Wyoming thrust belt (Fig. 11.22a). The upper part of the structure is constrained by surface geology and the wells but the deeper structure is open to question. Is the interpretation valid? The presence or absence of layer-parallel strain is unknown, as is the position of the lower detachment.

The data points from nine measurable horizons in the anticline fall reasonably close to the least-squares best-fit straight line $h = (1/0.51) S - 1.10$, with correlation coefficient 0.961 (Fig. 11.22b). This result validates the interpretation and indicates that the structure could be a locally balanced detachment fold (Sect. 11.4.3). The displacement on the lower detachment is 0.51 km. The depth of the basal detachment is 1.10 km below the reference level at the base of the Triassic. This places the lower detachment at the stratigraphic level of the top of the crystalline basement. The extrapolation of the section to the lower detachment in Fig. 11.22a is based on the area-depth line and shows the size of the structural closures predicted from the area-depth line for horizons 10–12 below the base of the Triassic.

11.5
Rigid-Body Displacement

This section begins the presentation of the kinematic-model-based restoration and prediction techniques. If all the deformation is by the displacement of rigid blocks, then a cross section can be restored by rigid-block translation and rotation. Rigid-body displacement preserves all the original lengths and angles within the blocks on the deformed-state cross section. This method is suitable for cross sections consisting of internally undeformed fault blocks, most commonly found in extensional structural styles. The sequential restoration of the Rhine Graben (Fig. 11.9) was done by rigid-body displacement.

11.5.1
Restoration

Rigid-body restoration can be done by cutting the cross section apart on the faults, removing the offsets, and reassembling it so that the reference horizon has the required restored shape. An analogous strategy works with computer-drafting programs. The restoration can be done on paper without cutting by tracing the blocks on a transparent overlay (Fig. 11.23). Draw the pin line, the restored-bed reference line, and the footwall on the transparent overlay. Move the overlay so that the first block is in its restored position and trace it onto the overlay. Then move the overlay so that the second block is in the restored position and trace it. Repeat this step until the section has been restored. If the blocks do not fit together perfectly, it is preferable to leave gaps between them, rather than to overlap the blocks.

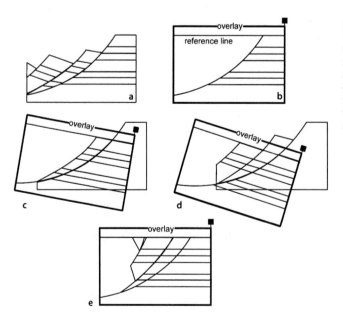

Fig. 11.23.
Restoration of rigid-body displacement by the overlay method. **a** Section to be restored. **b** Preparation of the overlay. **c** Restoration of the first block. **d** Restoration of the second block. **e** Complete restoration

11.5.2
Domino-Block Predictive Model

Dominoes are formed by parallel planar faults for which both bedding and the faults rotate (Fig. 11.24). Found typically in extensional and wrench regimes, the amount of rotation may be large, 30° to 60° or more. Such large rotations by other models require very large internal strains and seem not to be favored. This was one of the first kinematic models to be quantified (Thompson 1960; Morton and Black 1975). As the blocks extend, they rotate. Following the approach of Wernicke and Burchfiel (1982), the relationship between the extension and the geometry is

$$e = [\sin(\phi + \delta)/\sin\phi] - 1 \quad , \tag{11.37}$$

where ϕ = the final dip of the fault, δ = the final dip of bedding. The initial dip of the fault is $\phi_0 = \phi + \delta$ allowing Eq. 11.37 to be rewritten as

$$e = [\sin\phi_0/\sin(\phi_0 - \delta)] - 1 \quad . \tag{11.38}$$

The slip on the fault depends on the width of the fault block, the wider the block, the greater the slip on each block. The fault slip is (Axen 1988)

$$S_r = L_0 \sin\delta/\sin\phi \quad , \tag{11.39}$$

where S_r = rotational fault slip and L_0 = the original horizontal distance between two adjacent faults. The rotational slip is calculated using the L_0 value of the hangingwall block.

The correct frame of reference for calculating domino block strain is the median line through the blocks or, equivalently, the enveloping surfaces that touch the corners

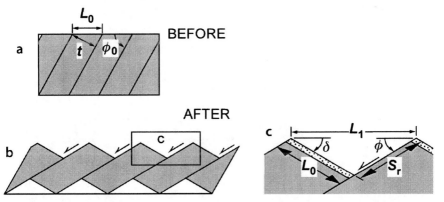

Fig. 11.24. Domino kinematics and strain (after Axen 1988). L_0: The original horizontal distance between two adjacent faults; L_1: final horizontal distance between two adjacent faults; ϕ_0: initial dip of the fault; ϕ: final dip of the fault; δ: final dip of bedding; S_f: slip on fault. **a** Before extension. **b** After extension. **c** Relationship among variables

of the blocks (Fig. 11.25a,b). Otherwise, external rotations can greatly change the amount of strain inferred from the domino geometry. To show the potential pitfall, an original set of domino blocks (Fig. 11.25a) is stretched 18.3% (Eq. 11.37, Fig. 11.25b), then rotated 15°. A counterclockwise external rotation of the stretched dominoes of 15° (Fig. 11.23c) produces horizontal bedding and faults having their original dips. According to Eq. 11.37, using a horizontal reference line, the domino strain would be zero because the bedding and faults are not rotated. From a median surface reference line, the strain remains 18.3%. A 15° clockwise rotation of the dominoes (Fig. 11.25d) increases the bedding dip and decreases the fault dip with respect to the horizontal

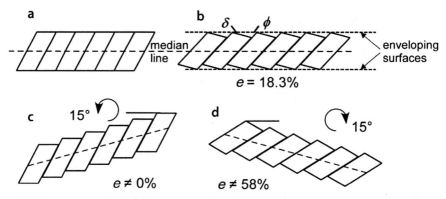

Fig. 11.25. Effect of external rotation on strain inferred from domino block geometry **a** Unrotated dominoes. **b** Internal rotation of domino blocks, without external rotation, 18.3% extension. **c** 15° counterclockwise rotation of **b**. **d** 15° clockwise rotation of **b**

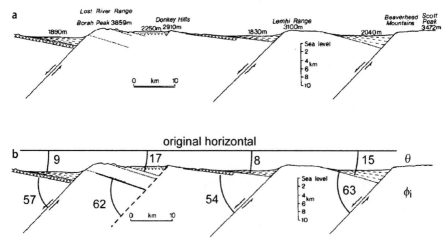

Fig. 11.26. Domino model applied to example in Basin and Range Province of the western U.S. **a** East-west cross section from the Wasatch fault to Bear Lake, about 100 km north of Salt Lake City, Utah (after Westaway 1991). **b** Measurements for domino-model strain calculation

and thereby greatly increases the amount of strain (to 58%) given by Eq. 11.37 using horizontal as the reference line. Again, using the median line as the reference, the strain is 18.3%.

Seismically active faults in the Basin and Range province of the western U.S. provide an example of domino-block deformation. The faults are planar and steeply dipping to depths of at least 10 km, breaking the crust into a series of dominoes (Fig. 11.26). How much crustal extension is implied by the geometry of the dominoes? The extension can be found from Eq. 11.38 and the measurements shown in Fig. 11.26b. Because the cutoff angles are different for each block, so are the extension strains. The horizontal extension is, from the left domino to the right domino, 12.9%, 24.9%, 12.5%, and 19.9%. Rigid dominoes should rotate the same amount and so the different amounts of rotation and the different strain magnitudes could be viewed as representing departures from perfect rigid-block deformation.

11.5.3
Circular-Fault Predictive Model

Circular-arc faults permit rigid rotation of the hangingwall block. Rarely are faults complete circular arcs, but the upper part of a listric thrust may be approximately circular. Quantitative kinematic models for block rotation on listric thrusts have been developed for and successfully applied to normal faults (Moretti et al. 1988) and basement-involved thrust faults (Erslev 1986, 1993; Seeber and Sorlien 2000). The model below is for a listric thrust fault. It allows the prediction of the location of the lower detachment and the displacement from the fault dip and the width of the structure.

The kinematic model (Fig. 11.27) is based on a fault that is a segment of a circular arc with radius R from a center C. The circular portion of the fault meets the planar

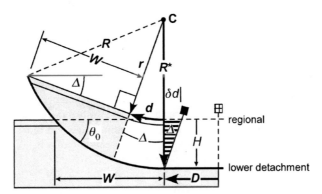

Fig. 11.27. Kinematic model of a rotated block above a circular-arc thrust. The hangingwall is displaced D along the lower detachment. *Light shading* is rigid; Δ: rotation angle; R: radius of curvature; C: *center of curvature*; R^*: *radius on a line or*thogonal to the regional; θ_0 is the dip at the fault at regional; d: displacement of a marker at a distance r from the center of curvature; W: width of the structure at regional; H: depth to detachment from the regional. The *dashed pin line* is the original trailing edge. The *solid-head pin* is the trailing edge of a constant-bed-length structure; δd is the length imbalance with respect to a vertical trailing edge

lower detachment along the radius R^*, orthogonal to the regional, located where the hangingwall marker bed in the anticline returns to regional dip. Rigid-block rotation is possible for the main part of the hangingwall (shaded, Fig. 11.27). The original dip of the fault at regional is θ_0, which, together with the width of the structure, W, is a function of the radius of curvature, r, of the marker bed

$$r = W / \tan \theta_0 \ . \tag{11.40}$$

The center of curvature of the fault is located a distance r above the regional along R^*. If the dip of the fault at regional is not visible, then it can be found from the dip of the fault at its tip, θ_1, from

$$\theta_0 = \theta_1 - \Delta \ , \tag{11.41}$$

where Δ dip of the fault block. Once r and W are known, the radius, R, to the detachment from the center of curvature is

$$R = (W^2 + r^2)^{1/2} \ . \tag{11.42}$$

The depth to the lower detachment from the regional is H, where

$$H = R - r \ . \tag{11.43}$$

The displacements are

$$d = r \, \Delta / 57.3 \ , \tag{11.44}$$

$$D = R \, \Delta / 57.3 \ , \tag{11.45}$$

where d = displacement at the level of the regional datum, D = displacement on the lower detachment, and Δ is in degrees.

Total rigid-block rotation is impossible because the trailing edge of the rotated block (unshaded, Fig. 11.27) moves from a planar lower detachment onto the curved ramp, requiring some form of internal deformation. If the original hangingwall remains rigid, the displacement increases downward toward the detachment. To maintain constant bed length and bed thickness behind the rigid block, flexural slip is required and would cause the trailing pin line to rotate (solid head, Fig. 11.27). The magnitude of the slip is

$$\delta d = D - d = H \tan \Delta \ . \tag{11.46}$$

The final geometry of the trailing pin line is of great importance. If the pin line rotates to maintain constant bed length and bed thickness, as shown by the solid-head pin in Fig. 11.27, then the rotated block will impose regional layer-parallel shear on the hinterland of the structure. Mechanically it seems much more likely that displacement of the hinterland block will be approximately constant, which requires that the trailing edge remain vertical in the deformed state. This leads to a second reason why perfect rigid-block rotation with the geometry of Fig. 11.27 is unlikely: it is not length or area balanced with a vertical pin line. The length imbalance is δd (Eq. 11.46). The area imbalance is equal to the area, A, of the triangle bounded by H and δd,

$$A = (H^2 \tan \Delta)/2 \ . \tag{11.47}$$

Thus it is unlikely that the hangingwall will maintain constant length and thickness, but instead it is likely to deform internally in some fashion in order to remain locally balanced.

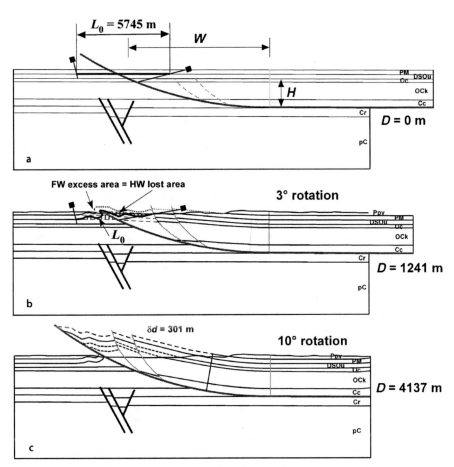

Fig. 11.28. Model-based sequential restoration of Wills Valley anticline. *D:* displacement on lower detachment. **a** Undeformed. **b** After 3° rotation of hangingwall, just before master fault breaks across PM. *Shading* shows area transferred from hangingwall to footwall. **c** After 10° rotation of hangingwall into present day configuration. Everything above the erosion surface is schematic, based on the model

The Wills Valley anticline provides an example of rigid-block rotation above a circular-arc thrust and of a model-based sequential restoration. The cross section is from the middle of an approximately 140 km long, straight, anticline located in the frontal part of the southern Appalachian fold-thrust belt. The cross section on which Fig. 11.28c is based is derived from outcrop geology and a seismic reflection profile (Fig. 1.46). This structure fits a circular-block model with best-fit measurements of $\Delta = 10°$, $W = 8.88$ km and $\theta_0 = 22°$. From Eqs. 11.40, 11.42, and 11.43, respectively, $r = 21.98$ km, $R = 23.7$ km, and $H = 1.73$ km. From Eq. 11.45, the displacement that produced the structure is $D = 4.137$ km. These values allow the structure to be restored to zero displacement (Fig. 11.28a), from which the bed lengths are measured and included in Fig. 11.28c. From Eqs. 11.46 and 11.47, respectively, $\delta d = 301$ m, and $A = 5.3 \times 10^5$ m^2. These values are included in Fig. 11.28c with the displacement accommodated by the hangingwall splay thrusts and the excess area accommodated by the uplift on the splays.

A footwall syncline is preserved beneath the master thrust indicating a fault-tip fold was once present, which is reconstructed in Fig. 11.28b. The outer limits of the region inferred to be affected by fault-tip folding are indicated by the pin lines shown in Fig. 11.28a,b. The outer limit of folding on the footwall is obtained directly from the cross section. The outer limit of folding on the hangingwall is somewhat arbitrary because the fold has been eroded from the hangingwall. The pin is placed at the hangingwall cutoff of the top OCk, which is appears not to be folded, and is sloped sharply outward to include comparable widths of PM in the hangingwall and footwall, giving a length L_0 of 5 745 m at the base of the PM. If L_0 is too short, the necessary area cannot be obtained on the hangingwall; if it is too long, an unrealistic amount of the hangingwall is folded. The reconstruction of Fig. 11.28b is constrained by the known geometry of the footwall folds, area balance, and maintaining constant bed length at the base of the PM. A 3° rotation of the hangingwall and attendant displacement causes the straight-line distance between the endpoints of L_0 to be reduced to 4 751 m. These constraints require folding on the hangingwall as well as the footwall. For a 3° rotation, δd at the top of the Cc is 117 m (Eq. 11.46) and the excess area required to maintain a vertical trailing pin line is $A = 1.6 \times 10^5$ m^2 (Eq. 11.47). These values are incorporated into the geometry of the hangingwall as displacement and thickening on the splay faults and their hangingwalls.

11.6
Flexural-Slip Deformation

Flexural-slip restoration is based on the model that bed lengths do not change during deformation (Chamberlin 1910; Dahlstrom 1969; Woodward et al. 1985, 1989). Internal deformation is assumed to occur mainly by layer-parallel simple shear (Fig. 11.3b). For the area to remain constant, the bed thicknesses must be unchanged by the deformation as well. This is the constant bed length, constant bed thickness (constant BLT) model. Flexural-slip restoration is particularly suitable where the beds are folded and structurally induced thickness changes are small, the style of deformation in many compressional structures. Flexural slip is also commonly used to restore the complex deformation above salt layers (Hossack 1996).

11.6.1
Restoration

Flexural-slip restoration consists of measuring bed lengths and straightening the lengths while preserving the thicknesses to produce the restored section. The section to be restored is bounded on one end by a pin line and on the other by a loose line (Fig. 11.29). The pin line is a straight line on the deformed-state cross section th it is required to remain straight and perpendicular to bedding on the restored section. The loose line is a straight line on the deformed-state section that may assume any configuration on the restored section, as required by the restoration. For the sake of clarity, pin lines are shown here as having solid heads and loose lines as having open heads.

A loose line that is straight and parallel to the pin line is the best indication of a valid restoration (Fig. 11.29b). Faults on the deformed-state cross section are shown in their restored positions on the restored section (Fig. 11.29b). The reasonableness of the restored shape of the fault is another criterion for the quality of the restoration. Note that the lengths of lines oblique to bedding and the cutoff angles between bedding and faults are changed by the deformation and so are different in the deformed-state and restored cross sections. A loose line that is highly irregular is the clearest indication of a section that is not valid (Fig. 11.30a). A straight but inclined loose line (Fig. 11.30b) indicates a systematic length difference that may represent an invalid cross section, a poor choice of the pin line or the loose line, or material introduced or removed by layer-parallel simple shear. Usually at least four to five beds must be tested to determine whether the section is correctly restorable or if some type of error is present.

In complex structures, the choice of pin line and loose line requires care. The goal is to choose the pin line and loose line in locations that can be expected to restore to perpendicular to bedding and consequently to have constant bed lengths between them. There are a number of possible choices. The best location for a pin line and loose line

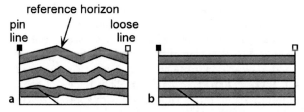

Fig. 11.29. Cross-section restoration. *Pin line*, solid-head; *loose line*, open-head. **a** Deformed-state cross section. **b** Restored section: equal bed lengths and a *straight loose line* indicate a satisfactory restoration

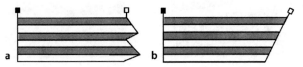

Fig. 11.30. Imperfect restorations. **a** *Irregular loose line* indicates an invalid section. **b** Systematic error: the *linear loose line* may indicate an invalid section or a poor choice of pin line or loose line

is in undeformed beds adjacent to the area of interest, but such locations may not be available. Often both the pin line and the loose line must be placed in deformed beds. Some of the better possibilities are

1. Along an axial surface. This is appropriate for fixed-hinge folds in which shear does not occur across the axial surfaces.
2. Perpendicular to horizontal beds. This is appropriate if the strain is zero or constant where beds are horizontal.
3. Pin line and loose line located where the dip is the same in amount and direction. This is appropriate where layer-parallel shear is proportional to dip.

The typical result of an incorrect choice of a pin line or loose line is the systematic length error shown by a straight but inclined loose line (Fig. 11.30b). A systematic error might be corrected by a better choice of pin line and/or loose line without any changes to the cross section itself.

To restore a structure, the pin line and loose line are chosen (Fig. 11.31a), and the bed lengths are measured between them, stretched out and placed on the restoration (Fig. 11.31b). Lengths may be measured by a variety of means. Chamberlin (1910) used a thin copper wire, curved to follow bedding, then straightened. A ruler or a straight piece of paper can be rotated along the contact, the lengths of many small segments marked, and the total measured at the end. Computer methods are especially quick and convenient (Groshong and Epard 1996). Thicknesses in the deformed-state are preserved on the restored cross sections. Faults are drawn on the restored cross section in the positions required by the restored bed lengths. The restored fault trajectories should match those expected for the structural style. Ordinarily faults should restore to planes, smooth curves, or ramp-flat geometries.

A question that must be addressed with any restoration is how straight must the loose line be in order to represent a valid cross section? The loose line in Fig. 11.31b is not perfectly straight, but the cross section can be considered valid for most purposes. Constant BLT line-length restoration is a robust technique for which small violations of the constant BLT assumption cause only small effects in the restoration. The lower, shorter part of the loose line in Fig. 11.31b may be caused by a small amount of layer-parallel shortening and thickening of the units in the inner arc of the physical model. Discrepancies that clearly indicate an invalid cross section are large (Fig. 11.32). Wilkerson and Dicken (2001) provide an excellent review of common problems revealed by the restoration of thrust faults along with appropriate corrective measures.

 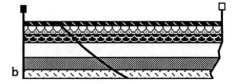

Fig. 11.31. Flexural-slip restoration of the cross section of a physical model. a Deformed-state section. b Restored section. (After Kligfield et al. 1986)

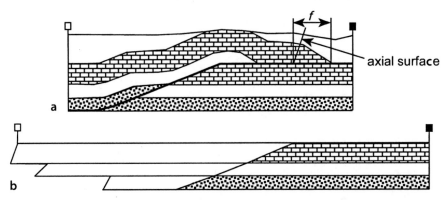

Fig. 11.32. Flexural-slip restoration of an invalid cross section. **a** Deformed-state cross section. Preserved beds maintain constant bed thickness (after a problem in Woodward et al. 1985). *f:* hangingwall fault ramp. **b** Flexural-slip restoration based on length measurements at the top and base of each unit

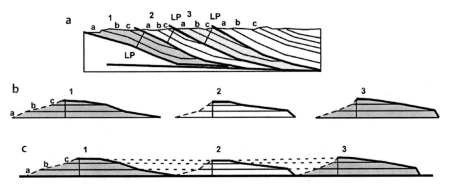

Fig. 11.33. Restoration using local pin lines (after Woodward et al. 1989). **a** Deformed-state cross section of eroded thrust sheets. *LP:* local pin line; *1–3:* thrust sheets; *a–c:* stratigraphic units. **b** Thrust sheets individually restored from local pin lines. **c** Thrust sheets restored to minimum displacement positions. *Dashed lines* are eroded beds

A simple measurement pitfall to avoid is including a fault length with the bed length. In Fig. 11.32a, a portion of the base of the limestone is a fault ramp (f) where the bedding is truncated against the upper detachment. The fault ramp segment is not part of the bed length of the base of the limestone. The fault ramp is separated from unfaulted bedding by an axial surface which, in the case of constant bed thickness, bisects the associated fold hinge (Sect. 6.4.1). Splitting the limestone bed by drawing many parallel beds within the unit would make the fault cutoffs, and hence the ramp, obvious.

Local pin lines (Fig. 11.33) should be used where erosion or lack of information renders the connection between hangingwall and footwall uncertain. A local pin line is placed in each block and the beds stretched out in both directions from the local pin. The individual restored blocks can then reassembled by bringing the individual blocks as close together as geologically appropriate.

A stratigraphic template (Woodward et al. 1989) greatly speeds up the reconstruction process for beds having uniform thickness or a constant thickness gradient. The template is a series of lines representing the restored bed geometry. For constant bed thickness the lines are parallel. A constant bed-length restoration can then be done by measuring bed lengths and marking the appropriate lengths on the template. This saves the time of drawing each bed segment separately. Regional thickness gradients can be shown on the template by diverging lines.

Units that have primary stratigraphic thicknesses changes may be restored by a modification of the constant line-length method. Begin with the uppermost unit that is to be restored, the top of which is the reference horizon. The thickness of this unit is measured at multiple points along the layer (Fig. 11.34a). The positions of the thickness measurements are recorded with respect to the pin line on the top of the unit. The top of the unit is restored and the distances of the thickness measurement points from the pin line are marked. The thicknesses are marked on the restoration in the direction perpendicular to the upper horizon (Fig. 11.34b). The lower horizon is drawn to maintain its original length but it may "slip" laterally past the points where the thickness was measured. The restoration preserves the lengths between segments on the base of the unit but not necessarily their positions. The total line length is preserved on the top and base. For a section containing faults (Fig. 11.34c,d), thickness measurements are taken at fault cutoffs and restored with respect to the upper cutoff if possible, otherwise with respect to the lower cutoff. Each stratigraphic unit on the cross section is measured and restored separately to the base of the unit above it until the restoration is complete.

Difficulties may be encountered in piecing the section back together across faults. The problems are usually related to uncertainty in the correct restoration of the fault shape. The best solution is to subdivide the stratigraphy into thinner units and then restore the thinner units as just described. In the limit, each unit being restored is a single line of variable thickness that is stretched out and placed along side the unit above it.

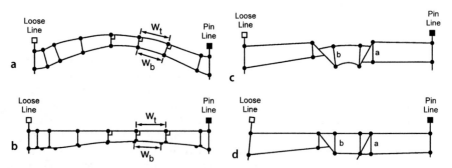

Fig. 11.34. Bed-length restoration of variable thickness units (after Brewer and Groshong 1993). *Dots represent points at which original thickness measurements are made.* W_t: segment width at top; W_b: segment width at base. **a** Deformed state fold. **b** Restored fold. **c** Deformed-state faulted fold. **d** Restored faulted fold

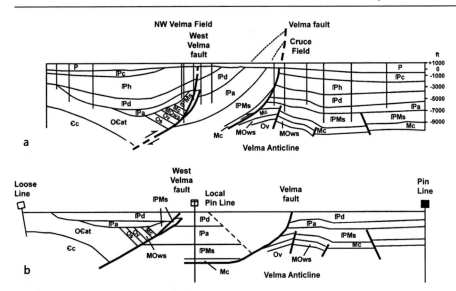

Fig. 11.35. Northwest Velma and Cruce oil fields. **a** Cross section based on wells and a seismic line (redrawn from Perry 1989). *P:* Permian red sand and shale; *Pc:* Cisco Group; *Ph:* Hoxbar sand, shale and limestone; *Pd:* Deese sand, shale and limestone; *Pa:* Atoka Formation; *PMs:* Springer sand and shale; *Mc:* Caney shale; *MOws:* Woodford shale, Hunton limestone and Sylvan shale; *Ov:* Viola limestone; *Os:* Simpson Gp. including Bromide and Oil Creek sands; *OCat:* Arbuckle Gp. limestone and Timbered Hills Gp. limestone and sandstone; *Cc:* Carlton Rhyolite. **b** Flexural-slip restoration

A cross section of a convergent-wrench structure from central Oklahoma provides an example of flexural-slip restoration where variable stratigraphic thicknesses are to be preserved (Fig. 11.35a). The section is located across the down plunge extension of the Wichita uplift, a basement cored anticline demonstrated by McConnell (1989) to be a convergent wrench-fault structure. The presence of folding as well as faulting suggests that the flexural-slip restoration technique is appropriate, as modified for variable original bed thickness. The pin line is placed in the horizontal beds on the right side of the cross section and the top of the Pd is restored to horizontal. The restoration proceeds to the left and downward from the pin line in the uppermost bed. The Velma fault is a major discontinuity in bed thickness and so the restoration from the pin line is stopped at the fault. A local pin line is placed perpendicular to bedding in the hangingwall of the Velma fault and the restoration continued to the left. The restoration (Fig. 11.35b) is good, indicating that the cross section is valid and, in addition, provides information about the geological evolution. The restoration is geometric, not palinspastic. The Velma fault marks a significant stratigraphic discontinuity on the restored section, as would be expected if it has an unrestored strike-slip component. Some amount of unrestored vertical-axis rotation of the cross section is also possible (c.f., Sect. 11.2.2). Restoration to the top Pd removes the displacement on the Velma fault but preserves the Velma anticline and the West Velma fault, showing that they formed earlier.

11.6.2
Fault Shape Prediction

Based on the assumption of constant BLT on a cross section bounded by vertical pin lines, there is a unique relationship between the hangingwall shape caused by movement on a fault and the shape of the fault itself. In the technique developed by Geiser et al. (1988), a complete deformed-state cross section is constructed while simultaneously producing a restored cross section. The method is based on: (1) constant bed length and bed thickness, (2) slip parallel to bedding, (3) fixed pin lines in the hangingwall and footwall of the fault, and (4) the hangingwall geometry is controlled by the fault shape. The pin lines are chosen to be perpendicular to bedding. The data required to use the method are the location of a reference bed and the hangingwall and footwall fault cutoff locations of the reference bed. The original regional of the reference surface is not required. The method produces a cross section that is length balanced, has constant bed thickness and has bedding-normal pin lines at both ends.

The technique is as follows:

1. Define the reference bed, its fault cutoffs, and the shape of the fault between the cutoffs (Fig. 11.36a, top).
2. Place the pin lines in the deformed-state cross section. Usually they are chosen to be perpendicular to bedding and beyond the limits of the structure of interest (Fig. 11.36a, top).
3. Measure the bed length of the reference horizon between the pin lines and draw the restored-state section (Fig. 11.36a, bottom).
4. Construct one or more constant thickness beds in the hangingwall between the hangingwall pin and the fault (Fig. 11.36b, top).

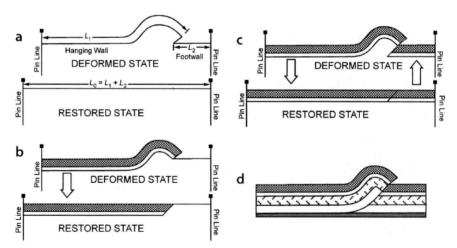

Fig. 11.36. Construction of fault shape from a hangingwall bed based on the constant BLT assumption (Geiser et al. 1988). **a** Initial geometry. **b** After first cycle of construction, hangingwall is restored. **c** After the first cycle of hangingwall restoration, the footwall is restored. **d** Completed cross section

5. Measure all the bed lengths in the hangingwall between the hangingwall pin and the fault and place them on the restored-state cross section (Fig. 11.36b, bottom). This defines the shape of the footwall beds in the restored state.
6. Draw the restored-state footwall on the deformed-state cross section (Fig. 11.36c, top). This gives a new hangingwall cutoff.
7. Continue the cycle of steps 4–6 until the section is complete (Fig. 11.36d).

This predictive technique is sensitive to the constant BLT assumption. Like the classical depth-to-detachment calculation (Sect. 11.4.2), small bed length changes will have a significant effect on the predicted depth to detachment. It is always best to confirm the predicted fault shape with additional information.

11.6.3
Flexural-Slip Kinematic Models

Numerous models have been developed that relate fold shape to fault shape for various flexural-slip structural styles. Many of these have adopted planar or ramp-flat fault shapes, leading to dip-domain cross-section styles (Sect. 6.4.1). Models like the fault-bend fold, fault-propagation fold and fault-tip fault-propagation fold of Fig. 11.37a–c are based on the assumption of constant bed length and constant bed thickness throughout and so yield unique relationships between the fold shape and the fault shape. Such models can be used to predict the entire structure from a very small amount of hard data. Other models, like the simple-shear, pure shear, and detachment folds in Fig. 11.37d–f, maintain constant BLT in the upper part of the structure but are area balanced in the decollement zone at the base. The latter style of model admits a wider range of geometries and so more must be known before the complete structure can be predicted. Other variants allow thickness changes in the steep limb, which creates even more degrees of freedom in the interpretation. A discussion of all the models and their variants is beyond the scope of this book. The objective here is to introduce the basic concepts and provide a guide to additional sources of information.

Fig. 11.37.
Fault-related fold models. *Unshaded:* constant-thickness units; *shaded:* variable-thickness unit. **a** Fault-bend fold (after Suppe 1983). **b** Fault-propagation fold (after Suppe 1985; Suppe and Medwedeff 1990). **c** Fault-propagation fold at the tip of a long ramp (after Chester and Chester 1990). **d** Simple-shear fault-bend fold (after Suppe et al. 2004). **e** Pure-shear fault-bend fold (after Suppe et al. 2004). **f** Detachment fold (after Jamison 1987)

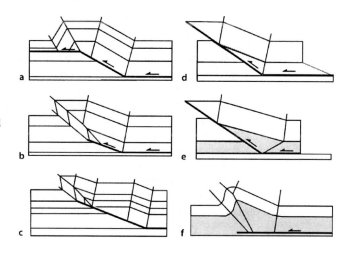

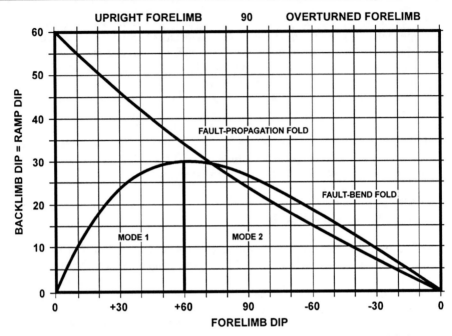

Fig. 11.38. Graph of forelimb vs. backlimb dips in constant bed thickness fault-bend and fault-propagation folds (Fig. 11.37a,b). The backlimb is always above the ramp

The starting point for a model-based interpretation is to determine which model is appropriate. Constant BLT fault-bend and fault-propagation folds have analytical relationships between forelimb and backlimb dips that can be expressed graphically. The graph (Fig. 11.38) is a plot of forelimb dip versus backlimb dip, with the dips being measured relative to the regional dip outside the fold. A field example can be plotted on Fig. 11.38 to see if falls on either the fault-bend or fault-propagation fold curve. If the point falls on one of the lines, then there is a high probability that the fold fits the model and the model can be used to predict the fold geometry.

As an example of the methodology, the deep structure of the fold in Fig. 11.39a is interpreted. The regional dip is inferred to be parallel to the planar domain between the two limbs and is confirmed because the same dip is seen outside the structure. The limb dips are measured with respect to regional Fig. 11.39a and plotted on Fig. 11.38, The points fall on the fault-bend fold curve, indicating the model to be used. In Fig. 11.39b the interlimb angles are measured and bisected to find the axial-surfaces. The backlimb axial surfaces (1 and 2) are parallel, as are the forelimb axial surfaces (3 and 4). Beds must return to regional dip outside the anticline, allowing the location where the forelimb flattens into regional dip (axial surface 4) to be determined. The fault ramp must be below and parallel to the backlimb (Fig. 11.39c). The exact location of the fault requires additional information about the position of the lower and upper detachments. Bed truncations in the forelimb of the hangingwall or against the ramp

Fig. 11.39.
Construction of a complete
fault-bend fold starting with a
single horizon. **a** Starting fold
geometry with dips measured
from regional dip. **b** Axial surfaces bisect hinges. **c** Location
of the fault. *Ud:* upper detachment; *f:* fault ramp; *Ld:* lower
detachment. **d** Final geometry

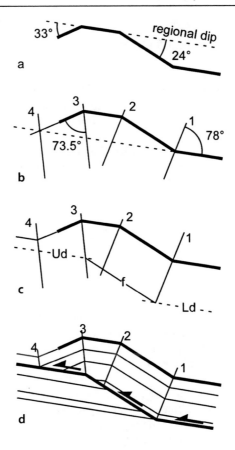

in the footwall may serve to locate the detachments. Alternatively, the stratigraphic
section may contain known detachment horizons that can be utilized in the interpretation. Additional controls from the model are that axial surfaces 1 and 2 always terminate at the upper detachment, and axial surface 1 always terminates at the base of the
ramp. Axial surface 3 can be on the ramp or at the top of the ramp. Having determined
the best location for the fault, the dip domains can be filled in with the appropriate
stratigraphy (Fig. 11.39d) and the interpretation is complete.

   Since the pioneering work by Suppe (1983), there have been numerous papers describing models for flexural-slip fault-related folds. The following is a representative
list of significant papers on the topic: Jamison 1987; Chester and Chester 1990; Mitra
1990; Suppe and Medwedeff 1990; Jordan and Noack 1992; Suppe et al. 1992; Mitra
1993; Epard and Groshong 1995; Homza and Wallace 1995; Wickham 1995; Poblet and
McClay 1996; Poblet et al. 1997; Medwedeff and Suppe 1997; Suppe et al. 2004. The
recent book by Shaw et al. (2005) is an excellent discussion of the application of models to compressional structures. Predictions based on 3-D axial-surface geometry are
described by Shaw et al. 2005b and Rowan and Linares 2005).

11.7
Simple-Shear Deformation

According to the simple-shear concept, a cross section deforms as if it were made up of an infinite number of planar slices that are free to slip past one another (Fig. 11.40). The shear direction is specified as having a dip α with respect to a regional that does not change during deformation. The special case for which the shear plane makes an angle of 90° to the regional is called vertical simple shear because the regional is usually horizontal (Fig. 11.40a). Oblique simple shear is simple shear along planes at some angle other than perpendicular to the regional (Fig. 11.40b, $\alpha \neq 90°$). (Some workers measure α from the vertical.) Simple-shear restoration is most widely used for extensional structures, in particular the hangingwall rollovers associated with half grabens.

11.7.1
Restoration

11.7.1.1
Vertical Simple Shear

Restoration by vertical simple shear involves the differential vertical displacement of vertical slices of a cross section to restore a reference horizon to a datum (Verrall 1982). This method has its origins in the well-log correlation technique in which the logs are slipped vertically until aligned side-by-side along an easily recognized marker horizon. The alignment makes the correlation of nearby markers much easier. Many computer programs designed for geological or geophysical data interpretation implement a function of this type. Folds and growth stratigraphy are easily restored by this method. The restoration of dipping beds by vertical simple shear preserves vertical (isocore) thicknesses but changes the stratigraphic thicknesses and the bed lengths.

Begin a vertical simple-shear restoration by picking the reference horizon to be flattened. In the example of Fig. 11.41a, the top of the sandstone is selected to be the reference horizon. Define the individual elements to be displaced (lithological logs in this example). Restore the cross section by shifting the elements vertically until the

Fig. 11.40.
Simple shear oblique to bedding. **a** Vertical simple shear. **b** Oblique simple shear

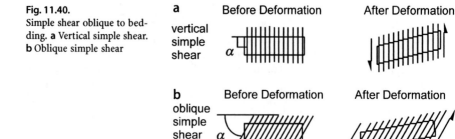

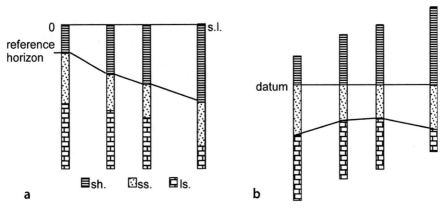

Fig. 11.41. Flattening to a datum by vertical simple shear. **a** Deformed-state cross section based on lithological logs. The zero elevation is sea level (*s.l.*); *sh:* shale; *ss:* sandstone; *ls:* limestone. **b** Cross section flattened to a datum at the top of the sandstone shows a paleo-anticline at the top of the limestone

reference horizon is horizontal (Fig. 11.41b). The operations required to restore the cross section may be easily performed by an overlay method. Draw a set of equally spaced vertical lines on the cross section to represent the positions of columns to be restored. The spacing of the vertical lines controls the level of detail in the horizontal direction that will be obtained in the restoration. Draw a horizontal line on a transparent overlay to represent the restored datum. Mark the horizontal position along the datum line of each column to be restored. Shift the overlay to bring each column to the restored datum and mark on the overlay the position of each stratigraphic unit. Complete the restoration by connecting the stratigraphic horizons to form a continuous cross section.

If the cross section includes faults, their displacements can be removed in the restoration. In the simple shear model, fault displacement parallel to the regional is constant and equal to the hangingwall block displacement, *D*. The steps in the restoration are given below (refer to Fig. 11.42 for the geometry).

1. Define the regional.
2. Find the block displacement by projecting a vertical line from the hangingwall cutoff of the reference bed to the regional. The block displacement is the distance *D* parallel to regional from the footwall cutoff of the marker bed.
3. Construct a series of vertical working lines, spaced *D* apart, starting at the fault.
4. To restore the hangingwall, each vertical segment along a working line in the hangingwall is moved up the fault until the horizontal component of displacement is equal to *D*. The base of each segment remains in contact with the fault and the top marks the position of the restored reference bed. Record the position of each marker horizon along the working line.
5. Repeat step 4 for as many vertical hangingwall segments as necessary to produce a smooth restoration of the hangingwall beds.

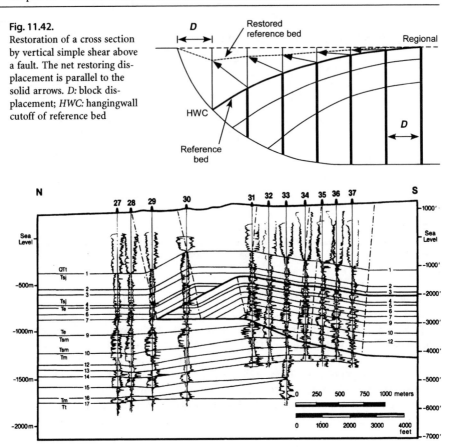

Fig. 11.42.
Restoration of a cross section by vertical simple shear above a fault. The net restoring displacement is parallel to the solid arrows. *D:* block displacement; *HWC:* hangingwall cutoff of reference bed

Fig. 11.43. Cross section of the thrust-cored Wheeler Ridge anticline, California (Medwedeff 1992). Wells show logs used for correlation of units

11.7.1.2
Pitfalls in Flattening

Restoration requires deciding whether a fold or fault is a pre-existing feature, to be passively deformed by the restoration, or is to control the restoration (an active feature). Fault-fold relationships are severely distorted if the hangingwalls of the active faults are not restored separately from the footwalls. As an example of the difference between restorations based on the active versus passive role of faults, consider the Wheeler Ridge anticline (Fig. 11.43). The cross section can be restored by treating the faults either as passive markers or as active elements with displacements that must be removed during restoration. Both alternatives are illustrated using the vertical simple shear technique. In both restorations the section is flattened to a datum at the top of the Tertiary San Joaquin Formation (Tsj). Assuming that the faults were passive markers, restoration (Fig. 11.44a) deforms the fault wedge but does not displace it horizon-

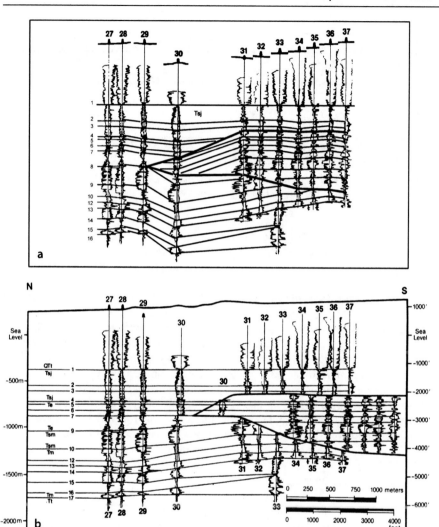

Fig. 11.44. Alternative restorations of Wheeler Ridge anticline (Fig. 11.43) by flattening to a datum at the top of the Tertiary San Joaquin Fm (Tsj, marker 1). Wells show logs used for correlation of units. **a** Flattening to a datum, disregarding possible slip on the faults. **b** Restoration by flattening to the Tsj datum with removal of fault slip (Medwedeff 1992)

tally. Assuming that the faults were active during the formation of the anticline, the fault wedge is restored by displacing the elements horizontally along the faults (Fig. 11.44b) so as to remove all of the hangingwall displacement. Which restoration is better is a geological decision. The restoration based on active faults (Fig. 11.44b) successfully flattens all the horizons down to horizon number 7, whereas the restoration based on passive faults (Fig. 11.44a) flattens only the original datum horizon. The

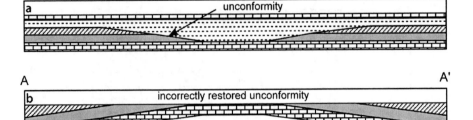

Fig. 11.45. Unconformity flattening pitfall. **a** Cross section of an undeformed unconformity and overlying valley fill. **b** Unconformity restored to horizontal showing an anticline that never existed

fault displacements in Fig. 11.44a must still be removed by a subsequent restoration. Based on the principles of simplicity and structural reasonableness, the restoration using active faults is a better representation of the geology at the time that the datum was horizontal.

Unconformities are not usually horizontal when they form, making the restoration of an unconformity a potential flattening pitfall (Calvert 1974). Flattening an unconformity to horizontal converts the original topography on an unconformity surface (Fig. 11.45a) into an anticline (Fig. 11.45b). The anticline indicated by the restoration was never present. An analogous problem occurs if the thickness variations of the unit directly above an unconformity are interpreted as representing a paleostructure when they, in fact, represent paleotopography. Interpretation of thickness variations in the shale overlying the unconformity in Fig. 11.45a as a paleostructure would lead to the interpretation of a growth syncline and the adjacent thinner areas as being growth anticlines, an incorrect interpretation of the true geometry in this example.

11.7.1.3
Oblique Simple Shear

The vertical simple shear concept can be generalized to shear in any direction (White et al. 1986). An oblique simple-shear restoration follows the same procedure as the vertical simple-shear restoration, except that the measurement lines are inclined to the regional at an angle other than 90°. The oblique lengths measured on the deformed-state cross section (Fig. 11.46a) are restored by translation in the shear direction to return the reference horizon to the regional (Fig. 11.46b). The spacing between any two measurement lines in the direction of the regional (S), and the shear angle (α), remains constant from the deformed state to the restored cross sections in both vertical and oblique simple-shear restorations.

The procedure for restoring a faulted section by oblique simple shear is analogous to the procedure for vertical simple shear except that the shear angle is less than 90°. The shear angle is measured downward from the regional. The fault displacement parallel to the regional is constant and equal to the hangingwall block displacement, D.

Fig. 11.46.
Restoration by oblique simple
shear. *Medium-weight solid
lines* are marker beds. *Dotted
lines* represent the shear direction and are spaced an arbitrary distance *S* apart. The
widest lines are the thicknesses
in the shear direction to be
restored. The shear angle is α.
a Deformed-state cross section.
b Restored cross section

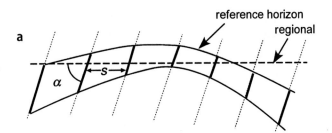

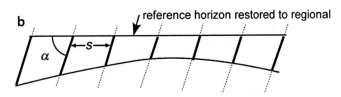

Fig. 11.47.
Restoration of the hangingwall
of a fault by oblique simple
shear. *FWC:* footwall cutoff of
reference bed; *HWC:* hangingwall cutoff of reference bed;
t_i: distance between reference
bed and fault, measured along
shear direction; α: shear angle;
D: block displacement

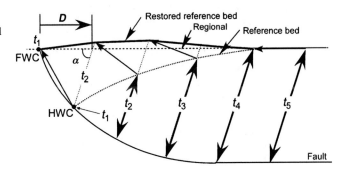

Note that for oblique simple shear, *D* is not equal to the fault heave on the ramp, but is
equal to the displacement on the lower detachment. The steps in the restoration are
given below (refer to Fig. 11.47 for the geometry).

1. Find the regional.
2. The block displacement is found by projecting a line parallel to the shear angle from
   the hangingwall cutoff of the reference bed to the regional. The block displacement
   is the distance *D* from the footwall cutoff of the marker bed.
3. Along the regional, mark off equal distances, *D*. Through each point draw a working
   line parallel to the shear direction, starting at the footwall cutoff of the reference bed.
4. Mark the position where each bed crosses a working line.
5. Each oblique segment of working line is moved up the fault a horizontal distance equal
   and opposite to *D*. Keeping the base of the oblique segment in contact with the fault,
   mark the location of the top, which is the restored position of the reference bed. Mark
   the restored position of all the other beds along this restored oblique segment.
6. Repeat steps 4 and 5 until the hangingwall is restored.

11.7.2
Fault Shape Prediction Techniques

11.7.2.1
Vertical Simple Shear

The vertical simple-shear method for predicting the fault shape from the hangingwall geometry was developed by Roger Alexander of Chevron and used successfully for many years in the Chevron Oil Company (Verrall 1982). The technique is sometimes referred to as the Chevron or Verrall method. If the data are vertically exaggerated, the inferred bed or fault geometries will have the same vertical exaggeration as the data used, which may differ from that in the region into which projections are made.

To use vertical simple shear to predict fault shape from the hangingwall fold, three pieces of geological information are necessary: (1) the shape of the fold in a hangingwall reference bed, (2) the regional for the hangingwall reference horizon, (3) the block displacement in the direction of the regional (D). In order to correctly determine D, the reference horizon must be correlated across the fault so that its hangingwall and footwall cutoffs are known. The technique is as follows (refer to Fig. 11.48 for the geometry):

1. Draw a straight line along the regional from footwall to hangingwall at the restored position of the reference bed.
2. Drop a perpendicular from the regional to the hangingwall cutoff of the reference bed.
3. Measure the fault separation parallel to the regional (D). Beginning at the footwall cutoff, mark equal distances along the regional with spacing equal to D.
4. Drop perpendiculars from the regional at the equal-D points to form working lines.
5. At a working line 1, measure the vertical distance from the regional to the fault.
6. Shift this measured length one D increment in the direction of transport (to line 2) and drop it down so that the top just touches the key bed in the rollover. The point marking the bottom of the vertical segment is the position of the fault.
7. Repeat steps 5 and 6, moving sequentially in the direction of transport (from line 2 to 11), until the section is complete.

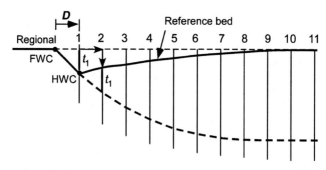

Fig. 11.48. Vertical simple-shear prediction of fault shape from rollover shape (after Verrall 1982). t_1 indicates the first vertical column that is shifted over D and down to the top of the reference bed. *FWC:* footwall cutoff of reference bed; *HWC:* hangingwall cutoff of reference bed; predicted fault is *heavy dashed line*

Fig. 11.49.
Determination of rollover
shape from fault shape using
vertical simple shear (modified from Verrall 1982).
Marker A has been used to
find the fault shape by the
method of Fig. 11.48. The
regional for bed B is obtained
from well 2. **a** Determination
of heave, **D**. **b** Construction of
position of marker horizon B;
predicted shape is *dashed*

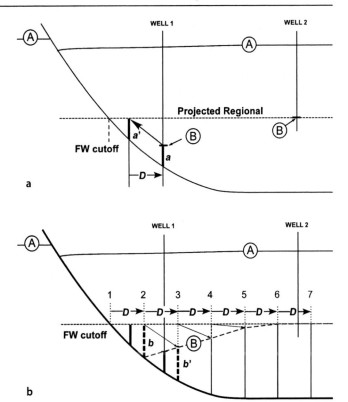

Once the shape of the fault is known, the geometry of any bed in the rollover can be
determined if its regional position and the location of at least one point in the rollover
is known (for example, B in Fig. 11.49a). The steps are as follows:

1. Find the regional by projecting a straight line along the regional from a known
   position of the regional in the hangingwall (B in well 2) to where it intersects the
   fault at the footwall cutoff. If the regional is known from the footwall, that level can
   be used for projection (Fig. 11.49a).
2. Drop a perpendicular from the regional, through point B in the rollover (in well 1)
   to the fault. The length *a* is the depth to the fault (Fig. 11.49a).
3. Move length *a* up the fault until it just touches the regional at *a'* (Fig. 11.49a).
4. The distance along the regional between *a* and *a'*, is the heave (*D*) of marker B on
   the fault (Fig. 11.49a).
5. Starting from the footwall cutoff, mark equal distances along the regional at a spacing equal to *D* and draw perpendiculars to form the working lines 2–7 (Fig. 11.49b).
6. Measure the distance between the regional and the fault along working line 2 (*b*)
   and shift it one heave increment down the fault in the direction of displacement to
   line 3 (*b'*) where the top of the measured line marks the position of marker B
   (Fig. 11.49b).

7. Repeat step 6 for the remaining working lines to find the shape of marker B in the rollover (Fig. 11.49b).

In both the restoration and prediction techniques, the starting locations of the working lines are arbitrary. The requirement is that a unit thickness marked off along a working line is always shifted laterally one displacement increment (D). The original working lines may be spaced as closely as desired to retain the detail of the cross section. This is also true for the oblique simple shear technique, discussed next.

11.7.2.2
Oblique Simple Shear

The oblique simple-shear technique for predicting the fault geometry from the shape of a hangingwall fold is the same as the vertical simple-shear method but with oblique working lines. The construction steps are as follows (refer to Fig. 11.50 for the geometry).

1. Draw the regional.
2. From the hangingwall cutoff of the reference bed, draw a line parallel to the direction of simple shear (α from the horizontal) to its intersection with the regional.
3. The length D is the displacement of the hangingwall block necessary to produce the observed geometry. Mark off equal distances D along the regional, starting from the footwall cutoff.
4. At each displacement increment, draw a working line parallel to the shear direction from the regional to the vicinity where the fault is expected to be (lines 1–6).
5. Measure the length t_1. Shift this length in the displacement direction one D increment and move it down the shear line until the top of the line just touches the reference bed in the rollover (location t_1'). The bottom of the line segment is the location of the fault.
6. Repeat step 6 for the next line segment ($t_3 = t_1' + t_2$), and continue to repeat in the direction of displacement until the section is complete.

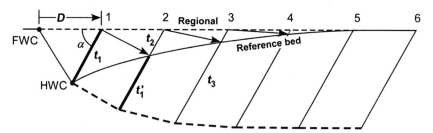

Fig. 11.50. Oblique simple-shear prediction of fault shape (*heavy dashed line*) from rollover shape. *D*: block displacement; *FWC*: footwall cutoff of reference bed; *HWC*: hangingwall cutoff of reference bed; predicted fault is *heavy dashed line*

11.7.3
Sensitivity of Prediction

For both vertical and oblique simple shear, the relationship between fault shape and the hangingwall rollover geometry is sensitive to (1) the shear angle, (2) fault dip between the hangingwall and footwall cutoffs of the reference horizon, (3) the stratigraphic correlation across the fault, and (4) the geometry of the reference bed near the fault (Withjack and Peterson 1993). Shape predictions are less sensitive to the exact location of the fault and to the rollover shape far from the fault.

Predictions are quite sensitive to small changes in the exact positions of the fault cutoffs. This is because small differences in the initial fault dip lead to errors that accumulate down the fault. From the opposite perspective, the agreement between observed and predicted fault trajectories is a sensitive indicator of the correct fault cutoff locations. Predictions are somewhat less sensitive to small changes in the correlation of the reference bed across the fault (Rowan and Kligfield 1989). Incorrect correlations also result in a mismatch between the predicted and observed fault trajectories. Additional discussions of simple-shear methods can be found in Wheeler 1987; Dula 1991; Withjack and Peterson 1993; Withjack et al. 1995; Hague and Gray 1996; Buddin et al. 1997; Shaw et al. 1997; Spang 1997; Hardy and McClay 1999; Novoa et al. 2000.

The major uncertainty in the simple-shear technique is in the choice of the shear angle. The next section describes the relationship between layer-parallel strain and the shear angle, followed by a discussion of how to choose the best shear angle.

11.7.4
Layer-Parallel Strain in Hangingwall

Simple shear oblique to bedding produces bed length and bed thickness changes (Sect. 11.3, Eqs. 11.14, 11.15) in beds above a fault. The relationships among the variables in Eq. 11.14 are illustrated by Fig. 11.51. For a normal fault, if the angle of simple shear is constant, the layer-parallel strain increases with the dip of the median surface of bedding, i.e., as a plane bed is rotated to steeper dips. Vertical simple shear produces much less strain for a given amount of bedding dip change than does oblique simple shear. Substantial dip changes without much strain imply vertical simple shear or that some other model is required. Synthetic simple shear (shear direction parallel to master fault) at angles lower than about 80° will cause layer-parallel contraction. Antithetic simple shear (shear direction dips opposite to master fault) produces layer-parallel extension. Small dip changes accompanied by readily visible second-order normal faults imply antithetic oblique simple shear, probably at angles of 60° or less.

The mechanism for the layer-parallel strain produced by oblique simple shear (Fig. 11.51) is usually small-scale faulting. The small-scale faults are probably not parallel to the shear direction, as logical as that may seem. The shear direction describes the geometry, not the mechanics. The faults produced by layer-parallel extension and the correlative thinning will be in response to the stress generated in the layer by the

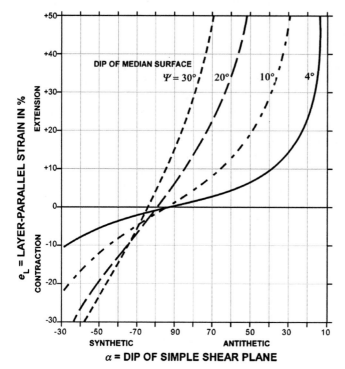

Fig. 11.51.
Layer-parallel strain in the hangingwall of a normal fault caused by simple shear oblique to bedding for various angles of dip of the median surface of bedding (Eq. 11.14, after Groshong 1990). All median surface dips shown are toward the master fault. Synthetic shear dips in the same direction as the master fault and antithetic shear dips opposite to the master fault dip

deformation. Thus conjugate faults or domino blocks are expected (Sect. 1.6.4) with orientations about 30° to σ_1. Because σ_1 is likely to be about normal to bedding, the expected faults should have initial dips close to 60°, regardless of the shear angle that connects the fault shape to the rollover geometry.

The structural styles that can be generated by small-scale faulting in a simple-shear rollover are illustrated with a series of forward models. Assume that the deformation mechanism in the rollover is the rotation of rigid dominoes with initial dips of 60°. The final geometry is controlled by the relative amounts and directions of (1) the external rotation of a median surface in the rollover, which is a function of the shear angle, and (2) the domino rotation, which is a function of the amount of layer-parallel extension (Sect. 11.5.2). The strain of the median surface is calculated from Eq. 11.14 and the domino rotation, given the layer-parallel strain, from Eq. 11.38.

Domino blocks bounded by faults that are precisely antithetic to the master fault at the beginning of deformation retain this geometry during extension (Fig. 11.52a). The domino-block rotation is exactly canceled by the rotation of the rollover. Dominoes that begin deformation with a steeper antithetic dip than the shear angle rotate away from the master fault during deformation (Fig. 11.52b). The net rotation of the domino blocks is relatively small. The geometries of Fig. 11.52a and 11.52b are common in the East African Rift valleys (Rosendahl 1987), the North Sea (Beach 1986), the Gulf of Suez (Colletta et al. 1988) and many other rifted environments. Dominoes that begin deformation dipping the same direction but at a lower angle than the shear angle rotate

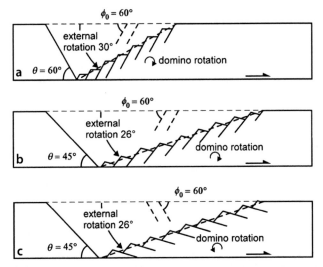

Fig. 11.52. Dominoes as the deformation mechanism in the rollover of a simple-shear half graben (modified from Groshong 1990). ϕ_0: initial dip of domino fault. The simple-shear direction in each rollover is exactly opposite to the dip of the master fault. **a** Dominoes begin with a dip exactly equal to the shear angle: domino rotation is equal and opposite to external rotation. **b** Dominoes begin with antithetic but steeper dip than the shear angle: domino rotation is opposite to and slightly greater than external rotation. **c** Dominoes begin synthetic to the master fault and with a steeper dip than the shear angle: domino rotation adds to the external rotation

a large amount toward the master fault because the external rotation and the domino rotation are in the same direction. The resulting low-angle normal fault geometry (Fig. 11.52c) is characteristic of the Basin and Range province of western North America. All the faults in the areas mentioned are not always parallel and the hangingwall extension may also be accommodated on conjugate normal faults.

11.7.5
Choosing the Best Shear Angle

It has been variously assumed that the shear angle is parallel to the dominant second-order fault trend (White et al. 1986), is exactly antithetic to the master fault (Groshong 1989), or is equal to the Coulomb failure angle of 30° to the maximum principal compressive stress direction (Xiao and Suppe 1992), but a variety of angles have been obtained in the laboratory and the field (Groshong 1990). The next sections examine methods for choosing the best angle.

11.7.5.1
Shear Angle by Trial and Error

One approach to finding the best shear angle is to find the angle that provides the best match between the shape of a bed in the rollover and the shape of the fault that caused it (White et al. 1986; Rowan and Kligfield 1989). This approach, and its pit-falls, is illustrated by relating the rollover to the fault shape in two different versions of the same seismic section of a thin-skinned normal fault from the Texas Gulf of Mexico. In the first version (Fig. 11.53) the vertical scale is time and thus contains an uncompensated vertical exaggeration. The oblique simple shear method applied by

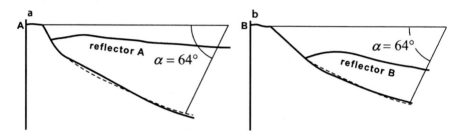

Fig. 11.53. Simple shear best fit to a normal-fault rollover on a seismic time section from Bruce (1973), modified from White et al. (1986). The *heavy solid lines* are drawn from the seismic line, the *dashed lines* are faults inferred from the model. **a** Best fit to reflector A, antithetic simple shear at 64° to the horizontal. **b** Best fit to reflector B, antithetic simple shear at 64° to the horizontal

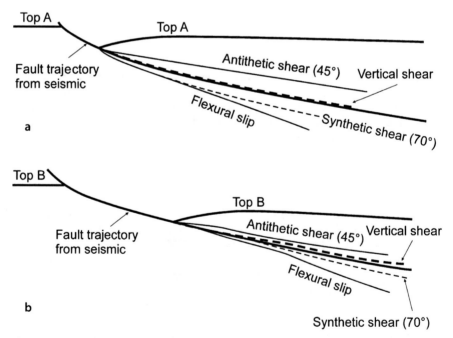

Fig. 11.54. Simple shear and flexural-slip best fits between rollover geometry and fault shape on depth-converted profile from Bruce (1973), redrawn from Rowan and Kligfield (1989). *Heavy solid lines* are from seismic line. **a** Marker A. **b** Marker B

White et al. (1986) gives a best-fit shear angle of 64° antithetic for two different reflectors on the same section (Fig. 11.53). Using a depth-migrated version (vertical exaggeration 1:1) of the same seismic line, Rowan and Kligfield (1989) predicted the rollover geometry using a variety of different shear angles (Fig. 11.54) and found that the angle that gives best match to the fault shape is 90° (vertical simple shear). This result demonstrates that an empirical best fit can be obtained in either time or depth but that finding the "true" shear angle requires a section without vertical exaggeration. The shear angle can compensate for an unknown vertical exaggeration but the strain related to the geometry and shear angle will only be correct if the shear angle is determined from an unexaggerated profile.

Additional confidence is obtained in the result if the same angle works for more than one bed as in both examples above. Statistical curve-fitting can be used to find the shear angle that minimizes the differences between the fault shapes predicted from multiple beds and to find the shear angle and shear azimuth in three dimensions (Kerr and White 1996).

11.7.5.2
Shear Angle from Axial Surface Orientation

The axial surfaces bounding dip domains in the pre-growth stratigraphy are parallel to the shear direction, a relationship that can be used to determine the shear angle (Fig. 11.55). Axial surfaces that are not parallel to the shear direction develop in growth sediments. See Xiao and Suppe (1992) and Spang and Dorobek (1998) for discussions of the geometry expressed by growth stratigraphy.

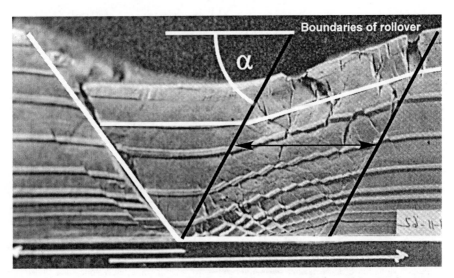

Fig. 11.55. Shear angle approximated as dip of axial surface (clay model from Cloos 1968, interpretation after Groshong 1990). The direction of oblique simple shear is given by the *heavy black lines*

11.7.5.3
Shear Angle from Strain

The layer-parallel strain is the basis of a third method to determine the best shear angle. The necessary relationship is derived by solving Eq. 11.14 for α (Groshong 1990). The result, Eq. 11.48, gives the appropriate shear angle from the layer-parallel strain and the dip of the median surface of bedding:

$$\alpha = \arctan\left[(e_L + 1)\sin\psi\right]/\left[(e_L + 1)\cos\psi - 1\right] \ , \tag{11.48}$$

where α is the angle of shear measured from the regional, ψ is the angle of rotation of the median surface of the bed from the regional and e_L is the layer-parallel strain as a fraction. The layer-parallel strain will usually be seen as second-order faults in the rollover. From measurements of visible bed length and the total extent of the horizon, e_L is determined with Eq. 11.2.

This relationship has been tested on experimental sand and clay models by Groshong (1990), including the model in Fig. 11.55, and yields predicted faults that are a close match to the actual faults. The Livingstone Basin (Fig. 11.56a) provides a large-scale field example illustrating the shear-angle calculation. It is a sub-basin at

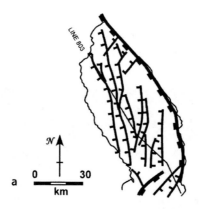

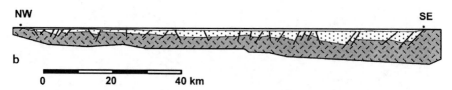

Fig. 11.56. Structure of Livingstone basin at the north end of Lake Malawi, East African Rift (after Wheeler and Rosendahl 1994). **a** Index map to Livingstone basin, showing the location of line 803. **b** Cross section along line 803, Livingstone basin, interpreted from depth-corrected seismic line (Wheeler and Rosendahl 1994). *Darker shaded unit* is basement, *lighter shaded unit* is sedimentary basin fill. No vertical exaggeration

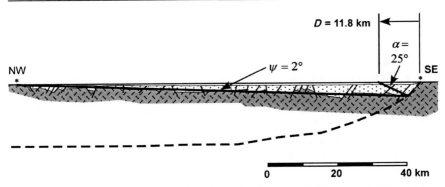

Fig. 11.57. Cross section in the transport direction of the Livingstone half graben (Fig. 11.56b), interpreted with the oblique simple shear model (Groshong 1995). *Darker shaded unit* is basement, *lighter shaded unit* is sedimentary basin fill. No vertical exaggeration. *Dashed line* is predicted fault

the north end of Lake Malawi (Nyasa), one of the East African Rift valleys. The master fault of the half graben is on the east side of the basin. The basin is broken by a number of second-order normal faults. The north-south trend of the faults within the basin seems to imply that the extension direction is east-west, however, the extension direction is interpreted to be northwest-southeast, oblique to the trend of the rift axis on the map (Scott et al. 1992; Wheeler and Rosendahl 1994). This interpretation is based on west-northwest to northwest plunging slickenlines on exposures of the master fault and on the presence of flower structures (wrench-fault indicators) along west-northwest trending faults in the basin. The cross section along the basin axis (Fig. 11.56b) is thus in the transport direction.

Using the relationship between layer-parallel extension and shear angle, it is possible to find the shape of the master fault, the depth to detachment, and the displacement that formed the graben. The amount of extension in the hangingwall is directly related to the dip change of the median surface and the angle of shear. Second-order normal faults are relatively evenly distributed throughout the hangingwall, as expected for the simple-shear model. The amount of extension along the median surface is determined by measuring the length of the top basement surface with and without the fault offsets and is $e_L = 0.08$ (Eq. 11.2). The dip of the median surface with respect to regional is $\psi = 2°$. From Eq. 11.48, $\alpha = 25°$. Given the shear angle, the shape of the master fault can be constructed by the oblique-shear method (Sect. 11.7.2.2) and is shown in Fig. 11.57. Measured from the figure, $H = 18.5$ km. By the oblique simple-shear model (e.g., Fig. 11.50), the displacement that formed the graben system is 11.8 km.

A depth of 18.5 km places the lower detachment of the Livingstone half graben in the middle of the crust. This depth is a reasonable possibility because it is the depth at which most seismic activity ceases below active cratonic rifts (Chen and Molnar 1983), even though the base of the crust is the expected location of the major strength minimum (Molnar 1988; Harry and Sawyer 1992) and rare deep earthquakes occur in the area (Shudofsky 1985; Jackson and Blenkinsop 1977).

11.8
Fault-Parallel Simple Shear

Fault-parallel displacement in the hangingwall is modeled with a slight modification of the vertical simple shear model (Williams and Vann 1987). The method uses a variable distance between working lines so as to maintain constant slip on the fault as it changes dip. This method approximates the hangingwall geometry developed above a listric normal fault in which there is nearly rigid rotation above the circular part of the fault, rigid-block displacement above the bedding-parallel fault segment, and a keystone graben between the rotated block and the translated block.

11.8.1
Restoration

The method for restoring a cross section by the constant dip separation model was developed by Chai (1994). The construction procedure is as follows (refer to Fig. 11.58 for the geometry):

1. Define the regional.
2. Draw a vertical working line through the hangingwall cutoff of the reference bed (point B).
3. Measure the straight-line dip separation on the fault, d (length AB).
4. Swing an arc from B of length d until it intersects the fault at C. This point locates the next vertical working line. Swing the next arc from C to find the location of D and so on. Draw vertical working lines through each marked point.
5. Measure the vertical distance from the reference bed to the fault, for example, t_1 the heavy line above point C, and shift it one working distance (W) horizontally and up the fault so that the base of the line is at the fault. The top of the line is the restored position of the reference bed. This is done for each vertical line to restore the section.

11.8.2
Fault-Shape Prediction

The construction of the fault geometry from the shape of a reference bed in the rollover is as follows (refer to Fig. 11.59 for geometry):

1. Define the regional.
2. Measure the straight-line dip separation, d, along the fault between the hangingwall cutoff and the footwall cutoff of the reference bed.
3. Draw a vertical working line through the hangingwall cutoff of the reference bed. Swing an arc with radius d from point A at the intersection of the working line and the regional. Where the arc hits the reference bed marks the position of the next vertical working line, which defines point B.
4. Repeat the swinging of arcs of constant length equal d from the intersection of each vertical working line with regional to the reference bed until the rollover returns to regional.

Fig. 11.58.
Restoration using the constant
fault slip model (after Chai
1994). *d:* dip separation on
fault; *W:* working distance
between two working lines

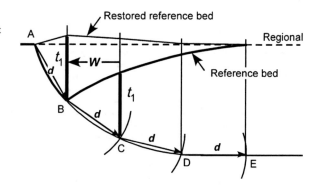

Fig. 11.59.
Fault shape prediction using
the constant fault slip model
(after Williams and Vann 1987).
d: dip separation on fault;
t_1: vertical distance between
fault and regional below *B;*
HWC: hangingwall cutoff;
FWC: footwall cutoff; *thick
dashed line:* predicted fault

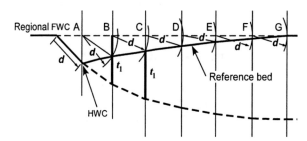

5. Sequentially from the fault cutoff, shift the vertical thickness between the regional and
   the fault over one working distance in the direction of fault displacement, then move the
   thickness down the working line until the top just touches the key bed (for example,
   the thickness t_1 below B moves to below C). The position of the bottom of the line corresponds to the location of the fault. Repeat this step until the cross section is complete.

The constant fault slip model predicts the fault shape from the rollover geometry in
a rigid-rotation experiment by Chai (1994). The hangingwall is homogeneous sand
with layers of different colors, the fault is a pre-cut shape in a rigid block, and the
extensional displacement is applied to the entire hangingwall by a flexible sheet on top of
the footwall block, a configuration developed by McClay and Ellis (1987). The critical
boundary condition in this experiment is that the flexible sheet forces the fault to maintain constant displacement and so the deformation mechanism is close to rigid rotation
above the listric part of the fault. Characteristic of the model is the formation of a
keystone graben between the rotated block above the listric portion of the master fault
and the translated block above the lower detachment. The graben (Fig. 11.60) represents the strain required between the two differentially displaced nearly rigid blocks.

The constant fault slip model produces a better match between the rollover shape
and the fault shape (Fig. 11.60) than any of the models described previously. The fit is
close, although not exact. The lack of a perfect fit between the predicted and observed
fault indicates that the best-fitting model does not perfectly duplicate the mechanics of
deformation. Kinematic models are simplifications of the mechanical processes and
so a close (but not necessarily perfect) match validates the interpretation.

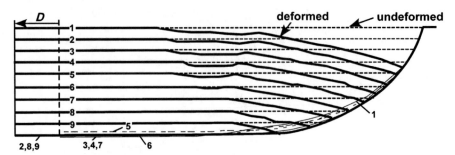

Fig. 11.60. Fault shapes predicted using the constant fault slip model for a sandbox experiment with constant fault displacement (Chai 1994). The predicted fault trajectories are numbered with the layer used to make the prediction. *D*: boundary displacement

11.9
Pure Shear Deformation

The final kinematic model to be discussed is pure shear deformation. The pure shear model is appropriate for the relatively uniform extension that forms a full graben. Significantly different predictions are produced depending on the dip of the boundary faults. The difference will be illustrated in a discussion of the Rhine graben.

11.9.1
Vertical-Sided Graben Model

Pure-shear extension of a vertical-sided graben (Fig. 11.61a) is the model proposed by McKenzie (1978) for the formation of an extensional basin. If this style of extension occurs above a rigid detachment (Fig. 11.61b), the subsidence of the graben is directly related to the amount of extension and the position of the detachment. If the extension occurs above a thick, deformable detachment unit (Fig. 11.61c), the amount of subsidence is controlled by isostacy and the temperature distribution (McKenzie 1978) in addition to the area balance of the graben. In this model the strain is uniform throughout the graben. The extension is usually calculated as the stretch, but based on the thinning, using Eq. 11.10.

11.9.2
Normal-Fault Bounded Graben Model

In this model the region of pure shear is bounded by normal faults and a planar detachment at the base (Fig. 11.62a). Extension causes the graben between the fault boundaries to stretch horizontally and thin vertically (Fig. 11.62b,c). The area-depth relationship provides a unique solution for the location of the lower detachment (Fig. 11.62d) and the strain is not uniformly distributed with depth.

The area-depth relationship, based on the deformed-state geometry of the 5 marker horizons in Fig. 11.62c, is a straight line (Fig. 11.62d) as expected for a locally balanced structure. The closeness of the area-depth points to the best-fit line through the data indicates the internal consistency of the cross section. The area-depth line goes to zero

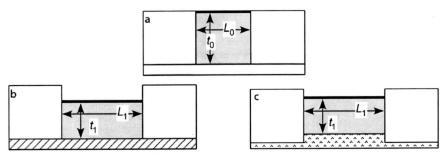

Fig. 11.61. Vertical-sided, area-balanced graben. **a** Before deformation. **b** After deformation, subsidence controlled by depth to the detachment. **c** After deformation, subsidence controlled by isostatic response of the system

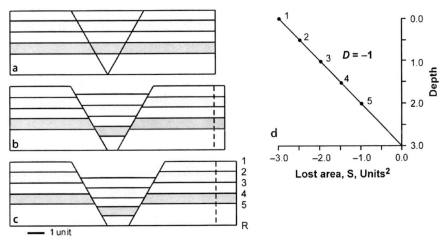

Fig. 11.62. Kinematic evolution of a pure shear full graben (after Groshong 1994). **a** Undeformed. **b** Initial stage of deformation. **c** Final stage of deformation. **d** Area-depth diagram of **c**. *Numbered points* correspond to *numbered horizons* in the cross section. Horizon 1 is the reference level

area at the lower detachment and the slope indicates a displacement of 1 unit. A model discussed previously (Fig. 11.18) is also a pure shear graben, but with growth sediments. Here the focus is on the internal strain in the graben.

Figure 11.63 shows the magnitudes of the layer-parallel strains for each horizon in the full graben model of Fig. 11.62c. The strain can be determined directly from the figure by comparing the bed lengths before and after the deformation (Eq. 11.2) as well as from the lost areas and depth to detachment as discussed in Sect. 11.4.3.2 (Eq. 11.26 or 11.27). Both methods give the same result. For this model the requisite strains are small in the upper part of the graben but increase rapidly toward the lower detachment and would approach infinity at the lower detachment because the original bed length there was zero. Quite a different strain magnitude is obtained if the total thickness change is used to calculate the extension (Eq. 11.10) as would be appropriate for a vertical-sided graben.

Fig. 11.63.
Strain in the graben of an
area-balanced model (after
Groshong 1994). The *dashed
lines* indicate the position of
the graben before deformation. Strain magnitudes are
in percent

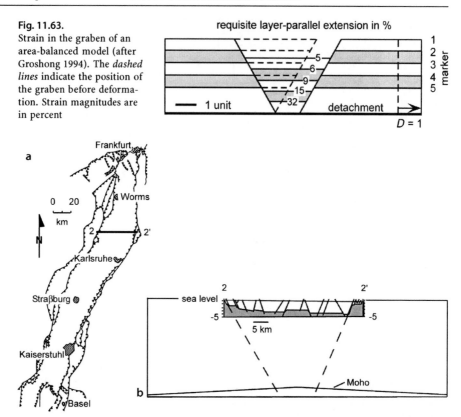

Fig. 11.64. Rhine Graben (after Groshong 1996). **a** Index map. **b** Cross section 2–2' inserted in a crustal profile. Triassic and younger sediments *unshaded*. The seismic refraction Moho is the base crust. (Illies 1977; Ziegler 1992; Rousset et al. 1993)

According to Eq. 11.11, the extension is $\beta_L = 1.34$ or $e_L = \beta_L - 1 = 34\%$ extension and is uniform with depth. This is clearly much different than the extension calculated directly from the model and shows the importance of the dip of the boundary faults.

The Rhine graben (Fig. 11.64a) provides an example of the dependence of the calculated extension on the model selected. The profile (Fig. 11.64b) is from the central area of the Rhine graben where it is relatively symmetrical, although the graben is asymmetrical to the north and south (Brun et al. 1992; Bois 1993; Rousset et al. 1993). The symmetry reverses across the profile area. The section to be interpreted was shown to be sequentially restorable by rigid block displacement in Fig. 11.9, thus it is valid. Length and area measurements were made on the preserved contact between the Permian and older basement and the Triassic Buntsandstein (shown in Fig. 11.9 and 11.64b).

Because shallowly buried rocks are both brittle and weak in extension, it is mechanically reasonable, although not certain, that nearly all of the layer-parallel extension within the sediments, including the Permian and older sediments of the shallow basement, occurs on the visible faults. The extension of the top-basement is 6.3% (Eq. 11.2), based on the bed length of L_0 = 38.3 km, and the straight line distance across the top basement in the graben of L_1 = 40.7 km. The Moho is at a depth of 27 km directly below the graben and is 31 km deep outside it (Fig. 11.64b). The β value for the graben is 1.29 from Eq. 11.10 (e_L = 29%), assuming the original thickness of the crust in the Rhine graben was 31 km and the final thickness is 24 km (27 km minus 3 km of post-Permian sediments in the graben). The dramatic difference in extension calculated from the two different equations is the result of the specific pure-shear model applied. The large value is for a vertical-sided graben and the small value is the expected result near the top of a normal-fault-bounded graben. Because the Rhine graben has normal-fault boundaries, it is reasonable to conclude that the model of Fig. 11.63 applies to the graben and that the small strain magnitude at the top of the basement does not imply a major discrepancy in the extension, but rather is the expected value.

11.10
Exercises

11.10.1
Cross-Section Validation and Interpretation 1

Validate (or invalidate) the cross section in Fig. 11.65. Is it length balanced? Is it area balanced? Apply the area-depth relationship to find the best-fitting lower detachment, displacement, and the strain in each layer.

11.10.2
Cross-Section Validation and Interpretation 2

Validate (or invalidate) the cross section in Fig. 11.66. Is it length balanced? Is it area balanced? Apply the area-depth relationship to find the best-fitting lower detachment, displacement, and the strain in each layer.

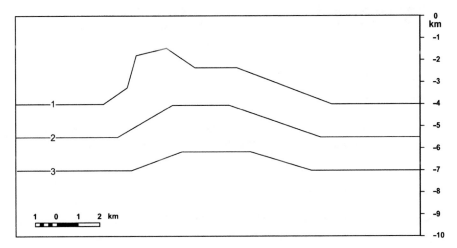

Fig. 11.65. Cross section of an anticline

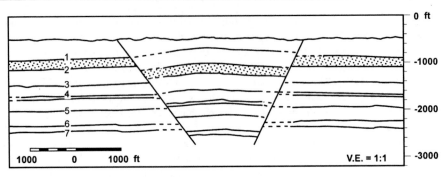

Fig. 11.66. Cross section of a full graben

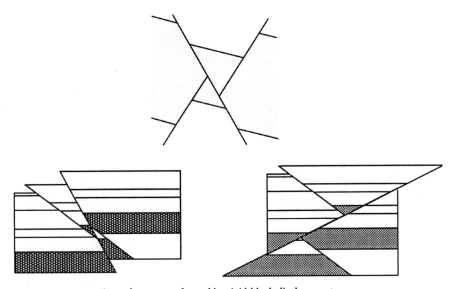

Fig. 11.67. Cross sections of structures formed by rigid-block displacement

11.10.3
Rigid-Body Restoration

Restore the cross sections in Fig. 11.67. Why is the rigid-body method appropriate? Are the cross sections valid? Show the structural evolution of each cross section.

11.10.4
Restoration of the Rhine Graben

Sequentially restore the cross section of the Rhine Graben in Fig. 11.68 to the top of the Bunte Niederroderner Schichten and the top of the Muschelkalk. What method is most appropriate? Is the cross section valid?

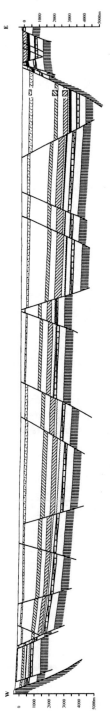

Fig. 11.68. Cross section of the Rhine Graben. (After Doebl and Teichmüller 1979)

11.10.5
Flexural-Slip Restoration 1

Restore the cross section of the Sequatchie anticline in Fig. 11.69. Why is the flexural-slip method appropriate? Discuss the effect of the choice of pin line and loose line on the result. Is the cross section valid?

11.10.6
Flexural-Slip Restoration 2

Restore the cross section of the Velma area in Fig. 11.70. Use the flexural slip technique and preserve the original stratigraphic thickness changes. Is the cross section valid? Discuss the origin of the major faults and their sequence of formation.

11.10.7
Flexural-Slip Restoration 3

Restore the cross section in Fig. 11.71 by flexural slip. First it is necessary to correlate units across the faults. Discuss the effect of the choice of pin line and loose line on the result. Could the interpretation be questioned or improved? Is the cross section valid?

11.10.8
Balance and Restoration

Restore the cross section of the Deer Park anticline (Fig. 11.72) by flexural slip or area balance, as appropriate. Construct an area-depth diagram for the entire anticline. What displacement caused the structure and what displacement is present on

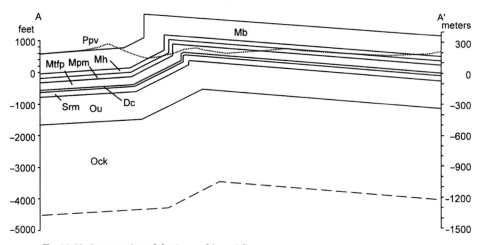

Fig. 11.69. Cross section of the Sequatchie anticline

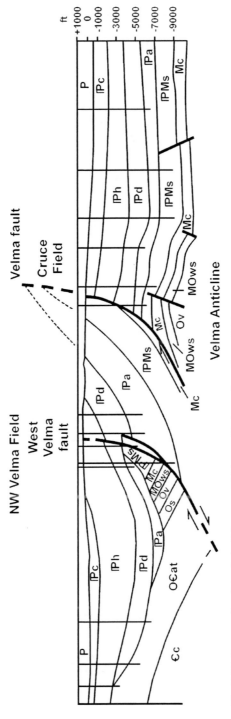

Fig. 11.70. Cross section through the Velma area, Oklahoma. (Redrawn from Perry 1989)

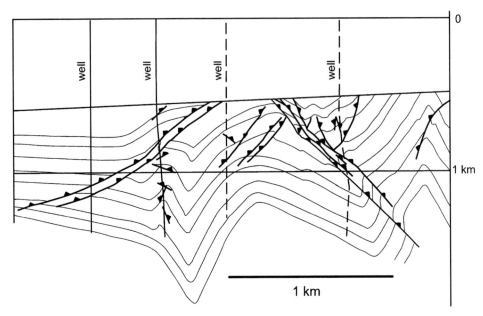

Fig. 11.71. Cross section of a portion of the Ruhr coal district, Germany. *Triangles* are located on the hangingwalls of the thrust faults. (After Drozdzewski 1983)

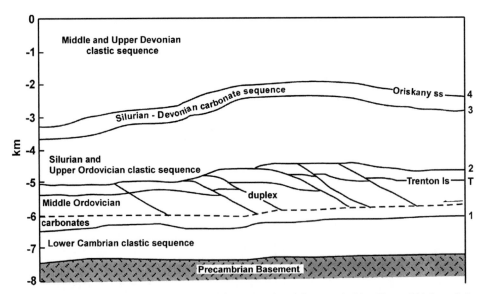

Fig. 11.72. Geological cross section across the Deer Park anticline, Appalachian Plateau fold-thrust belt, eastern U.S. The section is depth converted from Mitra (1986) using a velocity of 5 km s^{-1}. The inferred lower detachment is the *dotted line*. Epard and Groshong (1995) discuss the interpretation

the upper detachment, if any? Based on the results, is the structure locally balanced or regionally balanced? Compute the layer parallel strains for each layer. Is the cross section valid?

11.10.9
Predict Fault Geometry

The drape fold in horizons 1 and 2 in the South Hewett fault zone (Fig. 11.73) can be explained by an underlying rotated block. Apply the circular-arc fault model to predict the fault location and depth to detachment. The slip on some of the Zechstein normal faults has been reversed in the later deformation. Does the model explain which faults have reactivated?

11.10.10
Simple-Shear Restoration

Restore the growth normal fault in Fig. 11.74. This section contains growth stratigraphy and can be sequentially restored to the regional for horizons 2 and 3 to show the growth history. Why is the simple-shear method a reasonable choice? What is the appropriate choice of the regional? What is the most appropriate shear angle and how do you find it? Is the cross section valid?

11.10.11
Restoration and Prediction

Restore the cross section in Fig. 11.75 by either rigid-block displacement or flexural slip. Discuss the reason for your choice of method. Predict the deep geometry of the Schell Creek master fault using oblique simple shear. Find the shear angle from the strain in the rollover.

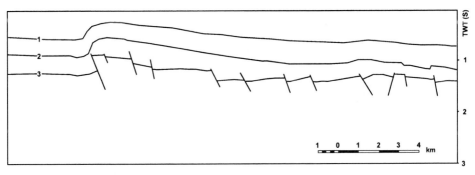

Fig. 11.73. The South Hewett fault zone in the North Sea. Interpreted and drawn from a seismic reflection profile in Badley et al. (1989). Assume the vertical exaggeration is approximately 1 : 1. Horizon *1:* top Creta- ceous Chalk; *2:* base-Cretaceous unconformity; *3:* top Zechstein; *TWT(S):* two-way travel time in seconds

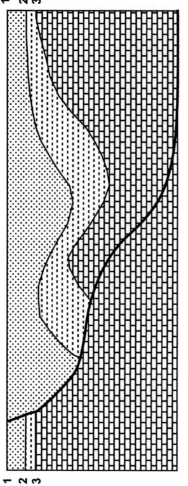

Fig. 11.74. Cross section of a ramp-flat normal fault

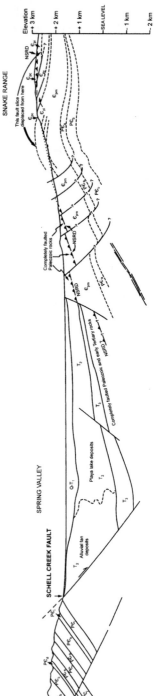

Fig. 11.75. Cross section of Schell Creek fault, U.S. Basin and Range province, Nevada from outcrop and seismic reflection profile. (After Gans et al. 1985). Groshong (1989) discusses the interpretation

Chapter 12

Direction Cosines and Vector Geometry

12.1
Introduction

The orientations of structural elements are most readily obtained analytically by 3-D vector geometry. A line is represented by a vector of unit length and a plane by its pole vector or its dip vector. The direction cosines of a line describe the orientation of the unit vector parallel to the line. Structural information such as bearing and plunge is converted into direction cosine form, the necessary operations performed, and then the values converted back to standard geological format. This chapter gives the basic relationships and outlines some of the ways vector geometry can be applied to structural problems involving lines and planes.

12.2
Direction Cosines of Lines

Direction cosines define the orientation of a vector in three dimensions (Fig. 12.1). The direction angles between the line OC and the positive coordinate axes x, y, z are α, β, γ, respectively. The direction cosines of the line OC are

$$\cos \alpha = \cos \text{EOC} \quad , \tag{12.1a}$$

$$\cos \beta = \cos \text{GOC} \quad , \tag{12.1b}$$

$$\cos \gamma = \cos (90 + \text{AOC}) \quad . \tag{12.1c}$$

In the geological sign convention, an azimuth of 0 and 360° represents north, 90° east, 180° south, and 270° west. Dip (of a plane) or plunge (of a line) is an angle from the horizontal between 0 and 90°, positive downward. From the geometry of Fig. 12.1:

$$\cos \delta = \text{OA} / \text{OC} \quad , \tag{12.2a}$$

$$\cos \theta = \text{OG} / \text{OA} \quad , \tag{12.2b}$$

$$\sin \theta = \text{OE} / \text{OA} \quad , \tag{12.2c}$$

$$\cos \alpha = \text{OE} / \text{OC} \quad , \tag{12.2d}$$

$$\cos \beta = \text{OG} / \text{OC} \quad , \tag{12.2e}$$

$$\cos \gamma = \cos (90 + \delta) = -\sin (\delta) \quad . \tag{12.2f}$$

Fig. 12.1.
Orientation of vector *OC* in right-handed *xyz* space; θ is the azimuth of the vector and δ is the plunge. The direction angles are α, β, γ

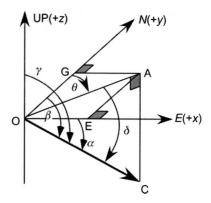

12.2.1
Direction Cosines of a Line from Azimuth and Plunge

The direction cosines of a line in terms of its azimuth and plunge are obtained from the relationships in Eq. 12.2:

$$\cos\delta\,\sin\theta = (OA\,/\,OC)\,(OE\,/\,OA) = OE\,/\,OC = \cos\alpha \quad , \tag{12.3a}$$

$$\cos\delta\,\cos\theta = (OA\,/\,OC)\,(OG\,/\,OA) = OG\,/\,OC = \cos\beta \quad , \tag{12.3b}$$

$$-\sin\delta = \cos\gamma \quad . \tag{12.3c}$$

12.2.2
Azimuth and Plunge from Direction Cosines

To reverse the procedure and find the azimuth and plunge of a line from its direction cosines, divide Eq. 12.3a by 12.3b and solve for θ, then solve Eq. 12.3c for δ:

$$\theta' = \arctan\left(\cos\alpha\,/\,\cos\beta\right) \quad , \tag{12.4}$$

$$\delta = \arcsin\left(-\cos\gamma\right) \quad . \tag{12.5}$$

The value θ' returned by Eq. 12.4 will be in the range of $\pm 90°$ and must be corrected to give the true azimuth over the range of 0 to 360°. The true azimuth, θ, of the line can be determined from the signs of $\cos\alpha$ and $\cos\beta$ (Table 12.1). The direction cosines give a directed vector. The vector so determined might point upward. If it is necessary to reverse its sense of direction, reverse the sign of all three direction cosines. Note that division by zero in Eq. 12.4 must be prevented.

An Excel equation that returns the azimuth in the correct quadrant has the form

$$=\text{IF(cell B<=0,180+Cell E,IF(cell A>=0,cell E,360+cell E))} \tag{12.6}$$

where cell A contains $\cos\alpha$, cell B contains $\cos\beta$, and cell E contains Eq. 12.4.

Table 12.1.	Azimuth	cos α	cos β	θ
Relationship between signs of the direction cosines and the quadrant of the azimuth of a line	000+ to 090	+	+	θ'
	090+ to 180	+	–	$180 + \theta'$
	180+ to 270	–	–	$180 + \theta'$
	270+ to 360	–	+	$360 + \theta'$

12.2.3
Direction Cosines of a Line on a Map

The direction cosines of a line defined by its horizontal, h, and vertical, v, dimensions on a map can be found by letting $\delta = \arctan(v/h)$ in Eqs. 12.3:

$$\cos \alpha = \cos(\arctan(v/h)) \sin \theta \ , \tag{12.7a}$$

$$\cos \beta = \cos(\arctan(v/h)) \cos \theta \ , \tag{12.7b}$$

$$\cos \gamma = -\sin(\arctan(v/h)) \ . \tag{12.7c}$$

Both v and h are taken as positive numbers in Eqs. 12.7 and the resulting dip is positive downward.

The direction cosines of a line defined by the xyz coordinates of its two end points are obtained by letting point 1 be at O and point 2 be at C in Fig. 12.1. Then

$$x_2 - x_1 = OE \ , \tag{12.8a}$$

$$y_2 - y_1 = OG \ , \tag{12.8b}$$

$$z_2 - z_1 = AC \ . \tag{12.8c}$$

Substitute Eqs. 12.8 into 12.3 to obtain

$$\cos \alpha = (x_2 - x_1)/OC \ , \tag{12.9a}$$

$$\cos \beta = (y_2 - y_1)/OC \ , \tag{12.9b}$$

$$\cos \gamma = -\sin \delta = -AC/OC = -(z_2 - z_1)/OC = (z_1 - z_2)/OC \ , \tag{12.9c}$$

where $OC = L$, given by

$$L = [(x_2 - x_1)^2 + (y_2 - y_1)^2 + (z_2 - z_1)^2]^{1/2} \ . \tag{12.10}$$

Using the convention that point 1 is higher and point 2 is lower, a downward-directed bearing is positive in sign.

12.2.4
Azimuth and Plunge of a Line from the End Points

To find the azimuth and plunge of a line from the coordinates of two points, substitute Eq. 12.9 into 12.4 and 12.5:

$$\theta' = \arctan((x_2 - x_1)/(y_2 - y_1)) \ ,$$
(12.11)

$$\delta = \arcsin((z_2 - z_1)/L) \ .$$
(12.12)

The value of L is given by Eq. 12.10. If $y_2 = y_1$, there is a division by zero in Eq. 12.13 which must be prevented. The value of θ is obtained from Table 12.1.

12.2.5
Pole to a Plane

The pole to a plane defined by its dip vector has the same azimuth as the dip vector, and a plunge of $\delta_p = \delta + 90$. Substitute this into Eqs. 12.3 to obtain the direction cosines of the pole, p, in terms of the azimuth, θ, and plunge, δ, of the bedding dip:

$$\cos \alpha_p = \cos(\delta + 90) \sin \theta = -\sin \delta \sin \theta \ ,$$
(12.13a)

$$\cos \beta_p = \cos(\delta + 90) \cos \theta = -\sin \delta \cos \theta \ ,$$
(12.13b)

$$\cos \gamma_p = -\sin(\delta + 90) = -\cos \delta \ .$$
(12.13c)

12.3
Attitude of a Plane from Three Points

The general equation of a plane (Foley and Van Dam 1983) is

$$Ax + By + Cz + D = 0 \ .$$
(12.14)

The equation of the plane from the xyz coordinates of three points is

$$\begin{vmatrix} x & y & z & 1 \\ x_1 & y_1 & z_1 & 1 \\ x_2 & y_2 & z_2 & 1 \\ x_3 & y_3 & z_3 & 1 \end{vmatrix} = 0 \ .$$
(12.15)

This expression is expanded by cofactors to find the coefficients A, B, C, and D:

$$A = y_1 z_2 + z_1 y_3 + y_2 z_3 - z_2 y_3 - z_3 y_1 - z_1 y_2 \ ,$$
(12.16a)

$$B = z_2 x_3 + z_3 x_1 + z_1 x_2 - x_1 z_2 - z_1 x_3 - x_2 z_3 \ ,$$
(12.16b)

$$C = x_1 y_2 + y_1 x_3 + x_2 y_3 - y_2 x_3 - y_3 x_1 - y_1 x_2 \quad , \tag{12.16c}$$

$$D = z_1 y_2 x_3 + z_2 y_3 x_1 + z_3 y_1 x_2 - x_1 y_2 z_3 - y_1 z_2 x_3 - z_1 x_2 y_3 \quad . \tag{12.16d}$$

The direction cosines of the pole to this plane are (Eves 1984)

$$\cos \alpha_p = A / E \quad , \tag{12.17a}$$

$$\cos \beta_p = B / E \quad , \tag{12.17b}$$

$$\cos \gamma_p = C / E \quad , \tag{12.17c}$$

where

$$E = (A^2 + B^2 + C^2)^{1/2} \quad , \tag{12.18}$$

and the sign of $E = -\text{sign}\, D$ if $D \neq 0$; $= \text{sign}\, C$ if $D = 0$ and $C \neq 0$; $= \text{sign}\, B$ if $C = D = 0$.

The direction cosines of the pole may be converted to the direction cosines of the dip vector. If $\cos \gamma_p$ is negative, the pole points downward. Reverse the signs on all three direction cosines to obtain the upward direction. If $\cos \gamma_p$ is positive, the pole points upward, and the dip azimuth is the same as that of the pole and the dip amount is equal to 90° plus the angle between the pole and the z axis:

$$\cos \alpha = A / E \quad , \tag{12.19a}$$

$$\cos \beta = B / E \quad , \tag{12.19b}$$

$$\cos \gamma = \cos (90 + \arccos (C / E)) \quad . \tag{12.19c}$$

To find the azimuth and plunge of the dip direction, use Eqs. 12.4 and 12.5 and Table 12.1 or the Excel equation 12.6.

12.4
Vector Geometry of Lines and Planes

Vector geometry (Thomas 1960) is an efficient method for the analytical computation of the relationships between lines and planes. A 3-D vector has the form

$$v = li + mj + nk \quad , \tag{12.20}$$

where i, j, k = unit vectors parallel to the x, y, z axes, respectively (Fig. 12.2) and the coefficients are the direction cosines l, m, n of the line with respect to the corresponding axes. The vector v is a unit vector if its length is equal to one, and gives the orientation of a line. The length of the vector is

$$|li + mj + nk| = (l^2 + m^2 + n^2)^{1/2} \quad , \tag{12.21}$$

Fig. 12.2.
Vector components in the
xyz coordinate system. Vector
components *i, j, k* each have
lengths equal to one

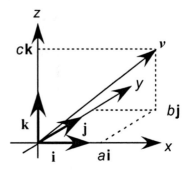

where the vertical bars = the absolute value of the expression between them. The direction cosines of any vector can be normalized to generate a unit vector by dividing each direction cosine (*l*, *m*, and *n*) by the right-hand side of Eq. 12.21.

12.4.1
Angle between Two Lines or Planes

The angle, Θ, between two lines, is given by the scalar or dot product of the two unit vectors with the same orientations as the lines. If the two vectors represent the poles to planes or the dip vectors of planes, then the dot product is the angle between the planes, normal to their line of intersection (Fig. 12.3). The *dot product* is defined as

$$v_1 \cdot v_2 = |v_1|\,|v_2|\cos\Theta \;, \tag{12.22}$$

where $|v| = 1$ for a unit vector and $v_1 \cdot v_2$ is evaluated as

$$v_1 \cdot v_2 = l_1 l_2 + m_1 m_2 + n_1 n_2 \;, \tag{12.23}$$

where the subscript "1" = the direction cosines of the first vector and subscript "2" = the direction cosines of the second vector. Equating 12.22 and 12.23 gives the angle between two unit vectors:

$$\cos\Theta = l_1 l_2 + m_1 m_2 + n_1 n_2 \;. \tag{12.24}$$

12.4.2
Line Perpendicular to Two Vectors

The orientation of a line perpendicular to two other vectors is given by the vector product, also called the cross product (Fig. 12.3). The *cross product* of two vectors is defined as

$$v_1 \times v_2 = n\,|v_1|\,|v_2|\sin\Theta \;, \tag{12.25}$$

where n = a unit vector perpendicular to the plane of v_1 and v_2. When the vectors are unit vectors given in terms of the direction cosines, the cross product is:

Fig. 12.3.
Geometry of the dot product and cross product. Θ is the angle between v_1 and v_2; v_3 is perpendicular to the plane of v_1 and v_2. The *shaded plane* is perpendicular to v_1 and the *ruled plane* is perpendicular to v_2

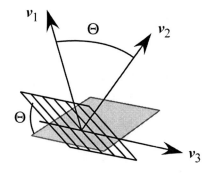

$$v_3 = \begin{vmatrix} i & j & k \\ l_1 & m_1 & n_1 \\ l_2 & m_2 & n_2 \end{vmatrix} = (m_1 n_2 - n_1 m_2)i + (n_1 l_2 - l_1 n_2)j + (l_1 m_2 - m_1 l_2)k \quad . \tag{12.26}$$

In Eq. 12.26, v_3 is the vector perpendicular to both v_1 and v_2. The order of multiplication changes the direction of v_3 but not the orientation of the line or its magnitude.

12.4.3
Line of Intersection between Two Planes

The properties of the cross product make it suitable for solving a number of problems in structural analysis. For example, two dip domains can be defined by their poles (or their dip vectors) and the cross product of these vectors will give the orientation of the hinge line between them. If the pole to a plane is defined by the azimuth (θ) and dip (δ) of a vector pointing in the dip direction, the orientation of this vector in terms of direction cosines is given by Eqs. 12.13, which are

$$l = \cos \alpha_p = -\sin \delta \, \sin \theta \quad , \tag{12.27a}$$

$$m = \cos \beta_p = -\sin \delta \, \cos \theta \quad , \tag{12.27b}$$

$$n = \cos \gamma_p = -\cos \delta \quad . \tag{12.27c}$$

The subscript "p" = the pole to a plane. Let the subscript "1" = the first domain dip, 2 = the second domain dip, and h = the hinge line, then substitute Eqs. 12.27 into 12.26 to obtain a vector parallel to the hinge line in terms of the direction cosines:

$$\cos \alpha = \sin \delta_1 \cos \theta_1 \cos \delta_2 - \cos \delta_1 \sin \delta_2 \cos \theta_2 \quad , \tag{12.28a}$$

$$\cos \beta = \cos \delta_1 \sin \delta_2 \sin \theta_2 - \sin \delta_1 \sin \theta_1 \cos \delta_2 \quad , \tag{12.28b}$$

$$\cos \gamma = \sin \delta_1 \sin \theta_1 \sin \delta_2 \cos \theta_2 - \sin \delta_1 \cos \theta_1 \sin \delta_2 \sin \theta_2 \quad . \tag{12.28c}$$

The direction cosines must be normalized to ensure that the sum of their squares equals 1, which is done by dividing through by N_c where

$$N_c = (\cos \alpha^2 + \cos \beta^2 + \cos \gamma^2)^{1/2} \ . \tag{12.29}$$

The final equation for the line of intersection between two planes is

$$\cos \alpha_h = (\cos \alpha) / N_c \ , \tag{12.30a}$$

$$\cos \beta_h = (\cos \beta) / N_c \ , \tag{12.30b}$$

$$\cos \gamma_h = (\cos \gamma) / N_c \ . \tag{12.30c}$$

The azimuth and dip of the line of intersection are given by Eqs. 12.4, 12.5 and Table 12.1.

12.4.4
Plane Bisecting Two Planes

The axial surface of a constant thickness fold hinge is a plane that bisects the angle between the two adjacent bedding planes and can be found as a vector sum. A vector sum is the diagonal of the parallelogram formed by two vectors. If the two vectors forming the sum have the same lengths, as do unit vectors, then the diagonal bisects the angle between them (Fig. 12.4). The sum of two vectors is the sum of their corresponding components:

$$v_1 + v_2 = (l_1 + l_2)i + (m_1 + m_2)j + (n_1 + n_2)k \ . \tag{12.31}$$

The unit vector that bisects the angle between the two vectors v_1 and v_2 is

$$v_3 = (1 / N_b)(l_1 + l_2)i + (1 / N_b)(m_1 + m_2)j + (1 / N_b)(n_1 + n_2)k \ , \tag{12.32}$$

where

$$N_b = ((l_1 + l_2)^2 + (m_1 + m_2)^2 + (n_1 + n_2)^2)^{1/2} \tag{12.33}$$

serves to normalize the length to make the resultant a unit vector. Depending on the directions of v_1 and v_2, this vector could bisect either the acute angle or the obtuse

Fig. 12.4.
Bisecting vector (*b*) of the angle between two vectors in the plane of the two vectors

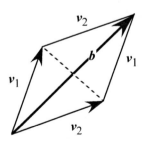

angle. If the two vectors are the poles to planes, the vector bisector is the pole to one of the two bisecting planes. The two bisecting planes are separated by 90°.

To bisect a given pair of planes, begin with the poles of the planes to be bisected (Eqs. 12.27). Substitute Eqs. 12.27 for both planes into 12.32 and use the same subscripting convention as above, with the subscript "B" = bisector, to obtain

$$\cos \alpha_{B1} = -(1/N_B)(\sin \delta_1 \sin \theta_1 + \sin \delta_2 \sin \theta_2) \ , \tag{12.34a}$$

$$\cos \beta_{B1} = -(1/N_B)(\sin \delta_1 \cos \theta_1 + \sin \delta_2 \cos \theta_2) \ , \tag{12.34b}$$

$$\cos \gamma_{B1} = -(1/N_B)(\cos \delta_1 + \cos \delta_2) \ , \tag{12.34c}$$

where, from Eq. 12.33:

$$N_B = [(\cos \delta_1 \sin \theta_1 + \cos \delta_2 \sin \theta_2)^2 + (\sin \delta_1 \cos \theta_1 + \sin \delta_2 \cos \theta_2)^2$$
$$+ (\cos \delta_1 + \cos \delta_2)^2]^{1/2} \ . \tag{12.35}$$

Equations 12.34 and 12.35 give the pole to one of the bisecting planes. The direction cosines of the pole should be converted to the dip vector. If $\cos \gamma_{B1}$ is negative, the pole points downward. Reverse the signs on all three direction cosines to obtain the upward direction. If $\cos \gamma_{B1}$ is positive, the pole points upward, and the dip azimuth is the same as that of the pole and the dip amount is equal to 90° plus the angle between the pole and the z axis:

$$\cos \alpha_1 = \cos \alpha_{B1} \ , \tag{12.36a}$$

$$\cos \beta_1 = \cos \beta_{B1} \ , \tag{12.36b}$$

$$\cos \gamma_1 = \cos (90 + \arccos (\cos \gamma_{B1})) \ . \tag{12.36c}$$

The procedure given above finds one of the two bisectors of the two planes. The other bisecting plane includes the line normal to both the line of intersection of the two planes (hinge line, Eqs. 12.30) and to the first bisecting line (Eqs. 12.36). Find the pole to the second plane by substituting Eqs. 12.30 and 12.36 into the cross product 12.26 to obtain

$$\cos \alpha_{b2} = (1/N_p)(\cos \beta_h \cos \gamma_{B1} - \cos \gamma_h \cos \beta_{B1}) \ , \tag{12.37a}$$

$$\cos \beta_{b2} = (1/N_p)(\cos \gamma_h \cos \alpha_{B1} - \cos \alpha_h \cos \gamma_{B1}) \ , \tag{12.37b}$$

$$\cos \gamma_{b2} = (1/N_p)(\cos \alpha_h \cos \beta_{B1} - \cos \beta_h \cos \alpha_{B1}) \ , \tag{12.37c}$$

where

$$N_p = [(\cos \beta_h \cos \gamma_{B1} - \cos \gamma_h \cos \beta_{B1})^2 + (\cos \gamma_h \cos \alpha_{B1} - \cos \alpha_h \cos \gamma_{B1})^2$$
$$+ (\cos \alpha_h \cos \beta_{B1} - \cos \beta_h \cos \alpha_{B1})^2]^{1/2} + 1 \ . \tag{12.38}$$

Fig. 12.5.
Geometry of conjugate faults.
δ_i: Dip vectors of fault planes;
σ_i: principal stress directions

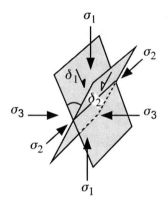

Equation 12.38 serves as a check on the calculation: it should always equal 1 because the starting vectors are orthogonal and of unit length. Convert this pole to the bisecting plane (Eqs. 12.37) into the dip vector of the plane as done above. First reverse the direction of the vector if it points downward ($\cos \gamma$ negative) by reversing the signs of all the direction cosines. Then

$$\cos \alpha_2 = \cos \alpha_{b2} \quad , \tag{12.39a}$$

$$\cos \beta_2 = \cos \beta_{b2} \quad , \tag{12.39b}$$

$$\cos \gamma_2 = \cos (90 + \arccos (\cos \gamma_{b2})) \quad . \tag{12.39c}$$

The azimuth and dip of the planes in Eqs. 12.36 and 12.39 are given by Eqs. 12.4, 12.5 and Table 12.1.

If the two planes being bisected are conjugate faults (Fig. 12.5), then the line of intersection of the faults is the intermediate principal compressive stress axis (σ_2), the line that bisects the acute angle is the maximum principal compressive stress axis (σ_1), and the line that bisects the obtuse angle is the least principal compressive stress axis (σ_3).

References Cited

Aitken FK (1994) Well locations and mapping considerations. Houston Geol Soc Bull Sept, 15–16

Allan US (1989) Model for hydrocarbon migration and entrapment within faulted structures. Am Assoc Pet Geol Bull 73:803–811

Anderson EM (1905) The dynamics of faulting. Edinburgh Geological Society, 8:387–402

Anderson EM (1951) The dynamics of faulting, 2nd edn. Oliver and Boyd, London, 206 pp

Asquith DO (1970) Depositional topography and major marine environments, Late Cretaceous, Wyoming. Am Assoc Pet Geol Bull 54:1184–1224

Asquith G, Krygowski D (2004) Basic well log analysis, Second Edition. AAPG Methods in Exploration Series, No. 16, 244 pp

Axen GJ (1988) The geometry of planar domino-style normal faults above a dipping detachment. J Struct Geol 10:405–411

Badgley PC (1959) Structural methods for the exploration geologist. Harper & Brothers, New York, 280 pp

Badley ME, Price JD, Backshall LC (1989) Inversion, reactivated faults and related structures: seismic examples from the southern North Sea. In: Cooper MA, Williams GD (eds) Inversion tectonics. Geological Society Special Publication No. 44, pp 201–219

Banks R (1991) Contouring algorithms. Geobyte 6:15–23

Banks R (1993) Computer stacking of multiple geologic surfaces. Petro Systems World, Winter, pp 14–16

Barnett JAM, Mortimer J, Rippon JH, Walsh JJ, Watterson J (1987) Displacement geometry in the volume containing a single normal fault. Am Assoc Pet Geol Bull 71:925–937

Bates RL, Jackson JA (1987) Glossary of geology, 3rd edn. American Geological Institute, Alexandria, Virginia, 788 pp

Beach A (1986) A deep seismic reflection profile across the northern North Sea. Nature 323:53–55

Becker A (1995) Conical drag folds as kinematic indicators for strike slip. J Struct Geol 17:1497–1506

Bengtson CA (1980) Structural uses of tangent diagrams. Geology 8:599–602

Bengtson CA (1981a) Statistical curvature analysis techniques for structural interpretation of dipmeter data. Am Assoc Pet Geol Bull 65:312–332

Bengtson CA (1981b) Structural uses of tangent diagrams; Discussion and reply. Geology 9(6): 242–243

Billings MP (1972) Structural geology, 3rd edn. Prentice Hall, Englewood Cliffs, 606 pp

Bishop MS (1960) Subsurface mapping. John Wiley, New York, 198 pp

Bois C (1993) Initiation and evolution of the Oligo-Miocene rift basins of southwestern Europe: Contribution of deep seismic reflection profiling. Tectonophysics 226:227–252

Bolstad P (2002) GIS fundamentals. Eider Press, White Bear Lake, MN, 411 pp

Bonham-Carter GF (1994) Geographic information systems for geoscientists: Modelling with GIS. Pergamon, Tarrytown, New York, 398 pp

Boyer S, Elliott D (1982) Thrust systems. Am Assoc Pet Geol Bull 66:1196–1230

Brewer RC, Groshong RH Jr (1993) Restoration of cross sections above intrusive salt domes. Am Assoc Pet Geol Bull 77:1769–1780

Brown W (1984) Working with folds. Am Assoc Pet Geol, Structural Geology School Course Notes, Tulsa, Oklahoma

Bruce CH (1973) Pressured shale and related deformation: Mechanism for development of regional contemporaneous faults. Am Assoc Pet Geol Bull 57:878–886

Brun JP, Gutscher MA, DEKORP-ECORS team (1992) Deep crustal structure of the Rhine Graben from DEKORP-ECORS seismic reflection data: A summary. Tectonophysics 208:139–147

Buddin TS, Kane SJ, Williams GD, Egan SS (1997) A sensitivity analysis of 3-dimensional restoration techniques using vertical and inclined shear constructions. Tectonophysics 269:33–50

Burchard EF, Andrews TG (1947) Iron ore outcrops of the Red Mountain Formation. Geological Survey of Alabama, Spec. Rep. 19. Geological Survey of Alabama, Tuscaloosa, 375 pp

Busk HG (1929) Earth flexures. Cambridge University Press, London, 106 pp

Butts C (1910) Geologic atlas of the United States, Birmingham folio, Alabama. United States Geological Survey, Washington, DC, 24 pp

Calvert WL (1974) Sub-Trenton structure of Ohio, with views on isopach maps and stratigraphic sections as basis for structural myths in Ohio, Illinois, New York, Pennsylvania, West Virginia, and Michigan. Am Assoc Pet Geol Bull 58: 957–972

Chai B (1994) Test of geometric models relating normal fault shape to rollover geometry in listric normal fault sand-box experiment. MS Thesis, University of Alabama, Tuscaloosa, 130 pp

Chamberlin RT (1910) The Appalachian folds of central Pennsylvania. J Geol 18:228–251

Charlesworth HAK, Kilby WE (1981) Calculating thickness from outcrop and drill-hole data. Bull Can Pet Geol 29:277–292

Chen WP, Molnar P (1983) Focal depths of intracontinental and intraplate earthquakes and their implications for the thermal and mechanical properties of the lithosphere. J Geophys Res 88: 4183–4214

Cherry BA (1990) Internal deformation and fold kinematics of part of the Sequatchie anticline, southern Appalachian fold and thrust belt, Blount County, Alabama. MS Thesis, University of Alabama, Tuscaloosa, 78 pp

Chester JS, Chester FM (1990) Fault-propagation folds above thrust with constant dip. J Struct Geol 12:903–910

Childs C, Easton SJ, Vendeville BC, Jackson MPA, Lin ST, Walsh JJ, Watterson J (1993) Kinematic analysis of faults in a physical model of growth faulting above a viscous salt analog. Tectonophysics 228: 313–329

Childs C, Watterson J, Walsh JJ (1995) Fault overlap zones within developing normal fault systems. J Geol Soc 152:535–549

Christensen AF (1983) An example of a major syndepositional listric fault. In: Bally AW (ed) Seismic expression of structural styles. Am Assoc Pet Geol, Stud. Geol. 15(2):2.3.1-36–2.3.1-40

Cloos E (1968) Experimental analysis of Gulf Coast fracture patterns. Am Assoc Pet Geol Bull 52:420–444

Coates J (1945) The construction of geological sections. Q J Geol Mining Metallurg Soc India 17:1–11

Colletta B, Le Quellec P, Letouzey J, Moretti I (1988) Longitudinal evolution of the Gulf of Suez structure (Egypt). Tectonophysics 153:221–233

Cooper MA (1983) The calculation of bulk strain in oblique and inclined balanced sections. J Struct Geol 5:161–165

Cooper MA (1984) The calculation of bulk strain in oblique and inclined balanced sections, reply. J Struct Geol 6:613–614

Cowie PA, Scholz CH (1992) Displacement-length scaling relationship for faults: Data synthesis and discussion. J Struct Geol 14:1149–1156

Crowell JC (1974) Origin of Late Cenozoic basins in southern California. Soc Econ Paleo Mineral Special Publication 22, pp 190–204

Cruden DM, Charlesworth HAK (1972) Observations on the numerical determination of axes of cylindrical and conical folds. Geol Soc Am Bull 83:2019–2024

Cruden DM, Charlesworth HAK (1976) Errors in strike and dip measurements. Geol Soc Am Bull 87: 977–980

Currie JB, Patnode HW, Trump RP (1962) Development of folds in sedimentary strata. Geol Soc Am Bull 73:655–673

Dahlstrom CDA (1969) Balanced cross sections. Can J Earth Sci 6:743–757

Dahlstrom CDA (1970) Structural geology in the eastern margin of the Canadian Rocky Mountains. Bull Can Pet Geol 18:332–406

Davis JC (1986) Statistics and data analysis in geology, 2nd edn. John Wiley & Sons, New York, 646 pp

Dawers NH, Anders MH, Scholz CH (1993) Growth of normal faults. Displacement-length scaling. Geology 21:1107–1110

De Paor DG (1988) Balanced section in thrust belts. Part I: Construction. Am Assoc Pet Geol Bull 72: 73–90

De Paor DG (1996) Bezier curves and geological design. In: De Paor DG (ed) Structural geology and personal computers. Pergamon, Tarrytown, New York, pp 389–417

Dennis JG (1967) International tectonic dictionary. Am Assoc Pet Geol, Mem. 7, 196 pp

Dennison JM (1968) Analysis of geologic structures. WW Norton, New York, 209 pp

Dickinson G (1954) Subsurface interpretation of intersecting faults and their effects upon stratigraphic horizons. Am Assoc Pet Geol Bull 38:854–877

Diegel FA (1986) Topological constraints on imbricate thrust networks, examples from the Mountain City window, Tennessee, USA. J Struct Geol 8:269–279

Dixon JM (1975) Finite strain and progressive deformation in models of diapiric structures. Tectonophysics 28:89–124

Doebl F, Teichmüller R (1979) Zur Geologie und heutigen Geothermik im mittleren Oberrheingraben. Fortschritte in der Geologie von Rheinland und Westfalen 27:1–17

Donath FA, Parker RB (1964) Folds and folding. Geol Soc Am Bull 75:45–62

Drozdzewski G (1983) Tectonics of the Ruhr district, illustrated by reflection seismic profiles. In: Bally AW (ed) Seismic expression of structural styles. Am Assoc Pet Geol, Stud. Geol. 15(3):3.4.1-1–3.4.1-7

Drucker DC (1967) Introduction to the mechanics of deformable materials. McGraw-Hill Book Co., New York, 445 pp

Dula WF (1991) Geometric models of listric normal faults and rollover folds. Am Assoc Pet Geol Bull 75:1609–1625

Elliott D (1976) The energy balance and deformation mechanisms in thrust sheets. Philos Trans R Soc London, Ser A, 283:289–312

Elliott D (1983) The construction of balanced cross sections. J Struct Geol 5:101

Elliott D, Johnson MRW (1980) Structural evolution in the northern part of the Moine thrust belt, N.W. Scotland. Trans Roy Soc Edinb Earth Sci 71:69–96

Elphick RY (1988) Program helps determine structure from dipmeter data. Geobyte 3:57–62

Epard J-L, Groshong RH Jr (1993) Excess area and depth to detachment. Am Assoc Pet Geol Bull 77: 1291–1302

Epard J-L, Groshong RH Jr (1995) Kinematic model of detachment folding including limb rotation, fixed hinges and layer-parallel strain. Tectonophysics 247:85–103

Erslev EA (1986) Basement balancing of Rocky Mountain foreland uplifts. Geology 14:259–262

Erslev EA (1993) Thrusts, back-thrusts and detachment of Rocky Mountain foreland arches. In: Schmidt CJ, Chase RB, Erslev EA (eds) Laramide basement deformation in the Rocky Mountain foreland of the western United States. Geol Soc Am Spec Pap 280:339–358

Eves H (1984) Analytic geometry. In: Beyer WH (ed) CRC standard mathematical tables. CRC Press, Boca Raton, Florida, pp 193–226

Faill RT (1969) Kink band structures in the Valley and Ridge Province, central Pennsylvania. Geol Soc Am Bull 80:2539–2550

Faill RT (1973a) Kink band folding, Valley and Ridge Province, central Pennsylvania. Geol Soc Am Bull 84:1289–1314

Faill RT (1973b) Tectonic development of the Triassic Newark-Gettysburg Basin in Pennsylvania. Geol Soc Am Bull 84:725–740

Fernández O (2005) Obtaining a best fitting plane through 3D georeferenced data. J Struct Geol 27: 855–858

Fleuty MJ (1964) The description of folds. Proc Geol Assoc 75:61–492

Foley JD, Van Dam A (1983) Fundamentals of computer graphics. Addison-Wesley, London, 664 pp

Fontaine DA (1985) Mapping techniques that pay. Oil Gas J 83(12):146–147

Fowler RA (1997) Map accuracy specifications. Earth Obs Mag 6:33–35

Freund R (1970) Rotation of strike-slip faults in Sistan, southeast Iran. J Geol 78:188–200

Galloway WE (1989) Genetic stratigraphic surfaces in basin analysis I: Architecture and genesis of flooding-surface bounded depositional units. Am Assoc Pet Geol Bull 73:125–142

Galloway WE, Ewing TE, Garrett CM, Tyler N, Bebout DG (1983) Atlas of major Texas oil reservoirs. Texas Bureau of Economic Geology, Austin, 139 pp

Gans PB, Miller EI, McCarthy J, Ouldcott ML (1985) Tertiary extensional faulting and evolving ductile-brittle transition zones in the Northern Snake Range and vicinity. Geology 13:189–193

Geiser PA (1988) The role of kinematics in the construction and analysis of geological cross sections in deformed terranes. In: Mitra G, Wojtal S (eds) Geometries and mechanisms of thrusting with special reference to the Appalachians. Geol Soc Am Spec Pap 222:47–76

Geiser J, Geiser PA, Kligfield R, Ratcliff R, Rowan M (1988) New applications of computer-based section construction: Strain analysis, local balancing and subsurface fault prediction. Mt Geol 25:47–59

Ghosh SK (1968) Experiments of buckling of multilayers which permit interlayer gliding. Tectonophysics 6:207–249

Gibbs AD (1989) Structural styles in basin formation. In: Tankard AJ, Balkwill HR (eds) Extensional tectonics and stratigraphy of the North Atlantic margins. Am Assoc Pet Geol Mem 46:81–93

Gill WD (1953) Construction of geological sections of folds with steep-limb attenuation. Am Assoc Pet Geol Bull 37:2389–2406

Greenhood D (1964) Mapping. University of Chicago Press, Chicago, 289 pp

Griggs D, Handin J (1960) Observations on fracture and a hypothesis of earthquakes. Geol Soc Am Mem 79:347–364

Groshong RH Jr (1975) Strain, fractures and pressure solution in natural single-layer folds. Geol Soc Am Bull 86:1363–1376

Groshong RH Jr (1988) Low-temperature deformation mechanisms and their interpretation. Geol Soc Am Bull 100:1329–1360

Groshong RH Jr (1989) Half-graben structures: Balanced models of extensional fault-bend folds. Geol Soc Am Bull 101:96–105

Groshong RH Jr (1990) Unique determination of normal fault shape from hanging-wall bed geometry in detached half grabens. Eclogae Geol Helv 83:455–471

Groshong RH Jr (1994) Area balance, depth to detachment and strain in extension. Tectonics 13: 1488–1497

Groshong RH Jr (1995) The meaning of extension in the hangingwalls of normal fault systems. Geol Soc Am, 1995 Abst. with Prog. 27:A-70

Groshong RH Jr (1996) Construction and validation of extensional cross sections using lost area and strain, with application to the Rhine Graben. In: Buchanan PG, Nieuwland DA (eds) Modern developments in structural interpretation, validation and modelling. Geol Soc SpecPub 99:79–87

Groshong RH Jr, Epard J-L (1994) The role of strain in area-constant detachment folding. J Struct Geol 16:613–618

Groshong RH Jr, Epard J-L (1996) Computerized cross section restoration and balance. In: De Paor DG (ed) Structural geology and personal computers. Pergamon, New York, pp 477–498

Groshong RH Jr, Usdansky SI (1988) Kinematic models of plane-roofed duplex styles. Geol Soc Am Spec Pap 222:97–206

Groshong RH Jr, Pfiffner OA, Pringle LR (1984) Strain partitioning in the Helvetic thrust belt of eastern Switzerland from the leading edge to the internal zone. J Struct Geol 6:5–18

Groshong RH Jr, Pashin JC, Chai B, Schneeflock RD (2003a) Predicting reservoir-scale faults with area balance: Application to growth stratigraphy. J Struct Geol 25:1645–1658

Groshong RH Jr, Pashin JC, McIntyre MR (2003b) Relationship between gas and water production and structure in southeastern Deerlick Creek coalbed methane field, Black Warrior basin, Alabama. Internat. Coalbed Methane Symp., Tuscaloosa, Alabama, Paper 0306, 12 pp

Gussow WC (1954) Differential entrapment of oil and gas. Am Assoc Pet Geol Bull 38:816–853

Gzovsky MV, Grigoryev AS, Gushehenko OI, Mikhailova AV, Nikonov AA, Osokina DN (1973) Problems of the tectonophysical characteristics of stresses, deformations, fractures, and deformation mechanisms of the earth's crust. Tectonophysics 18:167–205

Hague TA, Gray GG (1996) A critique of techniques for modelling normal-fault and rollover geometries. In: Buchanan PG, Niewland DA (eds) Modern developments in structural interpretation, validation and modelling. Geol Soc London Spec Pub 99:89–97

Hamblin WK (1965) Origin of "reverse" drag on the downthrown side of normal faults. Geol Soc Am Bull 76:1145–1164

Hamilton DE, Jones TA (1992) Algorithm comparison with cross sections. In: Hamilton DE, Jones TA (eds) Computer modeling of geologic surfaces and volumes. Am Assoc Petrol Geol, Tulsa, Oklahoma, pp 273–278

Handley EJ (1954) Contouring is important. World Oil 183(4):106–107

Hansen WR (1965) Effects of the earthquake of March 27, 1964, at Anchorage, Alaska. US Geol Surv Prof Pap 542-A:1–68

Hansen E (1971) Strain facies. Springer-Verlag, Berlin Heidelberg, 207 pp

Hardin FR, Hardin GC Jr (1961) Contemporaneous normal faults of Gulf Coast and their relation to flexures. Am Assoc Pet Geol Bull 48:238–248

Hardy S, McClay K (1999) Kinematic modeling of extensional fault-propagation folding. J Struct Geol 21:695–702

Harmon C (1991) Integration of geophysical data in subsurface mapping. In: Tearpock DJ, Bischke RE (eds) Applied subsurface geological mapping. Prentice Hall, Englewood Cliffs, pp 94–133

Harry DL, Sawyer DS (1992) A dynamic model of extension in the Baltimore Canyon trough region. Tectonics 11:420–436

Hawkins CD (1996) Thin-skinned extensional detachment in the Black Warrior basin: Areal extent. Masters Thesis, University of Alabama, Tuscaloosa, 45 pp

Hewett DF (1920) Measurements of folded beds. Econ Geol 15:367–385

Higgins CG (1962) Reconstruction of a flexure fold by concentric arc method. Am Assoc Pet Geol Bull 46:1737–1739

Hintze WH (1971) Depiction of faults on stratigraphic isopach maps. Am Assoc Pet Geol Bull 55:871–879

Hobson GD (1942) Calculating the true thickness of a folded bed. Am Assoc Pet Geol Bull 26:1827–1842

Homza TX, Wallace WK (1995) Geometric and kinematic models for detachment folds with fixed and variable detachment depths. J Struct Geol 17:575–588

Horsfield WT (1980) Contemporaneous movement along crossing conjugate normal faults. J Struct Geol 2:305–310

Hossack JR (1996) Geometric rules of section balancing for salt structures. In: Jackson MPA, Roberts DG, Snelson S (eds) Salt tectonics: A global perspective. Am Assoc Petrol Geol Mem 65:29–40

Hubbert MK (1931) Graphic solution of strike and dip from two angular components. Am Assoc Pet Geol Bull 15:283–286

Hurley NF (1994) Recognition of faults, unconformities, and sequence boundaries using cumulative dip plots. Am Assoc Pet Geol Bull 78:1173–1185

Illies JH (1977) Ancient and recent rifting in the Rhinegraben. Geologie en Mijbouw 56:329–350

Jackson MPA (1995) Retrospective salt tectonics. In: Jackson MPA, Roberts DG, Snelson S (eds) Salt tectonics: A global perspective. Am Assoc Pet Geol Mem 65:1–28

Jackson J, Blenkinsop T (1997) The Bilila-Mtakataka fault in Malawi: An active, 100-km-long, normal fault segment in thick seismogenic crust. Tectonics 16(1):137–150

Jamison WR (1987) Geometric analysis of fold development in overthrust terranes. J Struct Geol 9:207–219

Johnson HA, Bredeson DH (1971) Structural development of some shallow salt domes in Louisiana Miocene productive belt. Am Assoc Pet Geol Bull 55:204–226

Jones TA, Hamilton DE (1992) A philosophy of contour mapping with a computer. In: Hamilton DE, Jones TA (eds) Computer modeling of geologic surfaces and volumes. Am Assoc Pet Geol, Tulsa, Oklahoma, pp 1–7

Jones TA, Krum GL (1992) Pitfalls in computer contouring, Part II. Geobyte 7:31–37

Jones NL, Nelson J (1992) Geoscientific modeling with TINs. Geobyte 7:44–49

Jones TA, Hamilton DE, Johnson CR (1986) Contouring geologic surfaces with the computer. Van Nostrand Reinhold, New York, 314 pp

Jordan P, Noack T (1992) Hangingwall geometry of overthrusts emanating from ductile decollements. In: McClay KR (ed) Thrust tectonics. Chapman and Hall, London, pp 311–318

Jorden JR, Campbell FL (1986) Well logging II – Electric and acoustic logging. Soc Pet Eng, New York, 182 pp

Kelly VC (1979) Tectonics, middle Rio Grande Rift, New Mexico. In: Riecker RE (ed) Rio Grande Rift: Tectonics and magnetism. AGU, Washington, DC, pp 57–70

Kerr HH, White N (1996) Kinematic modelling of normal fault geometries using inverse theory. In: Buchanan PG, Nieuwland DA (eds) Modern developments in structural interpretation, validation and modelling. Geol Soc Spec Pub 99:179–188

Kidd JT (1979) Aerial geology of Jefferson County, Alabama, Atlas 15. Geological Survey of Alabama, Tuscaloosa, 89 pp

Kligfield R, Geiser P, Geiser J (1986) Construction of geologic cross sections using microcomputer systems. Geobyte 1:60–66, 85

Krajewski SA, Gibbs BL (1994) Computer contouring generates artifacts. Geotimes 39:15–19

Langstaff CS, Morrill D (1981) Geologic cross sections. International Human Resources Development Co, Boston, 108 pp

Levorsen AI (1967) Geology of petroleum. WH Freeman, San Francisco, 724 pp

Lisle RJ, Leyshon PR (2004) Stereographic projection techniques for geologists and civil engineers, 2nd edn. Cambridge University Press, Cambridge, 112 pp

Low JW (1951) Subsurface maps and illustrations. In: LeRoy LW (ed) Subsurface geological methods. Colorado School of Mines, Golden Colorado, pp 894–968

Mackin JH (1950) The down-structure method of viewing geologic maps. J Geol 58:55–72

Maher CA (2002) Structural deformation of the southern Appalachians in the vicinity of the Wills Valley anticline, northeast Alabama. MS Thesis, University of Alabama, Tuscaloosa, 45 pp

Mallet J-L (2002) Geomodeling. Oxford University Press, Oxford, 599 pp

Maltman A (1990) Geological maps: An introduction. Van Nostrand Reinhold, New York, 184 pp

Mansfield CS, Cartwright JA (1996) High resolution fault displacement mapping from three-dimensional seismic data: Evidence for dip linkage during fault growth. J Struct Geol 18:249–263

Marrett R, Allmendinger RW (1991) Estimates of strain due to brittle faulting: Sampling of fault populations. J Struct Geol 13:735–738

Marshak S, Woodward N (1988) Introduction to cross section balancing. In: Marshak S, Mitra G (eds) Basic methods of structural geology. Prentice Hall, Englewood Cliffs, New Jersey, pp 303–332

McClay KR (1992) Glossary of thrust tectonics terms. In: McClay KR (ed) Thrust tectonics. Chapman and Hall, London, pp 419–433

McClay KR, Ellis PG (1987) Analogue models of extensional fault geometries. In: Coward MP, Dewey JF, Hancock PL (eds) Continental extensional tectonics. Geol Soc London Spec Pub 28:109–125

McConnell DA (1989) Determination of offset across the northern margin of the Wichita uplift, southwest Oklahoma. Geol Soc Am Bull 101:1317–1332

McGlamery W (1956) Cuttings log of Shell Oil Company W. E. Drennan #1: records of Permit No 688. Alabama Oil and Gas Board, Tuscaloosa, Alabama, 15 pp

McKenzie D (1978) Some remarks on the development of sedimentary basins. Earth Planet Sci Lett 40: 25–32

Medwedeff DA (1992) Geometry and kinematics of an active, laterally propagating wedge thrust, California. In: Mitra S, Fisher GW (eds) Structural geology of fold and thrust belts. The Johns Hopkins University Press, Baltimore, pp 3–28

Medwedeff DA, Suppe J (1997) Multibend fault-bend folding. J Struct Geol 19:279–292

Mitchum RM (1977) Seismic stratigraphy and global changes of sea level, Part I. Glossary of terms used in seismic stratigraphy. In: Payton CE (ed) Seismic stratigraphy – Applications to hydrocarbon exploration. Am Assoc Pet Geol Mem 26:205–212

Mitra S (1986) Duplex structures and imbricate thrust systems: Geometry, structural position, and hydrocarbon potential. Am Assoc Pet Geol Bull 70:1087–1112

Mitra S (1990) Fault-propagation folds. Am Assoc Pet Geol Bull 74:921–945

Mitra S (1993) Geometry and kinematic evolution of inversion structures. Am Assoc Pet Geol Bull 77: 1159–1191

Mitra G, Marshak S (1988) Description of mesoscopic structures. In: Marshak S, Mitra G (eds) Basic methods of structural geology. Prentice Hall, Englewood Cliffs, New Jersey, pp 213–247

Mitra S, Namson J (1989) Equal-area balancing. Am J Sci 289:563–599

Molnar P (1988) Continental tectonics in the aftermath of plate tectonics. Nature 335:131–137

Moretti I, Colletta B, Vially R (1988) Theoretical model of block rotation along circular faults. Tectonophysics 153:313–320

Morley CK, Nelson RA, Patton TL, Munn SG (1990) Transfer zones in the East African rift system and their relevance to hydrocarbon in rifts. Am Assoc Pet Geol Bull 74:1234–1253

Morse PF, Purnell GW, Medwedeff DA (1991) Seismic modeling of fault-related structures. In: Fagan SW (ed) Seismic modeling of geologic structures. Society of Exploration Geophysicists, Tulsa, Oklahoma, pp 127–152

Morton WH, Black R (1975) Crustal attenuation in Afar. In: Pilger A, Rösler A (eds) Afar depression of Ethiopia, v. 1. Schweizrebart'sche Verlagsbuchlandlung, Stuttgart, pp 55–65

Mulvany PS (1992) A model for classifying and interpreting logs of boreholes that intersect faults in stratified rocks. Am Assoc Pet Geol Bull 76:895–903

Muraoka H, Kamata H (1983) Displacement distribution along minor fault traces. J Struct Geol 5:483–495

Nicol A, Watterson J, Walsh JJ, Childs C (1996) The shapes, major axis orientations and displacement patterns of fault surfaces. J Struct Geol 18:235–248

Nicolas A, Poirier JP (1976) Crystalline plasticity and solid state flow in metamorphic rocks. John Wiley, London, 444 pp

Norris DK (1958) Structural conditions in Canadian coal mines. Geol Surv Can Bull 44, 54 pp

Novoa E, Suppe J, Shaw JH (2000) Inclined-shear restoration of growth folds. Am Assoc Pet Geol Bull 84:787–804

Ocamb RD (1961) Growth faults of south Louisiana. Trans Gulf Coast Assoc Geol Soc 11:55–65

Oertel G (1965) The mechanism of faulting in clay experiments. Tectonophysics 2:343–393

Oliveros RB (1989) Correcting 2-D seismic mis-ties. Geobyte 4:43–47

Pashin JC, Raymond DE, Alabi GG, Groshong RH Jr, Jin G (2000) Revitalizing Gilbertown oil field: Characterization of fractured chalk and glauconitic sandstone reservoirs in an extensional fault system. Geol Surv Alabama Bull 168, 81 pp

Perry WJ Jr (1989) Tectonic evolution of the Anadarko basin region, Oklahoma: US Geol Surv Bull 1866-A, 19 pp

Poblet J, McClay K (1996) Geometry and kinematics of single-layer detachment folds. Am Assoc Pet Geol Bull 80:1085–1109

Poblet J, McClay K, Storti F, Muñoz JA (1997) Geometries of syntectonic sediments associated with single-layer detachment folds. J Struct Geol 19:369–381

Ragan DM (1985) Structural geology, 3rd edn. John Wiley, New York, 393 pp

Ramberg H (1963) Strain distribution and geometry of folds. Upps Univ Geol Inst Bull 42:1–20

Ramsay JG (1967) Folding and fracturing of rocks. McGraw-Hill, New York, 568 pp

Ramsay JG (1981) Tectonics of the Helvetic nappes. In: McClay KR, Price J (eds) Thrust and nappe tectonics. Blackwell, Boston, pp 293–309

Ramsay JG, Huber MI (1987) The techniques of modern structural geology, vol. 2: Folds and fractures. Academic Press, London, 700 pp

Ramsey JM, Chester FM (2004) Hybrid fracture and the transition from extension fracture to shear fracture. Nature 428:63–66

Reches Z, Eidelman A (1995) Drag along faults. Tectonophysics 247:145–156

Reches Z, Hoexter DF, Hirsch F (1981) The structure of a monocline in the Syrian arc system, Middle East – Surface and subsurface analysis. J Pet Geol 3:413–425

Redmond JL (1972) Null combination in fault interpretation. Am Assoc Pet Geol Bull 56:150–166

Rettger RE (1929) On specifying the type of structural contouring. Am Assoc Pet Geol Bull 13:1559–1561

Rippon JH (1985) Contoured patterns of the throw and hade of normal faults in the Coal Measures (Westphalian) of north-east Derbyshire. Proc Yorkshire Geol Soc 45:147–161

Robinson AH, Sale RD (1969) Elements of cartography, 3rd edn. John Wiley, New York, 415 pp

Rosendahl BR (1987) Architecture of continental rifts with special reference to East Africa. Ann Rev Earth Planet Sci 15:445–503

Rousset D, Bayer R, Guillon D, Edel B (1993) Structure of the southern Rhine Graben from gravity and reflection seismic data (ECORS-DEKORP program). Tectonophysics 221:135–153

Rowan MG, Kligfield R (1989) Cross section restoration and balancing as an aid to seismic interpretation in extensional terranes. Am Assoc Pet Geol Bull 73:955–966

Rowan MG, Linares R (2005) Medina anticline, Eastern Cordillera, Colombia. In: Shaw JH, Connors C, Suppe J (eds) Seismic interpretation of contractional fault-related folds. Am Assoc Pet Geol Seismic Atlas, Studies in Geology 53:77–82

Rowland SM, Duebendorfer EM (1994) Structural analysis and synthesis, 2nd edn. Blackwell, Oxford, 279 pp

Sanford AR (1959) Analytical and experimental study of simple geologic structures. Geol Soc Am Bull 70:19–51

Schlische RW, Young SS, Ackermann RV, Gupta A (1996) Geometry and scaling relations of a population of very small rift-related normal faults. Geology 24:683–686

Schlumberger (1986) Dipmeter interpretation. Schlumberger Limited, New York, 76 pp

Scott DL, Etheridge MA, Rosendahl BR (1992) Oblique-slip deformation in extensional terrains: A case study of the Lakes Tanganyika and Malawi rift zones. Tectonics 11:998–1009

Sebring L Jr (1958) Chief tool of the petroleum exploration geologist: The subsurface structural map. Am Assoc Pet Geol Bull 42:561–587

Seeber L, Sorlien CC (2000) Listric thrusts in the western Transverse Ranges, California. Geol Soc Am Bull 112:1067–1079

Shaw JH, Hook SC, Sitohang EP (1997) Extensional fault-bend folding and synrift deposition: An example from the Central Sumatra Basin, Indonesia. Am Assoc Pet Geol Bull 81:367–379

Shaw JH, Connors C, Suppe J (2005a) Seismic interpretation of contractional fault-related folds. Am Assoc Pet Geol Seismic Atlas, Studies in Geology 53, 156 pp

Shaw JH, Hook SC, Suppe J (2005b) Pitas Point anticline, California, U.S.A. In: Shaw JH, Connors C, Suppe J (eds) Seismic interpretation of contractional fault-related folds. Am Assoc Pet Geol Seismic Atlas, Studies in Geology 53:60–62

Shudofsky GN (1985) Source mechanism and focal depths of East African earthquakes using Raleigh wave dispersion and body-wave modelling. Geophys J Roy Astr Soc 83:563–614

Sibson RH (1977) Fault rocks and fault mechanisms. J Geol Soc 133:191–213

Snyder JP (1987) Map projections – A working manual. US Geol Surv Prof Pap 1395, 383 pp

Spang JH (1997) Use of synthetic and antithetic shear to model the development of Gulf Coast listric normal faults. Gulf Coast Assoc Geol Soc Trans 47:549–555

Spang JH, Dorobek SL (1998) Using antithetic normal faults to accurately map growth axial surfaces. Gulf Coast Assoc Geol Soc Trans 48:423–429

Stauffer MR (1973) New method for mapping fold axial surfaces. Geol Soc Am Bull 84:2307–2318

Stearns DW (1978) Faulting and forced folding in the Rocky Mountains foreland. Geol Soc Am Memoir 151:1–37

Stockmal GS, Spang JH (1982) A method for the distinction of circular conical from cylindrical folds. Can J Earth Sci 19:1101–1105

Stockwell CH (1950) The use of plunge in the construction of cross sections of folds. Proc Geol Assoc Can 3:97–121

Stone DS (1991) Analysis of scale exaggeration on seismic profiles. Am Assoc Pet Geol Bull 75:1161–1177

Suppe J (1983) Geometry and kinematics of fault-bend folding. Am J Sci 283:684–721

Suppe J (1985) Principles of structural geology. Prentice-Hall, Englewood Cliffs, 537 pp

Suppe J, Medwedeff DA (1990) Geometry and kinematics of fault-propagation folding. Eclogae Geol Helv 83:409–454

Suppe J, Chou GT, Hook SC (1992) Rates of folding and faulting determined from growth strata. In: McClay KR (ed) Thrust tectonics. Chapman and Hall, London, pp 105–121

Suppe J, Connors CS, Zhang Y (2004) Shear fault-bend folding. In: McClay KR (ed) Thrust tectonics and hydrocarbon systems. AAPG Mem 82:303–323

Tearpock DJ (1992) Contouring: Art or science? Geobyte 7:40–43

Tearpock DJ, Bischke RE (2003) Applied subsurface geological mapping with structural methods, 2nd Edn. Prentice Hall PTR, Upper Saddle River, 822 pp

Thibaut M, Gratier JP, Léger M, Morvan JM (1996) An inverse method for determining three-dimensional fault geometry with thread criterion: Application to strike-slip and thrust faults (Western Alps and California). J Struct Geol 18:1127–1138

Thomas GB (1960) Calculus and analytic geometry. Addison-Wesley, London, 1010 pp

Thomas WA (1986) Sequatchie anticline, the northwesternmost structure of the Appalachian fold-thrust belt in Alabama. Geol Soc Am centennial field guide – southeastern section. Geological Society of America, Boulder, Colorado, pp 177–180

Thompson GA (1960) Problem of Late Cenozoic structure of the basin ranges. Structure of the Earth's crust and deformation of rocks. International Geological Congress, Report of Session, Norden, 21:62–68

Thorsen CE (1963) Age of growth faulting in southeast Louisiana. Gulf Coast Assoc Geol Soc Trans 13: 103–110

Threet RL (1973) Down-structure method of viewing geologic maps to obtain sense of fault separation. Geol Soc Am Bull 84:4001–4004

Turner FJ, Weiss LE (1963) Structural analysis of metamorphic tectonites. McGraw Hill, New York, 545 pp

Van Wagoner JC, Posamentier HM, Mitchum RM, Vail PR, Sarg JF, Loutit TS, Hardenbol J (1988) An overview of the fundamentals of sequence stratigraphy and key definitions. In: Wilgus CK, Hastings BS, Posamentier H, Van Wagoner J, Ross CA, Kendall CG St C (eds) Sea-level changes: An integrated approach. Soc Econ Paleontol Mineral Spec Publ 42:39–45

Verrall P (1982) Structural interpretation with applications to North Sea problems. Course Notes No 3, Joint Association of Petroleum Exploration Courses (JAPEC), London, unpaginated

Walsh JJ, Watterson J (1991) Geometric and kinematic coherence and scale effects in normal fault systems. In: Roberts AM, Yielding G, Friedman B (eds) The geometry of normal faults. Geol Soc Spec Publ 6:193–203

Walsh JJ, Watterson J, Childs C, Nicol A (1996) Ductile strain effects in the analysis of seismic interpretations of normal faults. In: Buchanan PG, Nieuwland DA (eds) Modern developments in structural interpretation, validation and modelling. Geol Soc Spec Publ 99:27–40

Walsh JJ, Watterson J, Bailey WR, Childs C (1999) Fault relays, bends, and branch-lines. J Struct Geol 21:1019–1026

Walters RF (1969) Contouring by machine: A user's guide. Am Assoc Pet Geol Bull 53:2324–2340

Wang S (1994) Three-dimensional geometry of normal faults in the southeastern Deerlick creek coalbed methane field, Black Warrior basin, Alabama. MS Thesis, University of Alabama, Tuscaloosa, 77 pp

Washington PA, Washington RA (1984) The calculation of bulk strain in oblique and inclined balanced sections, discussion. J Struct Geol 6:613–614

Watson DF, Philip GM (1984) Triangle-based interpolation. Math Geol 16:779–795

Watterson J (1986) Fault dimensions, displacements, and growth. Pure Appl Geophys 124:365–373

Webel S (1987) Significance of backthrusting in the Rocky Mountain thrust belt. 38th Field Conf., 1987, Jackson Hole, Wyoming. Wyoming Geological Association Guidebook, pp 37–53

Weijermars R (1997) Structural geology and map interpretation. Alboran Science Publishing, Amsterdam, 378 pp

Wernicke B, Burchfiel BC (1982) Models of extensional tectonics. J Struct Geol 4:105–115

Westaway R (1991) Continental extension on sets of parallel faults: Observational evidence and theoretical models, In: Roberts AM, Yielding G, Freeman B (eds) The geometry of normal faults. Geol Soc London Spec Pub 56:143–169

Wheeler J (1987) Variable-heave models of deformation above listric normal faults: The importance of area conservation. J Struct Geol 9:1047–1049

Wheeler WH, Rosendahl BR (1994) Geometry of the Livingstone Mountains border fault, Nyasa (Malawi) rift, East Africa. Tectonics 13:303–312

White NJ, Jackson JA, McKenzie DP (1986) The relationship between the geometry of normal faults and that of the sedimentary layers in their hangingwalls. J Struct Geol 8:897–909

Wickham J (1995) Fault displacement-gradient folds and the structure at Lost Hills, California (U.S.A.). J Struct Geol 17:1293–1302

Wilkerson MS, Dicken CL (2001) Quick-look techniques for evaluating two-dimensional cross sections in detached contractional settings. Am Assoc Pet Geol Bull 85:1759–1770

Williams WD, Dixon JS (1985) Seismic interpretation of the Wyoming overthrust belt. In: Gries RR, Dyer RC (eds) Seismic exploration of the Rocky Mountain region. Rocky Mountain Association of Geologists and the Denver Geophysical Society, Denver, pp 13–22

Williams G, Vann I (1987) The geometry of listric normal faults and deformation in their hangingwalls. J Struct Geol 9:789–795

Wilson G (1967) The geometry of cylindrical and conical folds. Proc Geol Assoc 78:178–210

Wise DU (2005) Biemsderfer plotter for field recording of structural measurements on equal area nets. J Struct Geol 27:823–826

Withjack MO, Peterson ET (1993) Prediction of normal fault geometries – A sensitivity analysis. Am Assoc Pet Geol Bull 77:1860–1873

Withjack MO, Islam QT, La Pointe PR (1995) Normal faults and their hangingwall-deformation: An experimental study. Am Assoc Pet Geol Bull 79:1–18

Woodcock NH (1976) The accuracy of structural field measurements. J Geol 84:350–355

Woodcock NH, Fischer M (1986) Strike-slip duplexes. J Struct Geol 8:725–735

Woodward NB (1987) Stratigraphic separation diagrams and thrust belt structural analysis. 38th Field Conf., 1987, Jackson Hole, Wyoming. Wyoming Geological Association Guidebook, pp 69–77

Woodward NB, Boyer SE, Suppe J (1985) An outline of balanced cross sections. Studies in Geology 11, 2nd edn. University of Tennessee Knoxville, 170 pp

Woodward NB, Boyer SE, Suppe J (1989) Balanced geological cross sections: An essential technique in geological research and exploration. Short Course in Geology, vol 6, AGU, Washington, DC, 132 pp

Xiao H, Suppe J (1992) Origin of rollover. Am Assoc Pet Geol Bull 76:509–529

Ziegler PA (1992) European Cenozoic rift system. Tectonophysics 208:91–111

Index

CPSIA information can be obtained at www.ICGtesting.com
Printed in the USA
LVOW10*1209190114

370038LV00010B/402/P